Beatrix Pfleiderer
Gilles Bibeau (Eds.)

# Anthropologies of Medicine

curare

Zeitschrift für Ethnomedizin und transkulturelle Psychiatrie

Herausgegeben von der Arbeitsgemeinschaft Ethnomedizin e.V.

**Sonderband/Special Volume 7/1991**

For other volumes of this series see back cover.

# Anthropologies of Medicine

**A Colloquium on West European and North American Perspectives**

On behalf of the Arbeitsgemeinschaft Ethnomedizin
edited by
Beatrix Pfleiderer and Gilles Bibeau

vieweg

Arbeitsgemeinschaft Ethnomedizin e. V. (AGEM)
Völkerkundemuseum
Hauptstraße 235
D-6900 Heidelberg, Germany

Vieweg is a subsidiary company of the Bertelsmann Publishing Group International.



Printed by Lengericher Handelsdruckerei, Lengerich
Printed in Germany

ISBN 3-528-07820-0

# Contents

## Section 3: Local Cultures of Biomedicine



## Section 4: From Patients' Complaints to Cultural Narrative



## Section 5: Medical Systems and Cultural Change: An Analysis of Facts and Theories

# About the Origins of this Volume

**Beatrix Pfleiderer**

We, a group of medical anthropologists, met in Hamburg from December 4th to 8th, 1988 to discuss the various culture-specific (or nation-specific?) strands that had developed out of the new and - as we discovered - also old field of medical anthropology.

From all over Europe (except the then socialistic countries), Canada and the United States gathered participants whose contribution guaranteed the presence of the most important directions medical anthropology has produced in the recent past.

The need for this particular conference was felt and articulated on both sides of the Atlantic. As far as I remember, it was Arthur Kleinman from Harvard who had approached Thomas Maretzki from the University of Hawaií, a frequent traveller to Germany. The latter had worked on "die Kur" in Germany and used his contacts in order to set up a conference that helped Americans to have a look at the "sources". Having encorporated the work of Alfred Schütz, Foucault, Fleck, Mannheim, and of course, Umberto Eco into their recent theoretical development, American medical anthropologists wanted to establish contact to European traditions of medical anthropology, cultural psychiatry, ethnomedicine and perhaps even ethnopsychoanalysis.

But there were other reasons, too. A few years ago, while teaching a medical anthropology course, I instructed my class on the basic concepts of the field. When talking about the possible *Be-greifen* (perceiving) or *Be-schreiben* (describing) of a sickness in a subjective view as "illness" or in the (allegedly) objective view as "disease" as Kleinman (1980) has put it a medical student came up with the comment that he felt that the German founder of psychosomatic medicine Viktor von Weizsäcker (1987) had talked about the same thing, when he differentiated between "Krankheit" and "Kranksein". And that even long before him, Georg Groddeck (1983) who hated classifications where sick human beings were concerned and who wished medicine to be an art rather than a science, stated: "In Wirklichkeit gibt es gar keine Krankheiten, es gibt nur kranke Menschen." ("In reality there is no such thing as disease, there are only sick human beings."). A very brave sentence, indeed, in a time when Virchow and his cellpathology dominated the field (Groddeck 1983: 23-25 ). It was only that student's comment that made me and my fellow researchers aware of the fact that Groddeck's and von Weizsäcker's work are very little known outside of German-speaking countries. And that is in spite of the fact that for us German medical anthropologists the development of a "Cultural Medicine" is unseparable from the work of these two founders. In Germany medical anthropology made its entrance to the scene via the subdiscipline "Ethnomedizin", or Medicine in cultural comparison. Cut off from the mainstream by the fascistic scenario and its aftermath, cultural medicine did not develop in postwar-Germany as it did in other Western countries. This is also true for "Ethnomedizin". Although linked to the German philosophical tradition of the phenomenological school "Ethnomedizin " had to start from anew (cf. Sterly 1976/ 77). Caused by the postwar situation, academic "Ethnomedizin" followed the American formulation and articulation in teaching and research (Pfleiderer and Bichmann 1985). It did so without tracing its own roots, or at least by neglecting them.

The original idea of an American-German meeting was dropped soon, because there was another cause: European medical anthropologists started only in the eighties in small steps to organize themselves, that is to get to know each other. We wanted this conference to be a forum for the presentation of the culture-specific traditions of the respective medical anthropologies that would result in a more detailed view of the cultural context of these traditions. And second, we wished inner-European communicative activities to be encouraged as a consequence of this conference.

The planning committee for the conference met in May 1987 in Bad Homburg. We were sponsored by the Reimersstiftung. The members were A. Richters, W. Bonsels, S. v.d. Geest (The Netherlands), B. Pfleiderer, T. Hauschild, H.-J. Diesfeld, E. Seidler (Germany), C. MacCormack (U.K.), and T. Maretzki (U.S.A.). Herr v. Kroszigk of the Reimersstiftung helped us in a very hospitable atmosphere to develop the discourse we thought the conference should take. I want to thank the Reimersstiftung for funding our first meeting. In this pre-conference all participants gave papers on the state of art of our field in their own country.

The members of the committee parted with the task of composing a group of participants from his or her own country. I feel the result was a well-balanced group of scholars who first are in a wonderful way able to represent their discipline and second give us insight into their respective national tradition of anthropology. When the list of participants was established, Arthur Kleinman applied for funds from the Wenner-Gren-foundation. The application was successful and nine scholars from the other side of the Atlantic received travel funds. The European part of the conference was generously funded by the Stiftung Volkswagenwerk. I wish to express our gratitude to both of the funding agencies. I also want to thank Pfleiderer Industrie for their financial support which enabled us to organize a reception cum dinner party for the participants, observers, and guests from our University.

Without my students Peter Gössler, Elisabeth Bergner and Viola Langkusch we would not have had such good atmosphere during the meeting and a preparation with the fewest possible misunderstandings. With my thanks I would like to honour their talents as tough bargainers when it comes to hotel and meal prices, friendly hosts when it comes to catering and ticketing and "Mädchen für alles" when it came to replace lost xerox copies, among other things. And last but really not least I wish to thank Johannes Sommerfeld for his meticulous work, some extra day and night shifts and his special commitment while collecting these contributions and preparing this volume.

Most of the participants agreed to contribute to this volume with a final version of the paper read at the meeting. Some of the papers appeared also in a modified version in *Culture, Medicine and Psychiatry* (Vol. 14, 2, 1990). One paper was added later without having been read at the meeting.

The papers of this volume reflect the idea of the meeting. They show the state of art, tasks, and goals of medical anthropology. They also give us a taste of the country-specific flair of research. We, the editors, felt that the collection of papers suggests five sections for this volume. In the first section we arranged the papers that discuss the psychosomatizing process in their cultural specifity and context. We move from the original notion of "culture-boundness", a more static concept, to the dynamics and function of the illness narrative in the context of culture.

In the second section we continue the psychosomatic strand, but we move on to the form that the culturo-somatic dress furnishes (without labelling it as a A Bao A Qu [Pandolfi]). This form enables us to talk about "ethnicity" when it comes to the "symbolical and physiological" [Devisch], nerves [Lock], or the "extreme otherness of strange emotional worlds" [Pfleiderer]. In the third section we have four papers that deal with local cultures of biomedicine in three different worlds: America, Italy, and a rural area in Southwest-Germany. The fourth section starts out with an admonition that seems necessary to me at this point of the development of medical anthropology. I am sympathetic to this warning and not only because I lived and did fieldwork in a village called Nal-Damayantithal, named after two ardent lovers that serve Francis Zimmermann as a proper example for reminding us that by "focussing on illness, we might well miss that part of the ethnographic material that eludes the clinician". In the fifth section, finally, we start out with the interesting attempt to view the development of two medical traditions as schismogenic processes within their society. The second paper gives us a concise analysis of the recent developments and fads in medical anthropology. The look at the cultural constructivist perspectives is necessary for this volume, since it appeared too often as a tacitly assumed postulate. The grasp for the "beyond" is helpful. It is the interest in the "beyond" that it shares with the third - and epistemologically very special paper- of this section. Carol MacCormack suggests that we remove the boundaries of perception, action and practice which become obsolete when we grow tired of the Cartesian metaphysical dualism. She directs our look to the possible changes that a holistic reality view could cause for us and, of course, our field.

## References

GRODDECK G. 1983: *Krankheit als Symbol.* Frankfurt: Fischer.

KLEINMAN A. 1980: *Patients and Healers in the Context of Culture* . Berkeley: University of California Press.

PFLEIDERER B. and BICHMANN W. 1985: *Krankheit und Kultur.* Berlin: Reimer.

STERLY J. 1976/ 77: Versuch einer systematischen Ordnung der Ethnomedizin. *Ethnomedizin/ Ethnomedicine* Vol. 10 (3/4).

WEIZSÄCKER V. von 1987: *Gesammelte Schriften* Vol. 5. Frankfurt: Suhrkamp.

Beatrix Pfleiderer
August 1990

University of Hamburg
University of Hawaií at Hilo

# Section 1

# Culture-Specific Psychosomatic Paradigms

# Are Functional Syndromes Culture-Bound?

Thure von Uexküll

## What are Functional Syndromes?

In this article I will try to give an answer to two questions, firstly, "what are functional syndromes?" and, secondly, "what is their connection to social and cultural events?". Let me start off by summarizing the results of two fairly recent studies:

1. A nine-year follow-up-study conducted 1979 by Berkman and Syme in Almeda County (Cal.) compared the mortality of socially isolated individuals to the mortality of those with extensive social contacts. Their results, drawn from 6928 adults, show an *increase in the relative risk of mortality* by 2.3 in males, and 2.8 in females of the *socially isolated group*. And this increase was reported to be independent of life habits such as smoking and alcohol consumption, self reported physical health status, physical activity, obesity, socio-economic status and participation in preventive health services (Berkman and Syme 1979).

2. An epidemiological survey on recently divorced women by Chester in 1973 revealed an 85% incidence of health complaints. These complaints mainly consisted of 4-5 different symptoms, ranging from anorexia, heartache, stomachache and headache to insomnia, dizziness, skin-rash, abscess and hair-loss. 75% of patients went to see a physician. As a response to these symptoms psychotherapy or social counselling was however rarely practised. Therapy consisted almost exclusively in the prescription of psychopharmaca and sedatives. The duration of treatment thus exceeded two years in 25%, while only one remaining quarter of therapies came to an end within the first six months after initiation. Many cases of iatrogenic drug addiction were the consequence of these prolonged regimens (Chester 1973).

A combination of symptoms such as described by Chester in this study is by no means new in medicine. As early as 1681, Thomas Sydenham described "protean and chameleon-like" syndroms, capable of imitating any conceivable organic disease. Sydenham states the extraordinary frequency of these syndromes among his patients and estimates their percentage to amount to more than 50% in his afebrile cases (Wesiack 1974).

## Epidemiology

What about our present knowledge about the epidemiological relevance of these syndroms? In a survey conducted at the out-patient clinic in internal medicine at the University of Giessen in 1958, we screened 7825 patients with various symptoms, referred to us by their general practitioners for further investigation. Despite diagnostic scrutiny we were unable to elucidate any organic evidence in more than one quarter (25.5%) of these patients. Since we excluded statistical deviations unrelated to illness in this figure, our 1958 data remained in accordance with more recent results of similar studies.

*Today* the composition of patients in Western industrialized countries utilizing medical care in hospitals or private practises is thought to read as following:

- one third of the patients shows no explanatory evidence of organic disease;
- one third documents somatic *and* psychosocial problems of different degree of severity as the base of their difficulties;
- only the remaining third of patients shows pathological organic evidence sufficiently se-

vere to account for their complaints and their symptoms (v. Uexküll and Köhle 1986).

## Etiology and Pathogenesis

Considering the etiology and pathogenesis of functional syndromes, we must differentiate between the two basic questions of how these syndroms are *initiated* and how they are *maintained.*

To answer the first question it is important to realize that, as Sydenham observed, functional complaints have existed throughout history. As recent ethnomedical research has proven, they also exist throughout all human cultures (Good 1977, Kleinman 1986, Sich 1979). Accordingly, Chesters report suggests that *everybody* is capable of developing functional syndroms in situations of stress. On the other hand, functional syndromes appear quite differently in different cultures. Given this specific appearance of functional syndromes in specific cultures, the notion of *"culture bound syndromes"* has been put forward.

Applying this notion to Western industrialized culture involves a couple of *new* problems that are connected to the second question of how functional syndromes are *maintained.* Let me put this into three more questions: (1) Are the specific functional syndromes of industrialized countries typical examples of culture-bound syndromes? (2) How does modern biotechnical medicine handle and manipulate them and what is the influence of modern medicine on their clinical course? (3) And finally: could the assumption of a cultural impact on medicine be helpful in explaining the etiology and natural history of these syndromes?

The answer to all these questions inevitably leads to a bewildering phenomenon that - despite its enormous importance for our health-care system - is only very unwillingly realized by modern medicine: It is the fact that modern biotechnical medicine has become a first-degree risk-factor for patients with functional syndromes.

The modern physician, confronted with patients complaints, usually remains ignorant as to the psychosocial background of their histories. He has but learned to search for the *organic* causes of symptoms. Functional complaints however are capable of imitating the symptomatology of almost any organic disease. Hence a wide field of diagnostic considerations and techniques readily becomes involved in the physicians' search for the "real cause" of this or that complaint. It is no wonder that the medical history of patients with functional syndromes takes a radically *different course today* as compared to the era before the rise of modern biotechnical medicine.

Inasmuch as they are at all given the privileged status of distinct nosological entities, the very variety and plasticity of functional symptoms has led to various descriptions that perhaps reveal more about our present favourite myths and concepts of typical sources and sites of disease than about their actual etiology and pathogenesis.

I will give you just one example: ever since the introduction of the "DSM", terms like "somatization" and psychologization" dominate in theoretical discussions to such an extent that often enough the focus on *what* is actually somatized or psychologized is obscured. Clinical reality meanwhile remains widely untouched by even these limited discussions. Here physicians still aggressively engage in the attempt to fit functional symptoms into the schemes of organic disease. This is perhaps most drastically and instructively illustrated by the *surgical fate* of patients with functional syndromes. Epidemiological studies revealed quite early the astonishingly high percentage of surgical procedures performed on patients without organic evidence of disease: in a study of 1936 Bennet registered 244 surgical procedures performed in 150 patients of a psychiatric department within a general hospital, that had formerly been treated as organically ill.

A follow-up study in the Mayo-Clinic in 1934 investigated the fate of 235 patients formerly diagnosed with "chronic nervous exhaustion" in respect to possible organic causes of their complaints that might have been missed on primary examination. Six years after primary diagnosis of "chronic nervous exhaustion", 94% of patients still showed no signs of organic pathology, but nevertheless as many as 200 patients underwent 289 different surgical procedures (Macy and Allen 1949).

More recent data from West Germany (Reimer et al. 1979) suggests a time-span of *7 to 11 years* between the onset of functional symptoms and the reference of respective patients to psychosomatically trained physicians capable of supplying adequate diagnosis and therapy.

It is clear, though, that these patients, who suffered a long odyssey through all the various departments of organic medicine, including constantly repeated overall examinations and unfounded medical or surgical interventions, very rarely offer prospects for successful psychotherapy. One frequently has to be content with being able to prevent further deterioration of the problems or further useless interventions.

Weiner (1987) reports the present situation in the USA not to be any better. He emphasizes that most persons seeking health-care are "ill without having a disease" - disease here being defined as a state of organic alterations. Let me give you a couple of quotations from Weiner's report:

(1) "Persons in ill health constitute a major burden to themselves, their families and society. They are at risk for incorrect diagnoses, repeated, unnecessary surgery, and iatrogenic disease. They puzzle, frustrate and anger physicians." (Weiner 1987: 154)

(2) Weiner also quotes Smith et al. (1986), who calculated that those in chronic ill-health spend an average of seven days per month in bed, the general populations average being half a day per month. In 1989 the annual medical expenditures on this group of patients exceeded the figure for the average US-citizen by a factor of 9. To give an example: US patients hospitalized for functional gastrointestinal complaints numbered 115.000 per annum in 1976, accounting for 450.000 hospital-days and 250 million dollar costs. Some of the participants in this study looked back on patient careers of up to thirty years.

(3) In addition to examining the intrinsic stressor function of ill-health Weiner mentions stressors preceding changes in health status: "These stressors include mothers who give up their babies for adoption, widows and widowers, job loss and, in the case of functional bowel disorder marital discord, as well as concerns about children and job security. Threat of or actual job loss raises the risk or parasuicide and is associated with a 70% increase in illness episodes. A 150% increase in medical consultations and a 200% increase in attendance at out-patient clinics." (Weiner 1987: 156)

(4) Weiner concludes his statements by saying: "We must fill the blanks in our knowledge on the understanding of the psychobiology of stress - that is the manner in which stressors and anxiety are induced to alter bodily function without producing lesions. We must also learn to understand better how experience and personality interact." (Weiner 1987: 158). He suggests intensive investigation in the field of non-organic disease "because proper and timely intervention can radically reduce the cost of medical care and reduce the use of ill advised procedures" (Weiner 1987: 153).

In a 1988 study, Goldberg and Bridges investigated the onset of acute and subacute "somatization". They found no remarkable difference between "somatizers" and "psychologizers". Nevertheless they emphasize the adaptive advantages of somatization, especially its peculiar *"blame-avoiding function"*. They explain this concept as follows:

"Somatization is a great way for not seeing oneself as mentally ill and not seeing oneself as responsible for the life predicament that one happens to be in ... If anyone is to be responsible ... it is surely the doctor, who either has not made the correct physical diagnosis, or at any rate has not stopped his pain." (Goldberg and Bridges 1988: 140)

Now it is undoubtedly interesting to find out why some persons become "somatizers" and develop somatic symptoms while others become "psychologizers" and develop for example depressions. But this is not our main problem. We should rather proceed to asking ourselves, whether or not we tend to identify this blame-avoiding behavior in patients in order to avoid blaming our modern medicine and society for the living conditions that undoubtedly can be pathogenic for at least a significant minority of its members. Or, to put this into a question: Could somatization be connected to the failure of modern medicine to extend the scope of its interests beyond the limits of individual living conditions?

## The Functional Syndrome as a "Political Disease"?

We cannot give justice to patients with functional syndromes by merely making their individual and psychological constitution responsible for their inability to cope with the demands of modern society. I have quoted Berkman, Syme and Chester to show the importance of social environments for individual health and disease.

Maybe we can get closer to an understanding of functional syndromes by extending our study to the concepts of "somatization" and "psychologization". And it is here, where the concept of culture-bound syndromes, claiming the existence of specific diseases as related to specific cultures might offer some assistance. Shouldn't this apply to our industrialized culture, too? It does, but in a very special way. I will thus try to show that functional syndromes are produced by our society as well as by our medicine as a representative of our society.

The concept of culture-bound syndromes contains yet another claim the development of culturally specific therapeutic systems for specific cultural diseases. Why now, must we ask ourselves, has the specific type of medicine developed within modern industrial culture proven itself incapable of correctly treating patients with functional syndromes? And, moreover, how could it become a veritable threat to their lives?

This seemingly paradoxical constellation might be one of the factors capable of explaining another paradox, namely the increasing dissatisfaction and discomfort in the general population's reaction toward the spectacular successes of modern medicine in its fight against life-threatening diseases. To understand this we have to differentiate between those diseases that are affected by cultural factors in their course and those which are not - differentiation that can be made irrespectively of the different degrees of cultural impact on the origin of these diseases. If we want to identify certain cultural factors as "pathogenic" and others as "protective", we thus have to abandon the model of the body-machine that dominates biotechnical medicine. Instead, we have to conceive a new model that takes into account three fundamental experiences:

(1) The fact that every human being lives within his own *subjective reality*; this personal reality being the determining factor for the way in which he experiences and organizes his perceptions, his behavior and his bodyreactions - you could call this the *constructivistic message*.

(2) The fact that every human being has to integrate constantly his individual reality into his surrounding social reality if he wants to participate and feel at home in this *social reality* - you could call this the modern version of the Aristotelian *message of the social dependence of man*.

(3) The fact that social realities constitute a medium influencing health and disease of the individual - this you could call the *message of modern bio-psycho-social medicine*.

This new model is capable of extending our range of view beyond the individual patient and his organism. However it can only do this if it does not remain an abstract construction but develops into a realistic approach. Still it is a complicated approach. For we cannot understand its relevance unless we understand the role of "emergent" qualities in social (and other) systems. "Emergent" qualities of systems cannot, by definition, be derived from the qualities of their elements. They must instead be regarded as belonging to the next and more complex system level that gives rise to a "new" quality on a new level of system integration. Accordingly, social systems cannot be reconstructed from the individual properties of their participants. Instead the integration of individual human beings into social systems creates a new dimension of reality for them. All this shows that, in order to discuss these problems properly, we will have to introduce a number of terms derived from modern semiotics and modern system theory (Th.v. Uexküll 1980).

### The Model of "Bipersonality" as a Paradigm of Social Reality

Let me introduce a specific concept in some detail. It has been developed under the name of "bipersonality" by Christian and Haas as early as 1949 and can still serve as an example for the new model described above. Discussing this concept will supply us with *one example* of how the new systemic model might and should be conceived.

In their publication "Form and essence of bipersonality" with the programmatic subtitle "Foundations for a medical sociology", Christian and Haas present a subtle analysis of a seemingly simple human action: the cooperation of two human beings while sawing wood. The revolutionary character of this analysis remained widely unappreciated however in those early years - partly, I suppose - because system theory had not yet been developed. For it is the language of system theory that could most appropriately express their findings.

Christian and Haas define social realities as entities having an existence relevant for *all* participating subjects *independent* of their individualities. To repeat this in the terminology of system theory: human subjects can develop certain qualities and potentials within social systems that do not derive from their individual persons but from the system they are integrated in:

(1) "Hence relationships cannot be understood by adding up the potentials of their participants in a 1+1=2 manner. Individual participants are *integral parts* of the relationship, not *fractions*. "Bipersonal" is nothing additive and nothing that could be split up into fractions. Consequently, the relationship of three or more persons - "Tripersonality" and so on - does *not* represent a quantitative and calculable increase of bipersonality. It represents, on the contrary, a qualitatively new and different form of community and communication." (Christian and Haas 1949: 1).

(2) Social reality hence "cannot be generated by effort or acts of active empathy or sympathy of individuals. We can only accept it and experience its properties by changing it, driving it into crises or by destroying it. This means that we can only understand it from its limits. Its existence should not be made the object but the epistemiological basis of interpretations." (1949: 2).

(3) According to this analysis bipersonality is *defined* by the following five characteristics:

(a) "Bipersonality is a given fact in a relationship. Participants begin their cooperation without creating a contract and without specific effort for adaptation." (1949: 9)

(b) The *unifying principle* is the quality of "togetherness": "Both participants directly experience their harmonic cooperation or its deficiencies." (1949: 13)

(c) The *founding principle* is mutuality: "Anything A does for B, B also does for A: the game does not take place between A and B, it works for both of them." (1949: 10)

(d) As a consequence both subjects are not autonomous - today we would say, that their actions underly system-specific restrictions. In other words: the contribution of each participant may change from time to time, but "difference between both is constantly and imperceptibly compensated by actions of the counterpart" (1949: 10-11)

(e) "Solidarity is founded in self-concealment. Every participant plays his role as a part of the working-whole." (1949: 12)

Some of these statements we could put into semiotic terms, too. Participation in a social system could then be described as the dependence on a common code in which individual differences in the sending and reception of coded signs have to be continuously - and mostly unconsciously - compensated. G.H. Mead (1968) has spoken of "intelligent gestures", founding common activities. In contrast to "unintelligent gestures" only generating drill acts, these "intelligent gestures" use the reactions of a partner for orientation in a relationship by understanding them as the results of his interpretations of the signs received and perceived in their interaction. This is just what Christian and Haas have shown in their experiments. And now we are reaching a point where these theories become relevant for the discussion of what we have called the social factor in the generation of functional syndromes. Disturbances of social or common reality are thus perceived by both partners as a disturbance in their feeling of self-independence. For seemingly paradoxical, in social relations the feeling of being the sovereign master of ones own powers strongly depends on the supplementary but imperceptible counter play of ones social counterparts.

Every action has to be complemented by a counteraction. Hence the deficiency or absence of this complementary action can lead to an impediment or even a paralysis in the actor (Th.v. Uexküll 1980). It is, by the way, interesting to observe how non-actions can achieve the same or even superior importance as actions within social systems. This interdependence could be illustrated by some forms of Japanese prize-fighting. Christian and Haas emphasize this by stating:

"In the very moment of highest subjective self-independence during the cooperation, our analysis reveals the objective dependence of both partners in the stringent mutuality of their movements and actions. And still: the experience of self-independence and freedom can only be won if an objective mutuality of action is reached" (Christian and Haas 1949: 11).

In biological terms: Any environment, and especially a social environment must guarantee the function of an ecological niche for any living being supposed to survive in it. But this niche-function of the environment for its part depends on the ability of the respective organism to also fulfill a niche-function for its environment. We could call this a *"basic principle of ecology"*. Having developed this interactional model, I will now try to apply its implications to the role of cultural factors in the living conditions of modern industrial societies. Extending this to the general question of health and disease, I will also attempt to delineate potential "pathogenic social factors" in these societies.

## "The Failure of Success"

A recent report from the *New England Journal of Medicine* mentions an US study registering a markedly decreasing number of Americans who feel "healthy and bodily well despite undeniable progress in their objective state of health" (Barsky 1988). Trying to account for the ever widening gap between objective health status and subjective health experience the author calls this phenomenon a "failure of success". He blames the progressive "*medicalization*" of life as being responsible for the increasing tendency of Americans to interpret any disturbance in somatic or psychological functions as a disease. Our interactional model could help us to interpret these findings. It describes "downward effects" of social realities causing unconscious changes in the individual reality of people. As we have shown, a reduction of self-interdependence can lead to the feeling of lost autonomy, which in turn is interpreted as internal disturbance. This inner experience serves an important function within the social context: it acts as a sensitive *indicator* for unbalanced aspects of interaction and communication between partners. This indicator originally was interpreted as an unconscious *directive* toward harmonizing and rebalancing social behavior.

The medicalization of self-experience deprives these indicators (or signs) of their social quality and significance. They are no longer interpreted as behavioral *directives* but as *disturbances* in somatic or psychological functions. The unconscious attention is diviated from the goal of "togetherness" and mutual harmony to the own body and psyche. This shift in the focus of unconscious attention now is very rich in poor consequences.

*Medicalization in this sense should be seen as the transformation of somatic or bodily self-experience from a sensitive receptor of social signs into a sort of indicator screen for defects in biological mechanisms.* And nothing could be more welcome to very modern biotechnical medicine claiming to suffer from detrimental effects of medicalization. For the medical experts of biotechnical repair - and only they - are responsible for taking care of these defects.

This transformation of body-experience can serve as one example of a deficient social integration. But despite the undeniable variation in these deficiencies between different situations, individuals and cultures there still *is* a common denominator for social desintegration in all cultures: it is the deficient "translation" or interpretation of signs or indicators that normally serve the homeostasis of an environment by maintaining the mutual niche-functions we have spoken of.

Let me return to the semiotic theory of interaction for one more time. The example of two men sawing wood could serve as a paradigm for the semiotic process of communication: anything A does is perceived as an unconscious question or, more generally speaking, as a sign by B. And anything B does in reaction is again interpretable as an answer, another sign by A - and so on.

Using the terms coined by Jacob von Uexküll (1932) we could also say that the coding of a perception into a sign represents an assignment of significance", while signs of behavioral answers represent an "utilization of significance" - these signs mutually completing and correcting themselves in the process of communication. This form of active-passive interpretation of signs simply serves to generate informations by which the participants of a social system can secure themselves of a common code.

The feedback of these, as they have been called, semantic and pragmatic functions of signs is the base for any commonly shared social reality. And this base is destroyed if signs are perceived only as signals for disturbances in somatic or psychological functions of the body-machine.

To summarize: The interactional model developed here can help us to interpret functional syndromes as a consequence and the expression of disharmonies in social interaction. Disharmonies of this kind threaten the social balance in any culture. Hence every culture is obliged to develop a medicine capable of restoring the individual's ability of social resonance and communication once it is impaired. And this must be possible within and despite the specific limitations of the respective culture. "Bipersonality" as a model of interaction for and not between patient and physician would be one fundamental therapeutical principle within this medicine. It reminds us, by the way, of Balint's concept of the "physician as a drug" (Balint 1957).

In its ignorance of the supreme importance of psychosocial problems our modern biotechnical medicine fails to meet the requirements of a truly therapeutic medicine. Moreover, it partly takes the side of the pathogenic tendencies in our industrial societies that, regarding health and disease mainly as functions or disfunctions of biochemical body-machines.

## Closing Remarks

What do we have to do? Accusations are of no help whatsoever. They only end in fruitless polarizations, paralyzing any creative debate. They offer help neither to helpless physicians trying to understand patients with psychosocial problems, nor do they offer any assistance to patients feeling misunderstood by these physicians. What we rather need is a widening of the horizon of our modern medicine. Medicine finally has to acknowledge being an integral part of a specific, for example industrial culture - and it finally has to accept the sharing of some responsibility with and against this culture. Medicine has to accept the fact that many patients show symptoms containing hints and references towards the very cultural system they and their physicians live in. To put it rather provocatively into other words: Medicine has to accept the existence of "political diseases", diseases originating in deficiencies of our culture, our polis.

Physicians are "agents" of their societies and cultures. But relationships are never one-sided. They are mutual in the strict sense. Physicians are obliged to critically reflect the leading contemporary ideologies in medical research and clinical practice. These ideologies are not made sacrosanct by their utter predominance. Physicians also carry responsability for the potentially detrimental effects of the concepts of health and disease domination and determining the cultural climate of society. They may and they can change these concepts and thus exert considerable influence on the human climate of their cultures.

However more complex its recent models, modern medicine still defines the body as a biochemical machine and orientates its definition of health according to the principle of a technical surveillance service. Consequentely, medicine interprets the task posed to physicians by industrial culture as an ordeal of ever more intelligent repair. Within this concept it is absolutely impossible to assign psychological and social factors the same significance in their role for health and disease as to chemical, physical and microbiological influences. But as long as our modern medicine clings to this dangerously obsolete model, it will never be able to live up to the task of fully carrying the cultural and social responsability. The deplorable medical histories of patients with functional syndromes aren' t simply individual tragedies. They also are clinical symptoms of defects in our industrial society.

Let us keep in mind: functional syndromes have a function for the patient as well as for the social system. For the patient they create a solution easier to tolerate than the awareness of social frustration. And for the social system the call for harmonization of disturbed social interactions serves as a sign appealing to the quality of togetherness. Let me add one final comment: The problems confronting us cannot be solved by regression to pre-industrial eras or by reference to medical practices of ancient or distant cultures. They can only satisfactorily be solved by overcoming the fatal one-siddness of our reductionist concepts. We have to aim at a comprehensive revision of the mechanistic theory of body and disease. And we have to look out for a medicine truly deserving the by-word "humane".

One of the most urgent necessities of such a new theory would be to develop and experimentally investigate models that could allow us to diagnose functional syndromes in *statu nasciendi*. The invention and implementation of therapeutic or preventive measures must be founded in the understanding of the somatic and psychosocial evolution of disease. And this is true for all of medicine. For any true progress in medicine has its base in our ability and willingness to investigate *all* relevant factors in health and disease.

## References


BALINT M. 1957: *Der Arzt, sein Patient und die Krankheit*. Stuttgart: Klett.

BARSKY A.J. 1988: The Paradox of Health. *The New England Journal of Medicine* 318 (7): 414-418.

BENNET A.E. 1936: *Annals of the Medical Association* 130: 1203.

BERKMAN L.F. and SYME S.L.1979: Social Networks, Host Resistance and Residents. *American Journal of Epidemiology* 109 (2): 186-204.

CHESTER R. 1973: Health and Marital Break Down. *Journal of Psychosomatic Research* 17:317-321.

CHRISTIAN P. and HAAS R. 1949: Wesen und Formen der Bipersonalität. Grundlagen für eine medizinische Soziologie. *Beiträge aus der allgemeinen Medizin* 7 (Stuttgart: Enke).

GOLDBERG D.P. and BRIDGES K. 1988: Somatic Presentations of Psychiatric Illness in Primary Care Settings. *Journal of Psychosomatic Research* 32 (2): 137-144.

GOOD B. 1977 The Heart of What' s the Matter. The Semantics of Illness in Iran. *Culture, Medicine and Psychiatry* 11: 25-58.

KLEINMAN A. 1986: *Social Origins of Distress and Disease: Depression, Neurasthenia and Pain in Modern China.* New Haven, London: Yale University Press.

MACY I.W. and ALLEN E.V. 1934: *Annals of Internal Medicine* 7: 861.

MEAD G.H. 1968: *Geist, Identität und Gesellschaft aus der Sicht des Sozialbehaviorismus.* Frankfurt: Suhrkamp.

REIMER C.L., HEMPFING L. and DAHME D. 1979: Iatrogene Chronifizierung in der Vorbehandlung psychischer Erkrankungen. *Praxis der Psychotherapie und Psychosomatik* 24: 123-133.

SICH D. 1979: Naeng: Begegnung mit einer Volkskrankheit in der modernen frauenärztlichen Sprechstunde in Korea. *Curare* 2: 87-96.

SMITH G.R. Fr., MONSON R.A. and RAY D.C. 1986: Patients with Multiple Unexplained Symptoms: Their Characteristics, Functional Health and Health Care Utilization. *Archives of Internal Medicine* 146: 69-72.

UEXKÜLL J. v. 1932: *Bedeutungslehre*. Leipzig: Barth. (engl.: Jakob von Uexküll' s The Theory of Meaning" *Semiotica* 42 (1)).

UEXKÜLL Th.v. 1980: Das Problem der Entsprechung von Rollen und Gegenrollen bei Arzt und Patient. In: JAPPE G. and NEDELMANN C.(eds.) *Zur Psychoanalyse der Objektbeziehungen* (pp. 37-73) Stuttgart: Frommann & Holzboog.

-- 1982: Semiotics and Medicine. *Semiotica* 38: 205-215.

-- 1984: Zeichen und Realität als anthroposemiotisches Problem. In: OEHLER K. (ed.) *Zeichen und Realität* (pp. 61-72) Akten des 3. semiotischen Kolloquiums Hamburg. Tübingen: Stauffenberg.

UEXKÜLL Th.v. and KÖHLE 1986: Funktionelle Syndrome in der inneren Medizin. In: UEXKÜLL Th.v. (ed.) *Psychosomatische Medizin*. München: Urban und Schwarzenberg.

WEINER H. 1987: Some unexplored Regions of Psychosomatic Medicine. *Psychotherapeutica et Psychosomatica* 47: 153-159.

WESIACK W. 1974: *Grundzüge der psychosomatischen Medizin.* München: Beck.

# The Medical Anthropology of Viktor von Weizsäcker in the Present Clinical Context in Heidelberg

Peter Hahn

## The Medical Anthropology of Viktor von Weizsäcker

Clinical medicine in Heidelberg has always had its base in anthropology and philosophy to a certain extent, even when the impact of natural science was greatest, from the middle of the 19th to the beginning of the 20th century. Thus, highly respected Heidelberg clinicians and scientists such as J. W. Conradi (1780-1861), A. Kussmaul (1822-1907), W. B. Erb (1840-1921) among others, always remained open to psychological and social questions. In 1902 L. Krehl (1861-1937), who came to Heidelberg in 1907, had already begun to analyze possible relationships between Freud's psychoanalytic theories and functional diseases. The young Viktor von Weizsäcker (1886-1957) began working in Heidelberg in 1910. Although von Weizsäcker had been trained as an experimental physiologist in Freiburg by V. Kries, he was able to secure the support of his superior, L. Krehl, for his philosophical, psychological and even theological inclinations. Soon after his arrival he found access to a Heidelberg working group from which emerged philosophers such as K. Jaspers (1883-1969), scientists such as O. Meyerhof (1884-1951) and psychoanalysts such as A. Kronfeld (1886-1941), among others (cf. Henkelmann 1985, 1986). V. von Weizsäcker's interest in anthropological and philosophical questions was deepened by the experiences he and L. Krehl had in a field hospital during World War I. As a result of these experiences, they began to search for a new direction in a then explosively developing medical science. When V. von Weizsäcker became the head of Neurology at the Medical University Clinic in 1920, his interest also shifted from physiology of the heart and internal medicine to neurology.

Together with other colleagues who later became known as psychosomatic physicians (E. Wittkower, F. Mohr and others) and especially with R. Siebeck (1883-1965), von Weizsäcker developed a special style of anthropological medicine in Heidelberg. He formulated the first hypotheses of Gestaltkreis-theory [1]. The following diagram (Fig. 1) shows the relationship between V. von Weizsäcker's different scientific, clinical and philosophical interests in a chronological manner.

After a first period of "apprenticeship" (1906-1920) [2], came a period in which von Weizsäcker developed many of his productive ideas and after 1933, a "period of crises and resignation" during which he renovated his Gestaltkreis-theory and intensified his clinical activity as a neurologist. Between 1945 and 1953 von Weizsäcker deepened his anthropological thoughts and published his original critical-philosophical ideas on clinical medicine. With regard to content, von Weizsäcker based his own thinking on three principles which have been considered the link between medicine and philosophy ever since the romantic movement (Henkelmann and Hahn 1987).

### *Anthropological orientation*

This term refers to the effort to investigate the biological-physiological, psychological and cultural determining factors of human illness without any religious or philosophical prejudices (see also Bauer 1984) [3].

**Fig. 1: Main features of von Weizsäcker's scientific work (Hahn 1987)**

| **Born in 1896** | | | |
|---|---|---|---|
| **1906 - 1920** | Physiology of the heart | Internal Medicine | Philosophy of Nature Theology |
| **1920 - 1932** | Neurophysiology | Physician/Patient Neurology Social Medicine | |
| | | | Medical Anthropology |
| | | Psychoanalysis | |
| **1932 - 1945** | Gestaltkreis | | |
| **1945 - 1953** | | General Clinical Medicine | Pathosophy |
| | **Experiment** | **Clinic** | **Philosophy** |

*Stricter observation of methodological questions*

The great variety of interests in an anthropological sense requires meticulous observation of questions related to the adequacy and extent of the respective methodological approach chosen. The scientific paradigm highly developed by Cartesianism does not satisfy the needs of physicians or anthropologists. Phenomenological, hermeneutical as well as dialectical approaches must not only be included, but also must be specifically further developed (e.g., psychoanalysis as "depth-hermeneutics", see Habermas 1973).

*Interdisciplinary approach*

Von Weizsäcker emphasized the need for interdisciplinary approaches in contrast to the requirements of highly specialized individual sciences. The understanding of the specific limited view within each individual science encourages a deepening of interdisciplinary discussion, assuming Hegel's encyclopedic representation. This requires not only the transference of the principles recognized as important for the specific field, but also a new and changed style of scientific "behavior" [4]. Viktor von Weizsäcker's work as a physician principally determined his thoughts. Anthropological medicine was for him by no means just another medical discipline, but rather a new, complete, perhaps even holistic approach [5].

V. von Weizsäcker's concept of psychology was not a mere theory of interaction or of psychophysical parallelism, but rather a mutual representation of the body and the mind with the feature of reciprocal concealment. In other words: the psyche is the expression of the body and the body the expression of the psyche [6]. The biography of an individual is characterized by an interaction between body and mind interrupted by crises. The interpretation of at which point in time and under which circumstances a symptom appears or an illness develops leads to a personal understanding of the illness [7]. Von Weizsäcker used the methodological metaphor of the so-called "revolving door principle" in order to make it possible to focus attention on one aspect of the individual (e.g., the physical-biological) or the other (e.g., the psycho-social) without losing the general context [8]. These thoughts were developed or rather integrated in the early experiments of von Weizsäcker's working group on perceptual psychology. In order to analyze Menière symptoms better, a special experimental design was developed which permitted a more precise description of the interaction between perception and movement (Vogel 1931). While perceiving a rotating movement the subject's eyes not only became nystagmic, but he attempted to maintain coherence in perceiving the object by using correcting movements. If this did not succeed, a vital crisis occurred accompanied by the corresponding symptoms (rotation vertigo). The alternation of crisis and efforts to reestablish the coherence was thought to be a boundary experience of functional change and was considered to be the reason for the development of symptoms and pain. The reality of the subject can not only be understood ontologically-scientifically (the subject as "nystagmic being"), but rather reality can be defined in pathic terms (cf. von Weizsäcker 1956: 60-86) [9].

By making use of terms like mutual representation and concealment, von Weizsäcker succeeded in establishing a link to Freud's psychoanalytic ideas [10]. However, in contrast to Freud, the body as the "carrier of the senses" is for von Weizsäcker often wiser than consciousness and is not only a source of disturbance. In his two main publications, *Der Gestaltkreis* (1947) and *Pathosophie* (1956), von Weizsäcker discussed these questions from various viewpoints.

## The Present Clinical Context in Heidelberg

These basic principles of anthropological thinking found considerable overt and concealed resistance in conventional medicine. Nonetheless, their influence reached far beyond von Weizsäcker's field of activity in Heidelberg and Breslau. Nearly all authors who published medical-psychosomatic or medical anthropologies referred to him (Gadamer and Vogel 1972, Hartmann 1973, Schipperges 1972, Seidler 1983, Rothschuh 1975, von Uexküll and Wesiack 1979, Wyss 1977, Christian 1989). Since the post-war period and since the founding of the first "Psychosomatic Clinic" in Heidelberg von Weizsäcker has been considered one of the fathers of "Psychosomatic Medicine".

At this point I would like to describe the present mission of psychosomatic medicine in our department and how we have adopted and applied V. von Weizsäcker's ideas - based on the preliminary works of other Heidelberg internists like R. Siebeck, P. Christian and H. Plügge. Three aspects, based on von Weizsäcker's three main principles, will be discussed:

### *The anthropological orientation - the role of the physician and of science*

An appreciation of the physician's role and the "image of man" on which the medical healing and caring functions are based cannot be simply accepted as a given in a certain cultural setting. Like all fundamental questions, it must be analyzed in a complete anthropological context. Although V. von Weizsäcker's thinking was still deeply rooted in the religious tradition of Christianity, he had largely freed himself from the constraints of the Church. In a continuous philosophic dialogue with Catholic and Jewish friends such as J. Wittig, F. Rosenzweig, M. Buber, and other differently-minded friends, he tried to broaden his own knowledge and understanding and to encourage mutual empathy, tolerance and acceptance.

Man is a *Mängelwesen* (Gehlen 1958) whose life ends in death (von Weizsäcker 1956). Against this background the physician appears as marginal figure. He or she exists in the contradiction of being needed and at the same time of having to make him or herself superfluous. The physician's action is a reaction to suffering. His or her healing skills are limited. The paradoxes of the physician's profession require solutions which go far beyond the strict boundaries of the profession. We have therefore formulated three definitions of the physician' s profession:

1. The situation of the sufferer or ill patient is given
2. The physician's profession is a reactive
3. The physican's profession is paradox.

Particularly the last statement requires further explanation: "... The physician's task is not only to react to a given situation, but rather must have a basic intention to make his own task superfluous" (Hahn 1988). This requires solutions which counteract the conscious or unconscious misjudgement of the physician's role: the creative, self-affirming substance of medical activity must not only be derived from the "subject" or "activity": the goal of the action is always the patient who no longer needs the physician.

Since Paracelsus at the beginning of the Modern Age and E. Liek, who lived in the first decades of the 20th century, there has been an openly discussed conflict: Should the physician see himself as a scientist or helper? Where are these motivations complementary and where are they contradictory? Is there any sort of synthesis? Or is the notion historical? Is the development to a highly specialized medical technologist inevitable if there is to be progress in overcoming illness and suffering? The old reliable Greek term *techne*, which united science and practical skill, can no longer apply to the problems of the Modern Age. Today there is a deep gap between "practical skills" and "science." There is now an immense variety of practical skills of which the individual physician can only master a fragmentary part. How should a teaching institution be oriented? In this situation a lot of courage is needed for "terrible simplifications". There are formulations which are at the same time "right" and "wrong": right as a definition of intent, as a process of understanding, as a comparison; false as a supposed accurate description of living events.

Our solution for the problem of a "scientific approach" seems to be radically "anthropological and psychological": the scientific approach should no longer be defined according to a determination of the object (formulation of a question, hypothesis), the method or an interpretation of the results, but rather according to the attitude and motivation of the individual scientist. This means that a physician acts scientifically (whether his role is that of a researcher or healer) if his thoughts and actions are always ready for openness and criticism, for rational review, correction and change, and also if he accepts the assumption that that which is known and proved is only temporary (Hahn 1988: 72).

### *Stricter observation of methodological questions - The necessity of multidimensionality and flexibility in medical science*

For the physician's knowledge and action the methodological problem is necessarily multi-dimensional, at least if illness and suffering are to be treated by adequate means. Some older Heidelberg clinicians such as L. Krehl and R. Siebeck were interested in philosophical, psychological, theological and legal questions besides their scientific research. Krehl and Siebeck, for example held an honorary doctorate of theology, Krehl also a doctorate of law. As early as the early

1920's V. von Weizsäcker tried to solve the special medical methodological problem with its various somatic, psychological and social aspects by formulating the "revolving door principle" discussed above. According to this principle, physical, psychological and social phenomena should not be strictly perceived together at the same time, but rather each aspect should be observed alternatingly. In spite of mutual concealment, coherence and constraints can be recognized.

These basic hypotheses of the Gestaltkreis-theory permitted a first methodological approach to the "body-mind conflict." But today they are not sufficient, especially when one considers newer developments such as cybernetic and systemic theoretical models. The classical difference between methods of the natural sciences and the humanities, or rather nomothetic and ideographic approaches are no longer sufficient and result in misunderstandings. According to present-day knowledge in medicine, at least four different methods which differ with regard to procedure, justification and criteria must be taken into consideration: phenomenological, empirical-analytical, hermeneutical and dialectical approaches (see also Vogt 1979). Following the examples of Gestalt-kreis-theory and the "revolving door principle," these four possible approaches can be brought into

**Fig. 2: The Circle of Methods (Hahn 1988)**

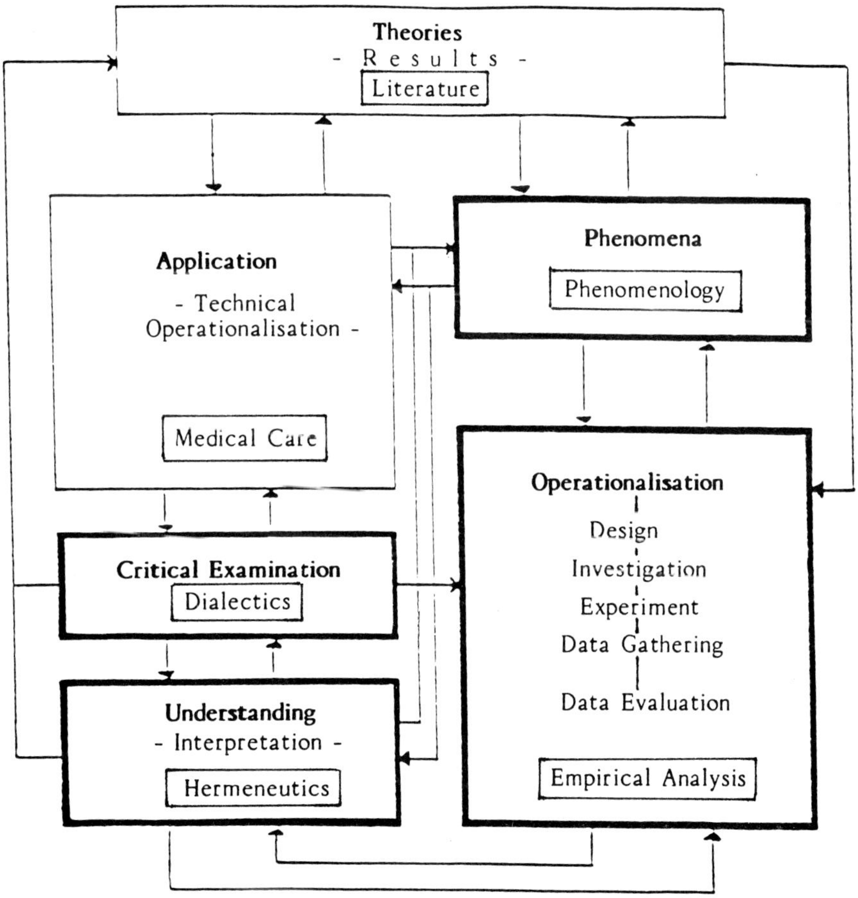

relation with each other. The result obtained - again following V. von Weizsäcker - is a circular model (with the "angular" variety of cybernetics). This model may be called a "circle of methods," in which the four methodological approaches as well as primary medical care and the results of scientific research are integrated.

When one places even more emphasis on the dynamic developmental ideas and the systemic aspects of this model, this "closed model" can be changed to an "open" one and the circle then becomes a spiral (Lolas-Stepke 1988, Klapp 1988). However, the graphic representation becomes more complicated.

Several further notions which result from the mutual interdependencies of this model are important for our clinical context: with regard to the chronological dimension, the principle of "successive simultaneousness" can be derived from the methodological interdependence, in the clinical context as so-called simultaneous diagnosis and therapy. In terms of method, it means the simultaneous perception of the three levels of content and the readiness for the most adequate direction of attention to the predominant problem area, i.e., a primarily patient-oriented procedure, in contrast to a methodologically centered approach.

With regard to the determination of method, other models which place emphasis on content and are relevant for clinical practice should not be omitted. Von Uexküll and Wesiack's "situational circle model" (1979, 1986), for example, which is characterized by its precise relationship with the point of departure of "here and now" can be understood by the phenomenological method of two-fold reduction.

**Fig. 3: The Situational Circle (von Uexküll and Wesiack 1979)**

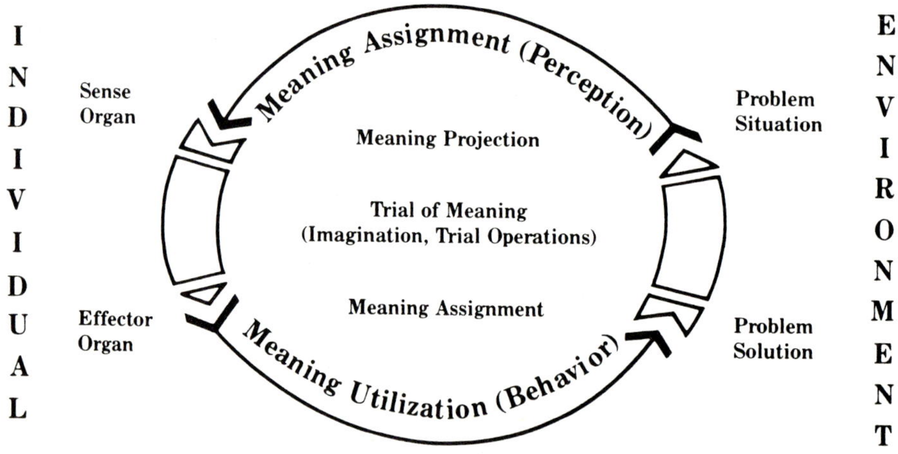

*Interdisciplinary requirements - "Scientific Behavior" and the problem of organizational forms*

In interdisciplinary work we must drop the old academic rivalry about the correct opinion or knowledge; rather the aim must be to become sensitive to the basic principles, ideas and methods of other academic disciplines so as to understand and eventually integrate them into one's own scientific thinking. At first this may appear an obstacle to the individual academic disciplines because attention is distracted - at least momentarily - from increasing one's knowledge in his or her own academic discipline. However, for medical science, whose object is "recognizing, evaluating and treating ill people" [11], it is an immanently necessary condition for adequate anthropological knowledge and activity.

The practical realization of interdisciplinary work is embodied in the various organizational forms in general and clinical medicine. Interdisciplinary work is practiced where empirical-analytical methods are used. Because of being used to clinical routine, physicians are generally not even conscious of working in an interdisciplinary manner when, for example, clinical-practical, radiological and biochemical results are considered together. Phenomenological and hermeneutical methods as well as dialectic approaches are, however, not so readily accepted; the hardliners who maintain the traditional scientific paradigms declare them to be "unscientific". The institutionalization of interdisciplinary work depends on the most varied factors. Viktor von Weizsäcker, for example, as the son of the former Prime Minister of the state of Baden-Württemberg, as a student and friend of Krehl and Siebeck, could depend on rather different support than G .R. Heyer in Munich, who worked in Friedrich von Müller's clinic without such connections. Political considerations, individual interests and sympathy or antipathy among colleagues continue to influence the realization of "adequate" or "ideal" concepts.

The unusual development of the present Department of Internal Medicine II (General Clinical and Psychosomatic Medicine) in Heidelberg within the Medical University Clinic, the third generation of a Department of Internal Medicine which sees itself as anthropologically based, results from the specific traditions of the University of Heidelberg and its medical departments among the departments of Philosophy, Romance Languages, and *Bildungsbürgertum* (educated classes) on the one hand and the inclination toward visions of the future and technical perfection on the other. Krehl's conflict between the "personal" approach of his medical thinking and action and an objectively subtile pathophysiology has remained until the present day and can still be seen in disputes among colleagues in our department.

The organization for patient care and intern training is a "Three Stage Model" of clinical psychosomatics, which has been used successfully with only minor variations since 1976. This model is based on the work of our predecessors: on the experiences of von Weizsäcker's "neurosis ward" in the 1920's, on Siebeck's "biological penetration" in Internal Medicine in the 1930's and 40's, on von Weizsäcker's and Kütemeyer's "anthropological ward" in the 1950's, and on Christian's "Institute for General Clinical Medicine" in the 1960's and 70's, which was oriented toward internal medicine.

*Therapeutic concepts*

Although von Weizsäcker was often extraordinarily successful both personally and as a physician, he always worked as a therapist - as an internist, neurologist, in rehabilitation medicine and as a psychotherapist [12]. However, he left behind no definite therapeutical concept, only some very general rules for biographical diagnostics [13], some practical comments on psychotherapy and psychoanalysis as well as remarks on how society sees the physician (cf. Hahn 1988). As a result, his co-workers often felt insecure about the problem of increasing psychotherapeutic knowledge and skills. The basic anthropological approach was of a relatively general, scientific-theoretical, cultural-anthropological nature, and could not offer a framework for the solution of therapeutic and psychosomatic problems [14].

**Fig. 4: The Heidelberg Three Stage Model of Clinical Psychosomatics**

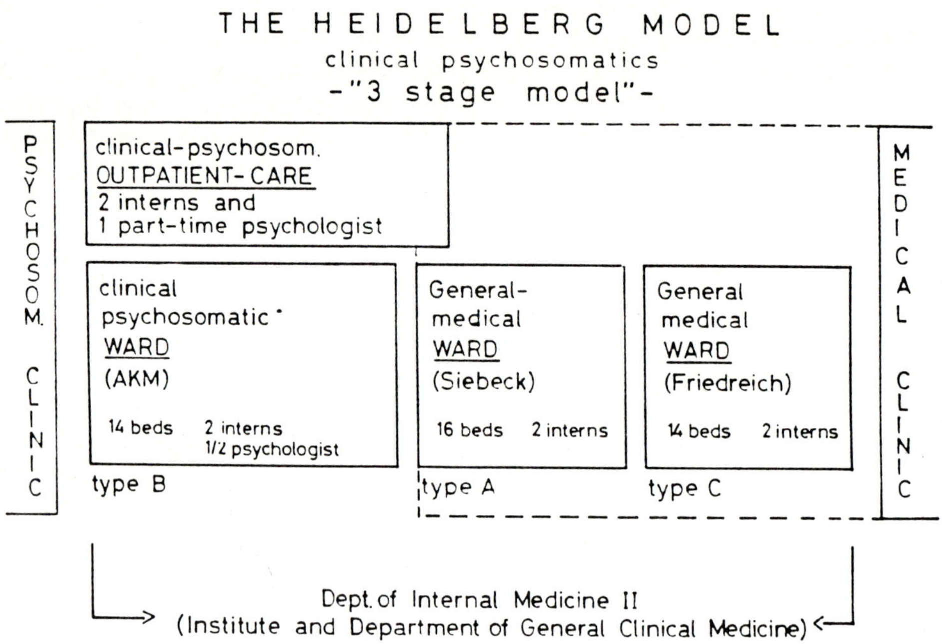


Our department, which sees itself as von Weizsäcker's legacy, has a variety of ways to train physicians as psychotherapists or psychoanalysts. These procedures, based on needs of practice and content, are largely identical with those used in diagnostic and therapeutic basic practice. Among these procedures is the introductory diagnosis, during which the colleagues in our department pay particular attention to their first impression of the patient, possibly their only "holistic" view of him or her. During the introductory diagnosis the physician takes down a complete history of the patient, which is a biographical, behavioral, and psychoanalytic analysis of the patient's living situation, illness and family. In-patient therapy in both general-internal wards includes further psycho-social diagnostics either by a colleague in the ward or a psychosomatic "liaison physician," extended conversation with the patient during daily rounds, conversations with the family, or the beginning of relaxation therapy or in-depth psychotherapy. For the staff team there is a weekly Balint session.

Initially, the patient is often not aware that the ward physician is a psychotherapist at the same time, which is often an advantage for us, since the patient might be prejudiced against psychotherapy. The physician often has difficulty evaluating the significance of his various roles, particularly at the beginning of his career. He or she must integrate the methodological approaches discussed above in his training as naturally as other methodological approaches such as clinical diagnostics and therapy.

## Summary and Final Remarks

In the further development of the anthropological and psychosomatic concepts of von Weizsäcker's Gestaltkreis-theory and Pathosophie our department has developed a diagnostic and therapeutic concept by means of which anthropological and clinical-psychosomatic viewpoints have been successfully integrated into internal medicine for over ten years. We have been able to find effective solutions for patient care as well as instruction and research and the training of staff members.

## Notes

[1] V. von Weizsäcker's Gestalt-theory should not be confused with the Berlin School's Gestalt-psychology (Wertheimer 1912, Köhler 1924, Metzger 1954 et al.) nor with the Gestalt-therapy of F. Perls and his co-workers (1974).

[2] As Th. Henkelmann described and characterized at an exhibition commemorating the 100th anniversary of V. von Weizsäcker's birth (1986) (cf. Hahn and Jacob 1987).

[3] The first use of the term *anthropo-logia* is found in M. Hunt (1501): man is the object of scientific research. The further history of this term takes us from authors like O. Cassmann (1594), an astronomer, physicist and theologian to E. Platner (1744-1818), a physician who designed an Anthropology for Physicians and World Journeys (1772, 1791) (cf. Bauer 1984).

[4] "Behavior" is for von Weizsäcker an anthropological term which deals with the autonomy of bipersonal (dyadic) and group dynamic processes while communicating or exchanging information of content. In a sense it anticipates the difference in communication theory between the levels of content and relationship (Watzlawick 1972).

[5] V. von Weizsäcker did not like the expression "holistic". He protested against the use: of the word *Ganzheit* 'wholeness', which for him is an idea not realizable scientifically. His methodological principle of "mutual concealment" of soma and psyche leads necessarily to *Teilheit* 'partial aspect' of that which is perceived.

[6] "Nichts Seelisches hat keinen Leib, nichts Organisches hat keinen Sinn." (There is no mind without body and no organic matter without sense) (Quoted in Jacob 1987).

[7] V. von Weizsäcker formulated three guiding questions which refer to the relationship between the events in a patient's life and illness to begin a patient's "biography": "Why just now?" "Why right here?" "Why just in this way?" (cf. von Weizsäcker 1956: 241 ff.)

[8] V. von Weizsäcker defined the methodological "revolving door principle" as follows: "Each act is perception and movement. But in perceiving one cannot perceive the enabling movement and in moving one cannot execute the conditioning perception. Therefore, moving is a not-perceiving-it, and perceiving is a not-moving-it. Both are in a relationship of mutual concealment." (von Weizsäcker 1947: 206). We use a didactic representation of this concept in our Department with a triangular field containing the notions "psycho", "socio" and "somato" ("bio"):

Fig. 5

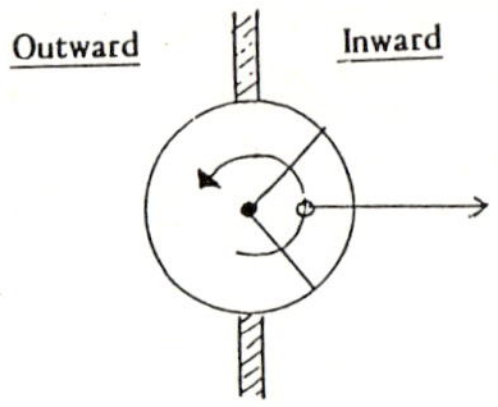


Fig. 6

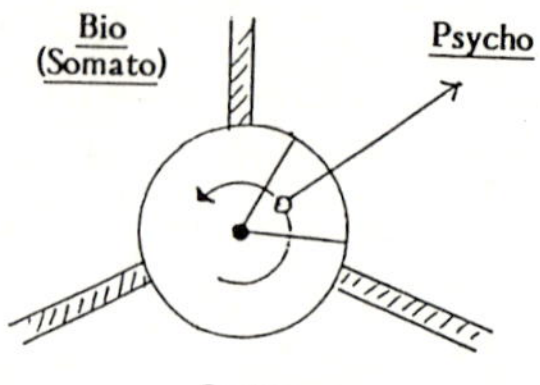


[9] The "pathische Kategorien" ("pathic categories") are: DÜRFEN (being allowed to do something), MÜSSEN (being obliged to do something), WOLLEN (wishing to do something), SOLLEN (ought to do something), and KÖNNEN (able to do something) (v. Weizsäcker 1956: 60-86).

[10] Von Weizsäcker describes his meeting with Freud (Nov. 1926 in Vienna) in "Natur und Geist" (*Gesammelte Schriften*, Vol. I, pp. 137-148). A synopsis of von Weizsäcker's attitude toward Freud's psychoanalysis can be found in Hahn (1987).

[11] Based on Krehl's and Siebeck's definitions, we define the "object" of medical science as "the recognition, evaluation and treatment of ill people" (Hahn 1988: 76).

[12] A further illustration of how V. von Weizsäcker applied psychotherapy in individual treatment is the description of "Case A" in "Körpergeschehen und Neurose" 1932 (Gesammelte Schriften, Vol. 6, pp. 121-251). For a view of his basic ideas on social rehabilitation, the "situational therapy" described in *Gesammelte Schriften*, Vol. 8, pp. 46-52 can be suggested. For a description of his personal style of deportment with patients, see M. Fuchs' report from 1939 in Buchheim et al. (1987: 23 ff.)

[13] See note 8.

[14] These problems were attempted to be solved by an approach using an "anthropological direction of psychotherapy" in the 1950's and 60's, until the dispute over the necessity of the training analysis (Bräutigam 1961, Gebsattel 1959, Sborowitz 1960, Wiesenhütter 1969, Zutt et al. 1963). A summary can be found in Hahn (1968).

## References

BAUER A. 1984: Bemerkungen zur Verwendung des Terminus "Anthropologie" in der Medizin der Neuzeit (16.-19. Jahrhundert). In: SEIDLER E. (ed.) *Medizinische Anthropologie.* Berlin, Heidelberg, New York, Tokio: Springer.

BRÄUTIGAM W. 1961: Psychotherapie in anthropologischer Sicht. In: WIESENHÜTTER E: (ed.) *Beiträge aus der Allgemeinen Medizin*, Vol. 15. Stuttgart: Enke.

BUBER M. 1954: *Die Schriften über das dialogische Prinzip.* Heidelberg: Lambert Schneider.

BUCHHEIM P., HAHN P. and SEIFERT Th. 1987: *Psychoanalyse und Psychotherapie in Vergangenheit und Gegenwart. In Erinnerung an den 100. Geburtstag Viktor von Weizsäckers im Jahre 1986*. Berlin, Heidelberg, New York, Toronto: Springer (=Sonderheft für die Lindauer Psychotherapiewochen 1987)

CHRISTIAN P. and HAAS R. 1969: Medizinische und philosophische Anthropologie. In: ALTMANN H.W. et al. (eds.) *Handbuch der allgemeinen Pathologie* Vol. 1 (pp. 232-278) Berlin, Heidelberg, New York: Springer.

CHRISTIAN P. 1989: *Anthropologische Medizin. Theoretische Pathologie und Klinik psychosomatischer Krankheitsbilder*. Berlin, Heidelberg, New York, London, Paris, Tokyo: Springer.

CONRADI J. and HEINRICH W. 1815: *Grundriß der medizinischen Enzyklopädie und Methodologie*. Zum Gebrauch bei seinen Vorlesungen entworfen von ... Prof. der Medizin zu Heidelberg. 2. ed. Marburg: Joh.Chr. Krieger.

DETER H.C. 1979: cited in HAHN (1980).

EICH W. and HAHN P. 1987: Viktor von Weizsäcker (1886-1957) and German Psychosomatic Medicine. In: CHRISTODOLOU G.N. (ed.) *Psychosomatic Medicine - Past and future*. New York, London: Plenum Press.

ERB W. 1910: *Gesammelte Abhandlungen 1864-1910*. 2 vol.

FAHRENBERG J. 1979: Das Komplementaritätsprinzip in der psychophysiologischen Forschung und psychosomatischen Medizin. *Zeitschrift für Psychologische Medizin* 27: 151-167.

FREUD S. 1941: *Gesammelte Werke Vol. 1-18*. London: Imago Publ. Co.

GADAMER H.G.1969: *Wahrheit und Methode*. Tübingen: Mohr.

-- 1967: *Apologie der Heilkunst. Kleine Schriften*. Tübingen: Mohr.

GADAMER H.G. and VOGLER P. (eds.) 1972: *Neue Anthropologie* Vol. 1-7. Stuttgart: Thieme.

GEBSATTEL V.E. von 1959: *Gedanken zu einer anthropologischen Psychotherapie*. Hbch. Neur. u. Psychother. III: 531-567. München: Urban und Schwarzenberg.

GEHLEN A. 1958: *Der Mensch, seine Natur und seine Stellung in der Welt*. 6. ed. Bonn: Athenäum.

HABERMAS J. 1973: *Erkenntnis und Interesse*. Frankfurt: Suhrkamp.

HAHN P. 1968: Die anthropologische (personale) Richtung in der Psychotherapie. *Hippokrates* 39: 65-69.

-- 1977: Die Entwicklung der psychosomatischen Medizin. In: *Enzyklopädie des 20. Jahrhunderts* Vol. 1: 932-952.

-- (ed.) 1979: *Die Psychologie des XX. Jahrhunderts* Vol IX: Ergebnisse für die Medizin (1). München: Kindler. Munich Newprint (1983) in 2 vols. Weinheim, Basel: Beltz.

-- 1980: Allgemeine Klinische und Psychosomatische Medizin. Entwicklung und Standort. *Heidelberger Jahrbücher* Vol. XXIV. Berlin, Heidelberg, New York, Tokyo: Springer.

-- 1988: *Ärztliche Propädeutik. Gespräch, Anamnese, Interview - Einführung in die anthropologische Medizin*. Berlin, Heidelberg, New York, London, Paris, Tokyo: Springer.

HAHN P., MAYER J. and STANEK B. 1973: Biometrische Befunde bei der Herzneurose. *Zeitschrift für Psychosomatische Medizin und Psychoanalyse* 19: 231-264.

HAHN P. and JACOB W. (eds.) 1987: *Viktor v. Weizsäcker zum 100. Geburtstag. Beiträge zum Symposium der Universität Heidelberg 1.-3.5.86*. Berlin, Heidelberg, New York, Tokyo: Springer.

HARTMANN F. 1973: *Ärztliche Anthropologie. Das Problem des Menschen in der Medizin der Neuzeit*. Bremen: Schünemann-Universitätsverlag.

-- 1984: *Patient, Arzt und Medizin. Beiträge zur ärztlichen Anthropologie*. Göttingen: Verlag für Medizinische Psychologie im Verlag Vandenhoeck & Ruprecht.

HENKELMANN Th. 1985: Zur Geschichte der Heidelberger Medizinischen Klinik im 19. Jahrhundert. *Heidelberger Jahrbücher* 29: 153-174.

-- 1986: *Viktor von Weizsäcker 1886-1957. Materialien zu Leben und Werk*. Berlin, Heidelberg, New York, Tokyo: Springer.

HENKELMANN Th. and HAHN P. 1987: Historical Insights of Psychosomatic Medicine in Heidelberg. In: CHRISTODOULOU G.N. (ed.) *Psychosomatic Medicine - Past and future*. New York, London: Plenum Press.

HEYER G.R. 1922: Psychische Faktoren bei organischen Krankheiten. *Münchner Medizinische Wochenschrift* 69: 1241.

JACOB W. 1967: *Medizinische Anthropologie im 19. Jahrhundert. Mensch - Natur - Gesellschaft. Beiträge zu einer theoretischen Pathologie*. Stuttgart: Enke.

JACOB W. 1978: *Kranksein und Krankheit. Anthropologische Grundlagen einer Theorie der Medizin.* Heidelberg: Hüthig.

JACOB W. 1987: Anthropologische Medizin. In: HAHN P. and JACOB W. (eds.) *Viktor v. Weizsäcker zum 100. Geburtstag*. Heidelberg, New York, London, Paris, Tokyo: Springer.

KÖHLER W. 1947: *Gestalt Psychology*. New York: Liveright.

KREHL L. 1902: Über die Entwicklung hysterischer Erscheinungen. In: BERGMANN E. von and WINCKEL F. (eds.) *Sammlung klinischer Vorträge* 330: 728-744 Leipzig: Breitkopf Härtel.

KREHL L. 1930: *Entstehung, Erkennung und Behandlung innerer Krankheiten*. 13. ed. Leipzig.

KLAPP B.E. 1988: Personal communication.

KÜTEMEYER W. 1947: Wandlungen medizinischer Anthropologie. In: *Beiträge aus der Allgemeinen Medizin.* Stuttgart: Enke.

KUßMAUL A. 1899. *Jugenderinnerungen eines alten Arztes.* Stuttgart: Bonz.

-- 1986: Seine aktuelle Bedeutung für die innere Medizin und Neurologie. In: KLUG E. (ed.) Stuttgart: Hippokrates Verlag.

LIEK E. 1927: *Der Arzt und seine Sendung. Gedanken eines Ketzers.* München: J.F. Lehmanns.

LOLAS-STEPKE F. 1988. Personal communication.

METZGER W. 1954: *Psychologie*. Darmstadt: Steinkopff.

PERLS F.S. 1974: *Gestalttherapie in Aktion*. Stuttgart.

PETZOLD E.; LUBAN-PLOZZA B.; MATTERN H. and BERGMANN G. (eds.) 1987: *Brücken von der Psychosomatik zur Allgemeinmedizin.* Berlin, Heidelberg, New York, Tokyo: Springer.

PETZOLD E.; HERZOG W. and KRÖGER F. 1988: Der simultandiagnostische Würfel und einige Überlegungen zur Familientherapie. In: *Die Entzauberung des Zauberbergs.* Dortmund. Verlag Modernes Lernen.

PLATNER E. 1772 (1791): *Neue Anthropologie für Ärzte und Weltweise*. Leipzig.

PLÜGGE H. 1957: Zur Phänomenologie des Leib-Erlebens, besonders bei inneren Krankheiten. *Jahrbuch für Psychologie und Psychotherapie* 5: 155-168.

PLÜGGE H. 1962: *Wohlbefinden und Mißbefinden. Eine phänomenologische Studie*. Tübingen: Niemeyer.

RAD M. von 1974: *Anthropologie als Thema von psychosomatischer Medizin und Theologie*. Stuttgart: Kohlhammer.

REINDELL A., PETZOLD E., DETER C., STINDL E. and HAHN P. 1977: Simultandiagnostik und Simultantherapie auf einer klinisch-psychosomatischen Station. *Zeitschrift für Psychosomatische Medizin* 23: 387-396.

ROSENZWEIG F. (ed.) 1926-28: *Die Kreatur*. Berlin.

ROTHSCHUH K.E. 1975: *Was ist Krankheit?* (= Wege der Forschung 362). Darmstadt: Wissenschaftliche Buchgesellschaft.

SBOROWITZ A. and MICHEL E. 1960: *Der leidende Mensch. Personale Psychotherapie in anthropologischer Sicht.* Düsseldorf.

SCHIPPERGES H. 1971: *Ausbildung zum Arzt von morgen*. Stuttgart: Thieme.

SCHIPPERGES H. 1972: Anthropologie in der Geschichte der Medizin. In: GADAMER H.G. (ed) *Biologische Anthropologie* Vol II. Stuttgart: Thieme.

SEIDLER E. 1983: *Medizinische Anthropologie*. Berlin, Heidelberg, New York, Tokyo: Springer.

SIEBECK R 1953: *Medizin in Bewegung*. Stuttgart: Thieme.

THOMÄ H. and KÄCHELE H. 1973: Wissenschaftstheoretische und methodologische Probleme der klinisch-psychoanalytischen Forschung. *Psyche* 27 (3/4): 205-236, 309-355.

UEXKÜLL Th. von 1984: Zeichen und Realität als anthroposemiotisches Problem. In: OEHLER K. (ed.) *Zeichen und Realität*. Tübingen: Stauffenberg.

UEXKÜLL Th. von et al. (eds.) 1986: *Psychosomatische Medizin*, 3. ed. München: Urban & Schwarzenberg.

UEXKÜLL Th. von and WESIACK W. 1988: *Theorie der Humanmedizin*. München: Urban & Schwarzenberg.

VOGEL P 1931: Über den Drehschwindel. *Pflügers Archiv* 228: 510, 632.

VOGT R. 1979: Wissenschaftstheoretische Leitlinien in ihrer Bedeutung für die psychosomatische Medizin. In: HAHN P. (ed.) *Enzyklopädie des XX Jahrhunderts* (Vol. IX) München, Zürich: Kindler.

WATZLAWICK P., BEAVIN J.H. and JACKSON D.D. 1972: *Menschliche Kommunikation*. Bern: Huber.

WEIZSÄCKER C.F. von 1956: Gestaltkreis und Komplementarität. In: VOGEL P. (ed) *Weizsäcker V. von - Arzt im Irrsaal der Zeit* (pp. 21-53) Göttingen: Vandenhoeck & Ruprecht.

WEIZSÄCKER V. von 1947: *Der Gestaltkreis. Theorie der Einheit von Wahrnehmungen und Bewegungen*. Stuttgart: Thieme.

-- 1956: *Pathosophie*. Göttingen: Vandenhoek & Ruprecht.

-- 1986: *Gesammelte Schriften* (eds. ACHILLES T., JANZ D., SCHRENK A. and WEIZSÄCKER C.F. von) Vol. 1-9 Frankfurt: Suhrkamp.

WERTHEIMER M. 1912: Experimentelle Studien über das Sehen von Bewegung. *Zeitschrift für Psychologie* 61.

WIESENHÜTTER E. 1960: *Einführung in die Neurosenlehre*. Stuttgart: Hippokrates.

WITTIG J. (ed.) 1926-1928: *Die Kreatur*. Berlin.

WYSS D. 1977: *Die tiefenpsychologischen Schulen von den Anfängen bis zur Gegenwart*. Göttingen: Vandenhoek & Ruprecht.

ZUTT J. 1963: Über verstehende Anthropologie. In: *Psychiatrie der Gegenwart* I/2 (pp.763-752). Heidelberg, Berlin, New York: Springer.

# Ethnomedical Fundaments in the Work of Viktor von Weizsäcker

Wolfgang Jacob

The general topic "Ethnomedical fundaments in the work of Viktor von Weizsäcker" might create the impression of Weizsäcker's work to be founded on concepts so specific as to serve as an immediate basis for ethno-anthropological research. But Weizsäcker has in fact not expressedly spoken of ethnomedical fundaments or tried to substantiate such phenomena.

In the following I shall concentrate upon a few predominant and pragmatical perspectives, i.e. upon some basic concepts and propositions of Weizsäcker's anthropological medicine. Thus my actual topic will read "Some basic concepts in Victor von Weizsäcker's anthropological medicine and their significance for ethnomedical research".

Peter Hahn has already shown the relation between Weizsäcker's anthropological medicine and psychosomatic clinical work. His subject could be enlarged by including not only more recent developments in American psychosomatics, but also Thure von Uexküll's and Wolfgang Wesiack's theory of Human Medicine. Within the field of ethnomedicine Weizsäcker's anthropological medicine is as yet little known, it might however prove important. The fact that it cannot be easily integrated into current scientific medical concepts needs extra mentioning for the following reasons:

1) Scientific medicine tends to develop ways of thinking comparable to those in modern physics.

2) The fascinating successes of scientific medical research and their enormous variety of operational possibilities have lead to a more and more limited view of "the actual situation of the sick person". It is this "actual situation of the sick person", however, upon which Weizsäcker's anthropological medicine is focused (Jacob 1978).

Let me give an example from everyday practice: A seven year old boy contracts scarlet fever and lice simultaneously; he brings both home from school. The scarlet fever is treated - as is usual today - with antibiotics and soon subsides. Three or four weeks later there is a recurrence among the school children - a well-known phenomenon. The scarlet fever recurs not only once but several times, and one does not hesitate to treat each recurrence as usual with antibiotics. Nor are there any qualms about totally ignoring the time-honoured rules of experience with the contagious nature of scarlet fever in terms of epidemic hygiene: in spite of the fact that the disease must be reported to the authorities, there is no health authority actively informing the public or, if necessary, taking dirigistic protective measures. The inevitable occurs: the extremely sensitive child contracts pneumonia not once but twice, reaching a state of near collapse which the mother correctly considers alarming. She calls for the pediatrician who comes once, because the boy shows a high temperature. She does the usual laboratory tests, sees the child in her practice after the fever has subsided, ascertains a cough, auscultates the chest, runs a blood test, and prescribes again antibiotics against the cough. Later on she calls the mother to tell her she need no longer administer the antibiotics because the blood test is normal and the blood sedimentation rate near the upper limit for the norm.

Had one taken the child's biography into account, as a physician one would have had to proceed quite differently. One would have been aware how highly sensitive this extraordinarily gifted boy reacted not only to mental and spiritual insults, but also to physical ones, so that for the course of the illness which he had to suffer and to endure the rules of formal somatic medicine do not suffice, they do not even apply. On the occasion of her first visit, the doctor should at least have realized that the little patient was in need of more frequent visits at his sick-bed. She should have been aware of his extremely sensitive and intelligent reactivity, i.e. this child's fundamental social condition responsible for the somatic course of the disease, for the recurrences and finally for the serious collapse symptoms.

This child is greatly appreciated by the other children at school for his sensitivity and giftedness: however, the ambivalent forces of social chaos raging within the school have found a peculiar form of expression towards this child, in that it splits friend and foe - although it would like to be friends with everyone: It polarizes the class, without realizing it. The signs that at least one party would like to destroy his sensitive nature are on the increase, culminating in the blunt announcement: "we are going to kill you", followed by the direct, frequently repeated action of suddenly pushing the child in the chest and making him fall. He is prepared to endure this procedure without complaint. At least he would not betray any of the ill-doers to an authority. Yet even from above, i.e. in terms of school supervision, the child remains unprotected. The teacher asks the members of the group threatening him: "Do you mean it seriously or just in fun?". When they claim that it is just fun, he simply turns away with the remark: "Well, it can't be too bad then".

At home the child has bad dreams about the terror at school: he screams in the night and during the acute phase of his sickness he has the convincing feeling that he must die.

The mother feels that the child is already beyond the limits of what he can tolerate and intends to keep him away from school to strengthen him as long as the scarlet fever recurrences are still going on. However, the doctor who had successfully treated the pneumonia with Euromycin announces the results of the blood tests to the mother with the remark: "Well, now he can soon go back to school".

I would like to finish this account by stating that this child, in spite of his young age, bears a quite considerable gift for poetry and acting, which - when he is feeling good - he uses freely and without prior reflection. Some of his poems show an astonishing degree of maturity and expressiveness, like the following:

*Heidelberg calls out again*
*while I compose songs of farewell.*
*I turn and wave, once more,*
*fighting the torment in my soul*
*and even in foreign lands*
*to which I may turn*
*from that old town will never depart*
*the piece it holds of my heart.*
*So much have I seen in you,*
*Never do I wish to leave, and yet,*
*farewell beautiful town!*

The special gifts of the child require special protection. However, I feel this necessary protection should not be limited to this particular child, but extended to all sick and sensitive children who need it, indeed not only to sick children but to all sick, old and frail people who require it to the largest extent possible.

In view of the case just presented, let us now look at some of the basic concepts in Viktor von Weizsäcker's anthropological medicine. First there are the "pathic categories" i.e. *to will, to be able to do, to be permitted to do, to be obliged to do, to be forced to do* (von Weizsäcker 1950: 183ff.). These pathic categories are dependent of and related with each other. We know which conflicts are bound to arise if we wish to do something but do not have the faculty to achieve it; or if we *are forced* to do something which we by no means *wish* to do. Due to their social dynamics - which would require a special discussion - the pathic categories are assigned to sociological dimensions. Initially we look at the pathic categories - which are based, as the term indicates, on a passionate intensity of emotion - merely intrapsychically, i.e. in terms of the meaning they have for the sick individual. But soon we realize that it does not suffice to look at the pathic categories just intrapsychically: they must be reflected on a larger scale, namely be extended to the social dimension.

There is not the least doubt that the pathic categories are of greatest importance for psychosomatic medicine. With regard to depth-psychologically oriented psychosomatic medicine, Peter Hahn has already mentioned in his contribution that they are associated or fundamentally related with the so-called 'unconscious' in man.

For ethnomedical research, these pathic categories might not only serve as a fundament, they are to be considered indispensable. They reveal coherences in the ways of man's thinking and acting which would remain inexplicable as cultural achievements unless we were willing to consider and investigate the anthropological-archaic interaction of these categories in the different cultural areas! The depth-psychological dimension and the discovery of the "unconscious" which we owe to S. Freud are insofar indispensable as the pathic categories and their dynamic interaction would be hardly understandable without them. This is especially true for the psycho-dynamic interpenetration of the pathic pentagram - i.e. the relatedness and dynamic interaction of "willing", "being able to do", "being permitted to do", "being obliged to do", and "being forced to do" (von Weizsäcker 1956: 61f.). We simply cannot escape the "passionate character" of the pathic categories, not even when subjecting them to scientific abstraction.

Another basic concept in Weizsäcker's anthropological medicine is that of the unity of perceiving and moving, derived from the Gestaltkreis (gestalt sphere) - research. In "Gestaltkreis" Weizsäcker speaks of "Prolepsis", in his late work of a "proleptic structure of biography" (von Weizsäcker 1956: 257f.).

Using once more the example of the formative process in intentional movement, the term "prolepsis" indicates that something which lies temporarily ahead becomes real in the process of action.

In intentional movement on an uneven plane, for example on a mountain path, or when a shepherd boy runs after an animal over hedge and ditch, one has neither the time nor the possibility to gauge or calculate even in a kind of geometrical flash-procedure where exactly to put one's foot when perceiving a stone. Rather, in taking the next step the "ground-relation" sets in as an anticipated unity of perceiving-movement. And this is exactly why I can take a next step without stumbling or falling but also without being able to figure it out beforehand.

This is but one example for the process of "prolepsis". Another even more mysterious process is that of the eye when it pursues two points of light moving in a darkfield and connects their movement. Under certain experimental conditions they appear as *circles* when they are directed towards one another with different velocities, although their actual movement is not circular. The eye perceives their movement as if they obeyed a natural law. Thus, as Weizsäcker puts it, in sensual perception one is bound to allow for an anticipation of "Kepler's laws" (von Weizsäcker 1950: 12).

Weizsäcker has mentioned a third example of "prolepsis" to which I have already briefly referred. He speaks of the proleptic structure of reality by which he ultimately means the proleptic structure of biography. This can however only be demonstrated in the light of concrete examples. What he aims at is elucidated by two sentences from chapters in his last work "Pathosophie":

"Dreams, prognoses, explanations, also an understanding that there is something concealed in a disease which but represents it, yet is a bad enough substitute; even the purging effect of suffering - inasmuch it alludes to future events - is no value in itself, at best a condition for other values which however need not be higher ones. So here we have a parity of proleptic interpretation, and nothing but parity." (von Weizsäcker 1956: 257)

Weizsäcker's statement that man as a suffering and sick being is on an inner way which furnishes the "presuppositions for other values" within the reality of life might be of importance for the understanding of some ethno-anthropological relations.

Weizsäcker continues: "Yet the irreality of the future as well as the reality of dreams. thoughts, prophecies reinforce one another and belong (paritetically) together. Although not in such a way that only something which appeared proleptically within the psyche would also come to reality; this seemed to happen at times, and such semblance calls to attention the proleptic structure as such, nothing else."

This "insight into the proleptic structure", however, is based upon something different, namely an area of life which appears to be of greater impact and easier to grasp for ethno-anthropological research than for Western scientific medicine: Weizsäcker calls it the category of the "impossible" in its significance for the reality of the healthy and the sick. "One can say the following: if the impossible becomes real one could also say that the impossible becomes possible ... the issue intended appeared impossible being erroneously projected upon a concept of reality which later proved wrong"; and "the impossible became possible inasmuch as it was understood that "the possible" looks different from what was earlier called impossible or possible ... Namely that the impossible is really possible" (von Weizsäcker 1956: 257).

The *unity of perceiving and moving* - the subtitle of "Gestaltkreis" - can now be acknowledged more clearly as a fundamental category of ethno-anthropological research!

Starting from notions in psycho-physical research, one used to consider emotional movements of expression as relative and at the same time as objective by interpreting them only psychologically. In view of the "unity of perceiving and moving" the psychological interpretation recedes more and more into the background. "Expression" means "existence", man's "being in existence" as such!

What Weizsäcker actually attempted in his anthropological medicine is now more distinctly discernible: form itself and the process of formation in perception and movement are not merely essential properties of every living being; moreover, we have to realize that the formative process of performing a purposeful, willful movement is not confined to the purposeful direction of the movement. More than that: it designs man's existence, the very reality of life appointed to him. It is a *unity* which cannot be described or depicted as a result of causal psycho-physical investigations. It is an original, *ever newly occurring* unity in nature and in human existence. It shapes our human and social reality incessantly anew, that is, it "expresses" it.

Let us finally consider the range of two more basic concepts in Weizsäcker's anthropological medicine: "Logophanie" and "Eidologie" (von Weizsäcker 1956: 158ff.). For classical psychosomatic research the introduction of these two concepts might signify a breaking of new ground.

As soon as we encounter transcultural phenomena in medicine - above all in ethnomedicine - there is a realm of reality which can only be termed as "origin" or "creation". "Logophanie" and "Eidologie" denote formative processes of this origin in the anthropological field.

Even pathology is a creative event! What is meant by this? First of all that "the nature of passions can be defective and morbid and thus pathogenous in effect." In order to be able to explain pathological effects in the future one has, according to Weizsäcker, to investigate the nature of logophane and eidological processes. It is not at all sufficient to explain the "affect", the most simple, usually socially intended emotion of man, in terms of the causal determinism of natural sciences. It is what results from "affect" which counts: "victory" or "defeat" are the decisive criteria for man's own and social actions proceeding from passion. The more trivial the explanation of such actions the more immediate is the effectiveness of "psychic states" "which either accompany causal thinking or replace it". Weizsäcker continues: "Such states are amongst others boredom, curiosity, envy and all forms of cunning".

He observes that the seemingly more intellectual actions "are still thoroughly tinted by emotions", as for example by mistrust, suspicion, blind trust or hope. One talks of "mystery", misfortune" or "luck" instead of analyzing one's own behavior intellectually. In human and social experience legend, myth and fable take the place of an intellectual analysis of the texture of reality.

The concept of "Logophanie" thus describes an experience of reality which can literally be destroyed through "causal explanation".

If one wishes to understand the ethno-anthropological reality of man's vital expressions, it must be mentioned here that the concept of "logophanie" can, may, must also be applied to the ways of experiencing time. In causal-analytical thinking, confined to the realm of natural sciences, the mathematical, i.e. homogenous concept of time is prevailing. This concept of time can be expanded, however. We then have to distinguish qualitatively between a physical, a biological, a historical and a psychological time, each with a "qualitatively different" past, present and future (von Weizsäcker 1956: 184f.). Weizsäcker even claims that reducing these different concepts of time to a single homogenous time-component would prove to be "an inadequate constraint", a sort of "violating restriction" (von Weizsäcker 1956: 186).

The concept of "Eidologie" may be considered the counterpart of "Logophanie". While the concept of "Logophanie" doubtlessly contains a destructive element which brings about chaos in man's thinking and consequently also in his deeds, the concept of "Eidologie" represents "a striving of eidos towards logos", the striving of "sensuality on its way to logical order" (von Weizsäcker 1956: 201).

In this context, the formative processes in language, in art, in philosophy or in movement appear to be of equal rank with the "logical order" represented by the natural sciences. Simultaneously with the disputes of scientific-logical discussion we encounter the forms of human deportment as a kind of "appeasement symptoms", e.g. in table manners, politeness and in the rules of social life "among nations, peoples, spouses, citizens etc." (von Weizsäcker 1956: 202).

Thus "Eidologie" means that in sensual perception as well as in man's social actions the formative organization of being prevails, "breaks through", so-to-speak. It is only thus that we can understand music, rhythm, language and grammar, that we can conceive of a "coherence of meaning".

One last question: Can there be anything more "meaningful" in ethnology - and thus in ethnomedicine - than the so-called "hermeneutics of meaning"?!

If not based on the awareness of the "unity of perceiving and moving", socio-cultural events, cultural changes - archaic or historical changes in peoples even - can only be described inadequately.

The concept of "gestalt" "(gestalt-sphere", "gestalt of time") is a continuous thread in Weizsäcker's anthropological medicine.

In this context "Eidologie" means that the formative organization of being establishes itself in sensual perception, "invades" it. Everything which has to do with "gestalt" and "form" requires categorical elucidation, a hermeneutical elucidation of human existence and its social dimensions.

## References

JACOB W. 1978: *Kranksein und Krankheit - Anthropologische Grundlagen einer Theorie der Medizin.* Heidelberg: Hüthig/ E. Fischer Verlag.

WEIZSÄCKER V. von 1950: *Der Gestaltkreis. Theorie der Einheit von Wahrnehmen und Bewegen.*Stuttgart:Thieme.

-- 1956: *Pathosophie* (2. edition). Göttingen: Vandenhoek und Ruprecht.

# Phenomenology of the Body
## The Subject-Object Problem in Psychosomatic Medicine and the Role of Traditional Medical Systems herein

Thomas Ots

### Preface

When in 1985 I returned from China and reported that traditional Chinese doctors manage to elicit emotional changes by analyzing the specific cluster of the patient's bodily symptoms (emotion-bodily symptom specificity) this message was met with disbelief by many of the psychosomatically oriented physicians, and vehemently rejected by almost all medical anthropologists I spoke to. This rejection of the Chinese model of psychosomatic specificity, and that of similar models, e.g., between bodily changes and psychodynamic conflicts (Alexander 1948, 1968), or personality traits (Dunbar 1954), was paired with a deep belief that many diseases bear symbolic meaning, thus constructing a very specific, symbolic relation between sociocultural events and the diseased body. This position is best expressed in the work of Geertz (1973).

In this symbolic mode of thought the body makes sense only as a culturally inscribed body, but not in itself, as a 'body sui generis', the 'lived-body' (Straus 1963; Leder 1984, 1990). This notion rests on the long-standing premise that the mind is superior to the body, as culture is superior to nature; or in other words, the human mind, and the world of culture are superior to any other organization of life. In modern scientific discourse this relation is expressed by the common assumption that body/nature - as opposed to mind/culture - cannot express intentions. This dichotomy reveals a specific subject-object relation: as the subject defines the object, the mind inscribes the body by giving meaning to the intentionless body; bodily symptoms are interpreted as symbolic utterances of the world of the mind (psyche), thus bodily symptoms are viewed as specific only in relation to their symbolic (cultural) meaning. I argue that in the scientific discourse of psychosomatic medicine, we do not try to understand bodily symptoms as what they present, but as what they represent. Thus, if we cannot detect symbolic or metaphorical meaning in bodily symptoms, in other words, if these symptoms make 'no sense' culturally, we declare them to be 'non-specific' and 'secondary', i.e., they are more or less neglected.

In this paper I try to show how to overcome this long-lasting subject-object relation between mind and body. One way is to open our senses to the multiform phenomenology of the body which is usually being neglected. I refer to the different views of bodily symptoms in Chinese culture. A second attempt is found in modern semiotics. I draw heavily on the thoughts that von Uexküll and Wesiack (1988) have put forth in their fascinating attempt to outline a 'theory of human medicine'. Taking Wittgenstein's assertion that "what we call descriptions are instruments of particular uses" as their point of departure, the authors write:

"Assuming that there is an intention behind every description we shall no longer be content with descriptions however fascinating they might be. We must seek the intentions that are hidden and disguised by the descriptions. As soon as we proceed in this manner the concept of nature as lacking intention falls apart. Because then it is revealed that behind our description of nature as intentionless, a mere mechanical play of forces, there lurks our intention of technical mastery and exploitation of nature. Our intention obscures our view of possible intentions of nature that could be contrary to our own intentions. Thus, it seems to be more feasible to negate even the possibility of the existence of intentions in nature." (v. Uexküll and Wesiack 1988: 64; my translation).

### The Neglect of Symptoms and the Inclination towards Mono-Symptomatology: the "etcetera"

Modern Western science has developed and - in part - has been developed by the need and capacity for quantification. Whereas it is only a question of method to quantify matter, it is next to impossible to quantify functions. In cellular and organically oriented biomedicine this led to a disregard of functional symptoms. This tendency is supported by technical innovation. Let us take the case of differential diagnosis between angina pectoris and heart infarction. A physician who has no access to modern technology relies on the patient's 'subjective' symptoms and his own observations, e.g., chest pain, fright, shortness of breath, perioral pallor, arhythmia, cold sweat on the face, cyanosis of the lips, cold extremities. A physician in a modern clinic merely needs to perform an ECG and maybe have a blood sample analyzed for specific enzymes. His diagnosis relies more on 'objective signs'. He can dismiss almost all the subjective symptoms and most of his own observations. The awareness of poly-symptomatology has given way to a reasoning within oligosymptomatology. The tendency clearly points towards mono-symptomatology and finally mono-signology. Biomedicine's ideal assumption would be: *one* cause - *one* sign - *one* disease - *one* therapeutic method. Those diseases that show poly-symptomatology are organized around the "leading sign", the "leitmotiv" (Dahmer 1983: 39). Often this sign is interpreted as the actual disease. This is the essential reductionistic mode of biomedical thinking.

Biomedicine's semiotic differentiation between 'subjective', 'non-specific', 'epiphenomenal' and thus more or less neglectable symptoms and 'objective' 'specific', and therefore 'reliable' bodily signs is surpassed by psychoanalytically oriented psychiatry, and is paralleled by current main-stream psychosomatics. When physical symptoms are incorporated into the disease construct they are usually given only secondary importance: "A number of 'vegetative' signs involving sleep, appetite, libido, etc., are important indicators of serious depression in psychiatric patients." (Derogatis 1987). Although classified as "signs", this author and many others paradoxically lump them together as 'various somatic complaints' or simply as the "etc.".

Wittgenstein (1979: 49e) noted "that the limits of my language mean the limits of my world". Fleck (1980) showed convincingly that the reality (facts) we seem to 'find', 'detect', or 'uncover' must be understood as reality (facts) that we create. Any relation we set up between our self-made constructs, e.g., stress and disease, depends on how we define these constructs as ontological and nosological entities. Let us take the heavily disputed question whether there exists a specificity between certain emotions and certain diseases. Let us assume that a group of researchers agrees on the definition of anger. Statistical research reveals that patients who - regardless of the reason - became angry and had to repress their anger suffer from a multitude of symptoms: gastric hyperacidity, duodenal ulcer, abdominal distention, constipation, globus hystericus, belching, hypertension, headaches, vertigo, dizziness, and migraines. The patients show varying numbers and patterns of these symptoms. Those researchers who define an individual symptom, the 'leitmotiv', as disease can convincingly demonstrate that there exists no specificity, because the identical emotion 'caused' different diseases. Yet, others who employ a phenomenologically oriented disease concept which defines disease as a pattern of symptoms are equally able to demonstrate that there exists specificity between repressed anger and a specific symptom pattern. Thus, the 'facts' we 'discover' depend on the ontological definitions of our disease constructs. It is precisely because I take cultural constructivism seriously that I feel the need to analyze and deconstruct medical anthropology's well-established symbolic mode of interpretation as applied to the realm of the body.

## Causal-Linear Thought : The 'Cultural Germ' Paradigm

In the epistemology of biomedicine's germ theory the patient is understood as an object that is 'hit' by the disease-producing germ. The relation between the germ and the disease is causal-linear. The interest lies in the isolation of the germ rather than in the eliciting of the factors that allowed the germ to become pathogenic. This is paralleled by research in psycho-somatic medicine, in psychology and related social sciences, which also try to identify etiological sociocultural factors; furthermore, it tries to quantify their disease-inducing qualities. One of the best-known examples of this research is the Life Event Scale by Holmes and Rahe (1967). Parallel to biomedicine's concept - and maybe even borrowed from it - the patient is viewed as hit by such sociocultural factors as 'life event' and 'stress'. We might call this the 'Cultural Germ' Paradigm. This paradigm is expressed most characteristically in the works of many so-called 'critical medical anthropologists' where diseases are not only linked to a specific political system but also serve as evidence of people's resistance against that particular system (see also Gaines' critique in this volume). In this specific context we might call it the 'Social Germ' Paradigm.

## The Subject - Object Problem

Sixty years ago, Viktor v. Weizsäcker (1987 [1926]: 20) called for a view in medicine that sees the human individual as subject. He argued that relying on the existing epistemologies and methods, biomedicine could successfully do research for another 2000 years without ever bringing the human subject into view. The social sciences made an important step by broadening biomedicine's reductionist view of disease causation and including socio-cultural factors into the etiology of disease. But as long as the basic paradigm remained that of causal-linearity, social science did not get much closer to humanly generated reality. This criticism includes a wide array of social research on the impact on health of such environmental factors as 'social support', 'social relations', 'social networks', and 'increased contact with significant others'. Whereas most researchers like House, Landis, and Umberson (1988) claim to have found "empirical evidence for a causal impact of social relationship on health", there is a growing number of critical reports of this causal-linear relation. Cohler and Liebermann (1980) were surprised by their own findings, which "show little relationship between availability of significant others and mental health, or that increased availability of others from the same ethnic group is associated with mental health." Differences were found in different ethnic groups, and between genders in the identical ethnic group, e.g., the "impact of ethnic 'embeddedness' on mental health appears to be negative for women within the Italian and Polish groups", whereas it was positive for men. The authors carefully reworded their previous expectations and omitted the notion of causal influence of the sociocultural factors on health. Seeman and Berkman (1988), and Boyce, Kay and Uitti (1988) reported equally ambiguous research outcomes. These outcomes call for a totally different kind of approach, an approach that views the human individual as subject.

In psychosomatic theory, this subject-object relation is reproduced by locating etiological factors in the individual's conflict with the environment: this conflict leads to psychic reactions which then lead to somatic changes. Although psychosomatic medicine tries to overcome the mind-body split, the individual remains dichotomized between a mind responsible for somatic changes and an intentionless body. Later in this paper, we will encounter in Chinese medicine a different cultural view in which emotional and somatic changes are understood as identical.

## The Phenomenology of the Body

In recent years, medical anthropologists and scholars of related social sciences have found new interest in the body. Kirmayer (1988) argues that this concern for the body seems to be a response to a dualism in contemporary Western thought that valued the activity of the mind over and against the life of the body. Does the renewed interest of anthropology in the body correlate with a renewed interest in the human individual as subject? To answer this question we must analyze the way in which the body is constructed by medical anthropology: as a mere representation of societal and cultural forces, or as a presentation of itself.

Nancy Scheper-Hughes' and Margaret Lock's (1987) programmatic article "The Mindful Body: A Prolegomenon to Future Work in Medical Anthropology" might serve as a valuable battle ground for our discussion on the new literature of "embodiment". The authors refer to the heuristic concept of three realms of the body: - the 'individual body', understood in the phenomenological sense of the lived experience of the body self (Mauss 1985); - the 'social body', referring to the representational uses of the body as a natural symbol with which to think about nature, society, and culture, as Mary Douglas (1980) suggested; - and the 'body politic', referring to the regulation, surveillance, domination and control of bodies (Foucault 1973, Turner 1980). The authors (1987: 8) argue that "these three bodies are not only three separate and overlapping units of analysis, but also three different theoretical approaches and epistemologies: phenomenology (individual body, the body self), structuralism and symbolism (the social body), and poststructuralism (the body politic)."

When we think of the phenomenology of the body, we might think of such experiences like hunger and thirst, empty and full stomach; lustful or painful defecation and urination; the multitude shivers of libidinous arousal; feeling cold or warm; sensing and becoming aware of my body: waking up and trying to feel my toes, feet, elbows, or any part of the body; lying in bed and trying to locate the 'I' (must I always sense myself in the head, must I always experience my body as distal of my head, or can I put my 'I' somewhere else, into a toe, into the solar plexus? Can I feel my body as stretching out in two directions from my 'I'?); touching myself and being touched; listening to my intestines grumble; feeling my heartbeat inside; seeing the oscillations of my pulse inside my closed eyes; smelling my body, different smells of sweat while exercising, when being anxious, when in sexual arousal; tasting myself, tasting the interior of my mouth, the coating of my tongue; looking at myself in the mirror and knowing how I look, and how I feel; breathing deeply; stretching my body; moving it slowly in *taijiquan*, moving differently than usual and feeling a different body; dancing, dancing myself into a kind of trance, experiencing a sheer endless flow of happiness and lightness; swimming and playing with myself in the water, swimming naked and being astonished about a different body; jumping, jumping into the water from a high board; feeling blocked, almost paralyzed, and lumped right under the heart - as if time stood still - when being scolded; being taken by surprise by that internal flush of heat that rushes up the body from the depth of the pelvis, constricting my heart, taking away my breath, and suddenly becoming frightened, starting to think "of what?"; ceasing to breathe, mouth and eyes wide open and chills on the arms and the back when suddenly being frightened by something or someone; feeling that shiver on the back along the spine when getting angry; or that big lump of heat in the upper part of the chest when in rage, followed by goose skin on the back; getting terribly frightened and feeling the cold run down the spine, right into the pelvis, everything there opening up, becoming wide, 'shitting into my pants'; feeling that biting sourness in my stomach when getting internally angry; how different when being happy, laughing, feeling light, almost elevated, the urge to scream out loud, to jump up, the eyes wide open, and blood pouring into the cheeks, feeling warm all over, the heart laughing and jumping; feeling slow, heavy and tired when being sad; how do I feel when grieving, when anxious, when being humiliated, when melancholy? Feeling the strain in the back and neck when sitting at the typewriter; the pains of blocked thought; nervousness before speaking in public: getting a dry mouth, wetting the lips with the tongue, stuttering, sweating and getting cold fingers; feeling different in different seasons: feeling the sun pleasantly warming the body, feeling uneasy when wrapped by clothes; experiencing the multiple symptoms of illness, such as dizziness, vertigo, nausea, abdominal distention, tinnitus, feeling sour all over, feeling stuffed and blocked, feeling bloated, feeling pressure in or around the chest, choking, feeling numb or paresthetic, feeling frail, aches and pains here and there; experiencing my chronic ailments getting worse or getting better; the feeling of relief; feeling empty; feeling guilt-ridden or ashamed; feeling lonesome and feeling good in company; feeling different from the others; feeling ex- or included; feeling another body, falling in love; loving and feeling what cannot be expressed; being astonished, being embarrassed, getting excited, feeling touched, and being moved.

What a colorful, wonderful, sometimes devastating, but nevertheless exciting world! But this world of bodily experiences is difficult to find in psychosomatic and medical anthropological texts. Who is to be blamed? When writing down my immediate body experiences I realize that I miss words. I have to take refuge in abstract and general concepts such as 'feeling like', 'feeling of', and 'as if'. There are obviously more 'things" happening inside me than I can communicate properly. The lively, sometimes stormy and wild sensations inside my body cannot even be described in what Bernstein called the 'restrictive code' , they are transformed into the rather lifeless concepts of the 'elaborate code'. Again I am reminded of Wittgenstein's observation that the limits of our language are the limits of our world. How can we look for what we cannot even express linguistically? But I 'feel' that there is something beyond my language, and also prior to cognition. In experiencing my own bodily sensations I am no longer the outside observer who tries to 'detect' an objective reality. I am the observer and also the observed. And I realize that man's world of culture and language is of little help, because it has stayed outside of this world of phenomenology in its preoccupation with the mind. When Michelle Rosaldo (1981: 143) says that "emotions are thoughts somehow felt" in the body, that they are "embodied thoughts", then I should be able to describe these thoughts. When Geertz (1964: 47) argues that "without the guiding patterns of human culture ... man would, quite literally, not know how to feel", then this statement is acceptable only when I stress the 'know'. Even if I would not 'know' how to feel, then I had felt already, my body would have sensed me, even if I had not understood 'the message'. Perception is prior to cognition, and as Maurice Merleau-Ponty (1962: 238 ff.) put it, it starts right in the 'lived-body'. He offers the concept of the 'pre-objective' and proposes that we should break with the attitude that mistakenly starts with the objects. Tom Csordas (1990: 9) taking up the point of Merleau-Ponty writes:

"Phenomenology is a descriptive science of existential beginnings, not of already constituted cultural products. If our perception 'ends in objects', the goal of a phenomenological anthropology of perception is to capture that moment of transcendence in which perception begins..."

The body cannot merely be thought of; it cannot merely be viewed as "the first and most natural tool of man" (Mauss 1989: 206). We must take notice of the 'lived-body's' experiences (*Erleben*). This calls for new research tools. Participant observation will no longer do.

### Emotions smell!

Emotions, affects, sensations, and feelings have both a cognized and an uncognized (pre-objective) aspect. Of course, we could say that my perception of today has been constructed by previous cognition, and also by previous socio-cultural experience such as child-rearing, education etc. This is only partly true. Yes, I am a biographical being, and I am influenced and shaped by society and culture, but not constructed. The term 'culturally constructed' bears a somewhat totalitarian taste, it tries to deny man his quality of subject, i.e., it sees the world as a world of objects. In the Hegelian (Hegel 1949) sense 'subject' is characterized by man's quality of self-organization and motion, of being a living body.

This phenomenology of the 'lived-body' has been widely neglected by the social sciences. Why are we so overengaged with the cognitive and socio-cultural aspects of man and his emotions? And when taking notice of the body, why have we focused our attention almost exclusively on the morphological, topographic-anatomical, i.e., the structural aspects of the body and neglected the functional aspects of the 'lived-body'? Why don't we accept that emotions smell!

Kirmayer (1988) noted that there exists a fourth semiotically distinct bodily realm and refers to the finding of temperature elevated on a thermometer that is "neither strictly bodily-felt, social-symbolical nor political, but indexical". This fourth body should actually be the first body. We could call it the 'body self'. In part it is a 'body pre-objective', and sometimes it is a 'body silent-working' that exists outside our awareness of it. A phenomenological approach towards the body should also concentrate on this body. Social scientists often believe that the biological aspects of the body are taken care of by biomedicine or the natural sciences. As I have shown, biomedicine can be criticized not only for its lack of attention to society and culture but also because of its reductionistic and particularistic approach towards the phenomenology of the body.

We have been overly concerned with the mind, and lately with the 'mindful body'. The 'mindful body' approach does not overcome the problems related to the mind-body dichotomy, because it stays inside the traditional subject-object relation: the body is reductionistically conceptualized as inscribed by the mind; it is understood as an object of the mind. Before all, the body is the body. It is time also to investigate the 'bodyful body' and the 'bodyful mind', i.e., the way the mind and our world of culture are shaped and constituted by the 'lived-body', i.e., culture as emergent objectification of bodily experience. Let me give you an example:

In 1981, shortly after my first clinical field research in China, I saw a 60 year old female cancer patient who said that she tolerated the chemotherapeutic treatment quite well, but there was something like a "blockage in the liver" (*Leberstau*). In Germany, I had never heard of this complaint before, it surely was no part of current biomedical nosology, there existed no semantic network for it, and the patient herself had no idea why she had referred to the metaphor of "blockage". Responding to my experience with Chinese medicine where 'liver blockage' (*ganqi yujie* 肝气郁结 ) is synonymous to repressed anger I inquired about her emotional situation. Almost immediately, she burst into tears and told me that the kids in her apartment building teased her because she had lost almost all her hair. And after she had bought a wig, one naughty kid tried to pull it down. "How I wish to hit that boy, but of course, I should not do so. I am so angry that I could throw all my china against the wall." This experience made me think of crossculturally identical bodily expressions of emotions. In Chinese medical as well as in German popular idioms the liver serves as a metaphoric expression which corresponds with anger (Ots 1990).

## Man's Hegemony over the World of Nature

After posing many questions and outlining present paradigmatic fallacies in social research, I move on to an analysis of why the social sciences show such contempt for biology and the somatic aspects of the body. Ever since the physics of Newton there has been a debate whether nature acts according to intentions or not. The former belief in divine intentions was counteracted by the image of nature without such intentions. The belief in intentionless nature expresses man's longing for almighty power, his self-image as being equal to God that is expressed in the Old Testament's assumption that God formed man after his own image. Man rules (t)his world with his mind. His mind is rational and logical, and it creates culture. Culture is thus the only world of real interest. Nature is rough, uncontrolled (but needing to be controlled) and random, and in dealing with nature, man shapes it according to his own image. The social sciences approach the body as a representation of man's world of culture. In the body we detect metaphors and symbolic expressions of man's world of culture. We speak of culturally and socially constructed diseases and illnesses and forget about nature's part in this process. Mary Douglas (1980: XIX f.; 1970) pointed out

"that systems of symbols, though based on bodily processes, get their meaning from social experience. They are coded by a community with a shared history. ... (T)hey are culturally learned and culturally transmitted. ... (T)here are no natural symbols, they are all social."

To say that bodily processes get their symbolic meaning from social experience is only half of the story. Our languages are full of metaphors which are deeply rooted in bodily sensations and take their meanings from these sensations (Lakoff and Johnson 1980). The contempt of the social sciences for biology, for the somatic body, its changes and symptoms, and the refusal of biomedicine to deal with culture are but two sides of the same coin: They are deeply rooted in the hubris of man who views himself as different from and superior to nature. If he deals with nature, then nature is made slave to culture, or it is seen as inanimate and machine-like, thus again slave to man who alone knows the laws of physics. This thinking is not Cartesian, it has been with us ever since Adam and Eve. The difficulty we face is not the Cartesian mind-body dichotomy but to overcome our reductionist view of the body in that relation. I believe that in the future there will be more obstacles to embodying the mind than to seeing the mind in the body.

## Modern Semiotics - Intention, Program, Information

In September 1988, John Cairns and his colleagues of the Harvard School of Public Health reported on an extremely surprising experiment outcome. They observed that when E.coli mutants lacking the ability to ferment lactose (Lac-) were grown on lactose medium that allowed them to survive but not to grow, new variants (Lac+) capable of processing sugar arose at a much higher rate than would be expected from a random process (Cairns et al. 1988). This work suggests that bacteria can influence their own genes to produce adaptive mutations and thus counteracts the idea that spontaneous mutation occurs randomly which has long been one of the underlying tenets of the theory of evolution by natural selection.

We might ask: How do the bacteria know how to change their own genes so they can make use of a changed environment? Again, we are reminded of Wittgenstein that our language makes up our world. How different would the problem be, if we said that the bacteria had a certain program that allowed them to read certain information. If they could not, then they would have acquired this new capability at random, i.e., much slower. The concept of program and information redefines and limits the role of the mind as necessary prerequisite for intention-oriented action. We obviously need new modes of thought to grasp this problem. But - as von Uexkuell and Wesiack (1988:65) point out - the above outlined concept of information and program that deeply changes the relation between object and subject, is not new at all. They refer to the Scottish physician John Brown who in 1780 laid the foundation for a conceptual framework that later was adopted by the German physician Andreas Roeschlaub who based his philosophical thinking on Kant and Fichte. Roeschlaub translated the term *incitabilitas* that Brown had introduced as *Erregbarkeit* and compared it with the prior concept of *irritabilitas* (*Reizbarkeit*). *Incitabilitas* reversed the subject-object relation of the concept of *irritabilitas*. Whereas the latter concept referred to a reaction provoked by exterior stimuli (the reacting matter thus being qualified as object) *incitabilitas* meant the capacity of (living organisms) for action by means of their own reality. This concept would thus not see the diseased body (object) as affected by the bacteria (subject), but see the bacteria (object) as chosen (or allowed to act) by the living body (subject). It is the living body that 'decides' over the action of the germ. The necessary precondition is the ability of the living body to 'read' the information of the object it is dealing with. This concept differentiates between living and non-living (or organic and inorganic) matter. The latter is only capable of reaction (representation) that can be defined by the laws of physics. E.g., a round stone does not move (effect) unless it is being hit by a certain exterior force (cause), and identical round stones react in the same predictable way to the identical exterior force. Only organic matter is capable of autonomous action (presentation); the success of exterior forces is mediated and determined by the living organism. The presentation cannot be defined by laws of physics. The living organism is never the same; it is creative and thus capable of learning, and its actions vary from time to time. In living organisms, it thus becomes useless to differentiate between cause (subject) and effect (object). Von Bertalanffy (1968) coined the term of 'primary active system'. In this paper I will give only a short overview of these concepts and refer the reader to von Uexkuell and Wesiack (1988). System theory simply means that interaction takes place as translations between different communication systems and not via cause and effect. A system must be able to read the other system's information and must be able to answer it. It contains codes and programs that are unique. The evolving action is more than the sum of its constituent parts. It takes place via qualitative leaps and generates new qualities (*Emergenz*). This concept is laid down in the feedback loop (Wiener 1948), a concept that has made its way into biomedicine, yet, without having altered biomedicine's general paradigm of cause and effect (see also MacCormack in this volume, especially p. 232)

Modern semiotic theory is concerned with the signs of communication. It differentiates between the syntactics (the actual signs and their combinations), the semantics (the meaning and interpretation of the signs), and the pragmatics (the clues for action or non-action). Von Uexküll and Wesiack (1988: 131) argue that

"semiotics is of importance for the physician insofar as it helps to overcome the dichotomy of mind and body and the impeding rupture of language into different systems of language. From a semiotic point of view there is no principal difference between the patient's 'subjective' complaints and 'objective findings'. This does not mean, however, that all signs are of identical importance. It simply means that the differentiation between signs does not follow the lines of *subjective* versus *objective* ..., but that every sign and every combination of signs must be analyzed according to its syntactic, semantic and pragmatic values." (my translation).

For centuries, semiotics has played a leading role in medicine. But in ancient and medieval times, the authors argue, only the semantic aspects of the signs were considered important, whereas their syntactic and pragmatic aspects were neglected. Disease was interpreted as a representation of societal or supranatural forces. This indicates that semiotic thought alone does not guarantee a framework that sees man as subject. To value man as subject means to drop the preconceived, merely semantic and symbolic subject-object framework and start off with a *syntactic phenomenology*. A subject-oriented phenomenology must include the bodily symptoms as well as man's biography.

### The Bodily Signs of Emotions

Walter B. Cannon (1922) could demonstrate in dogs that certain emotions (hunger, fear, anxiety, pain, and aggressiveness) were accompanied by certain bodily phenomena. He defined them as 'emergency states' that enabled the animals to get ready for fight or flight. Later, Wolff and Wolff (1943), and Engel, Reichsman, and Segal (1957) were fortunate to conduct research on humans with a fistula of the stomach and could prove identical correlations between emotional and bodily changes. These and many other findings led to the formulation of a psychosomatic theory that is distinctly different from the conversion model and free of symbolic interpretations. If physiological changes in emotional changes can be understood as integrated organismic responses to exterior stimuli that create an emergency state for a certain action, then psychosomatic disease will eventually occur when this action does not take place. The 'motive' (von Uexküll 1963) of action is not carried out. E.g.: In case of attempted aggression the body raises its blood pressure. After the aggressive act, blood pressure returns to normal. But if the aggressive act is not carried out, if the emergency state continues, the blood pressure stays elevated for a longer period of time. If this situation takes place repeatedly over a long period of time the blood pressure will eventually stay elevated. This concept does not primarily differentiate between states of health and disease as ontologically defined constructs. Normal physiological changes become pathological symptoms over a period of time along a health-disease continuum. This is a process prior to symbolic interpretation. The symbolic interpretation begins with the construction of illness. In psychosomatic disorders - as different from conversion disorders which are limited to willful innervation - it is not the individual symptom that carries symbolic (semantic) meaning. The symbolic meaning (symbolic of socio-cultural conflicts) begins with the mind's analysis of the symptoms. The body becomes object only in the cultural construction of illness, in the way the body's reality is reflected, labelled, and categorized, and put into a meaningful cultural context.

### The Role of Traditional Medical Systems

I have given outlines of a modern approach to psychosomatic disorders. Let me sum up: By viewing the patient as subject, we rely on eliciting biography as well as on bodily complaints. We take into account all of the patient's complaints and do not discard symptoms in favor of signs. We try to introduce as few preconceived definitional constructs as possible, i.e., we use an open disease classification that does not limit its insights by constructing disease entities that are based on an morphological-topographical framework of orientation. An ideal system would try to define the interface between body and mind as an organismic relation between emotional and bodily changes and would use an emotional disease classification.

None of the existing psychosomatic schools fulfills all of these conditions. Most rely on anatomical disease classifications. Few have a modern semiotic approach to the patient's psychological and bodily symptoms. In this situation we might ask whether traditional medical systems might be helpful in determining whether there exists a certain specificity in bodily responses to emotional changes.

Referring to Allan Young's (1976, 1983) analysis of 'internalizing' and 'externalizing' traditional medical systems, we might assume that internalizing medical systems fulfill the above given requirements better. Externalizing systems usually employ a semantic-symbolic view of supra-natural disease causation. Traditional Chinese medicine is an example of an internalizing medical system. Emotional disorders are seen as induced by exterior factors, but the emotions themselves are interpreted as interior pathological factors that lead to bodily symptoms. The disease begins with the lost equilibrium of the interior emotional factors. As Chinese medicine took little interest in anatomy and physiology, it became necessary to observe the phenomenology of the body, i.e., the functional relation between bodily symptoms and emotional changes. The disease classification knows few nosological entities but usually is made up of one to many pathodynamic statements. These statements may also employ an organ-related classification system that has often been misinterpreted as an example of 'organ language'. These organs, however, serve as metaphoric expressions of emotional changes and represent specific symptom patterns: liver stands for anger, heart for anxiety, and spleen for depression. (Ots 1990: 50). As traditional Chinese medicine does not emphasize the investigation of the patient's biography, and does not use psychological methods in therapy it cannot be classified as a truly psychosomatic approach. But its diagnostic view concerning bodily symptoms fulfills modern semiotic requirements to a wider degree than any other medical school. Subjective complaints and objective signs are given equal value and have been incorporated in the theory of differential diagnosis. One of the leading textbooks (Zhao 1984) of traditional Chinese medicine in mainland China lists 500 equally valued symptoms and signs. In medical practice the patients' complaints are often valued to a higher degree than assumed objective signs. E.g.: When pulse- and tongue diagnosis do not relate to the patient's subjective complaints, usually the pulse and the tongue are neglected (Ots 1987: 151).

Chinese traditional doctors are able to elicit the patient's emotional status by inquiring about the pattern of bodily changes. This specificity approach is different from Alexander's (1968). Alexander and his colleagues psychoanalyzed their patients and to a high degree were correct in predicting the related somatic diseases. The Chinese approach is simpler and more practically oriented, because the patients usually do not talk about their conflicts, but about their bodily symptoms. It is also much quicker than the reverse approach. It focuses on the phenomenology of the body whereas Alexander relied on arbitrary disease constructs such as hypertension, rheumatoid arthritis, ulcerative colitis, and peptic ulcer. In Chinese medicine these are not regarded as disease entities but as symptoms of a bigger syndrome.

The tables on the following pages represent the correlations between bodily and emotional changes of 240 patients who consulted a physician in the out-patient room in the department of internal medicine of a hospital of traditional Chinese medicine in Nanjing in 1984/85 (Ots 1990). 106 of these patients were diagnosed as suffering from a psychosomatic disorder.

The tables document only a short and by no means complete overview of the correlations of emotional and bodily changes in traditional Chinese medicine. This overview is limited due to the relatively small number of patients. The mean of symptoms was 6.3.

Certainly, the perception and labelling of symptoms bore sociocultural traits. But it posed no difficulty to retranslate semantic meanings into syntactic generics. Working with this diagnostic scheme in general practice, in Germany, I was equally able to diagnose emotional disorders.

## Table I:

## Correspondences between bodily symptoms and perceived emotional changes of 106 patients

| symptoms | number of cases | in % of total | anger | anxiety | feeling depressed |
|---|---|---|---|---|---|
| 1. vertigo | 38 | 36% | 20 | 10 | 8 |
| 2. insomnia | 29 | 27% | 11 | 16 | 2 |
| 3. belching | 27 | 25% | 25 | | 16 |
| 4. abdominal discomfort | 24 | 22% | 18 | | 16 |
| 5. constipation | 22 | 21% | 21 | 2 | |
| 6. palpitation | 20 | 19% | 8 | 14 | |
| 7. thoracic depression | 20 | 18% | 17 | | 9 |
| 8. loss of appetite | 19 | 20% | 11 | | 13 |
| 9. blurred vision | 19 | 20% | 11 | 7 | |
| 10. frigophobia | 18 | 16% | 14 | | 10 |
| 12. nausea | 17 | 16% | 15 | | 6 |
| 13. lumbago | 15 | 14% | 5 | | |
| 14. abdominal distention | 14 | 13% | 12 | | 9 |
| 15. increased dreaming | 14 | 13% | 5 | 8 | |
| 16. headaches (incl. 3 cases of migraine) | 14 | 15% | 9 | 6 | |
| 17. epigastr./right subcost. pain | 13 | 12% | 13 | | |
| 18. frailness | 10 | 9% | 2 | 1 | 7 |
| 19. distressed heart | 10 | 9% | | 7 | |
| 20. tinnitus | 9 | 8% | 2 | | 2 |
| 21. diarrhea | 9 | 8% | 2 | | 7 |
| 22. globus hystericus | 8 | 8% | 7 | | |
| 23. sweating | 8 | 8% | | 5 | |
| 24. retrosternal globe | 8 | 8% | 3 | | 5 |
| 25. flank pain | 8 | 8% | 8 | | |
| 26. dry mouth | 7 | 7% | | | 4 |
| 27. sighing | 7 | 7% | 6 | | |
| 28. hypertension | 6 | 6% | 4 | 2 | |
| 29. numbness of four extremities | 4 | 4% | | | 3 |
| 30. neck pain | 2 | 2% | | 2 | |

Evaluation of Table I : The correspondences between bodily symptoms and emotional changes show three characteristics:

A) Symptoms characteristic of a single emotion.

B) Symptoms characteristic of two or more emotions.

C) Symptoms clearly corresponding with a particular emotion, but which can also become the symptom of another emotion in a specific emotional dynamic, e.g., continuous anxiety leading to depression, or repressed anger leading to depression.

**Table II**

**Group A (symptoms characteristic of a single emotion)**

| anger: | anxiety: | feeling depressed: |
|---|---|---|
| belching (25) | insomnia (16) | loss of appetite (13) |
| constipation (21) | palpitation (14) | frigophobia (10) |
| thoracic depression (18) | increased dreaming (8) | diarrhea (7) |
| nausea (16) | distressed heart (7) | frailness (7) |
| epigastric/right subcost. pain (13) | sweating (5) | dry mouth (4) |
| headaches incl. migraine (9) | neck pain (2) | numbness of four extrem. (3) |
| flank pain (8) | | |
| globus hystericus (7) | | |
| sighing (6) | | |
| hypertension (4) | | |

**Group B (symptoms characteristic of more than one emotion)**

| anger: | anxiety: | feeling depressed: |
|---|---|---|
| vertigo (20) | vertigo (10) | |
| abd. discomfort (16) | | abd. discomfort (18) |
| vomiting (14) | | vomiting (10) |
| abd. distention (12) | | abd. distention (9) |
| blurred vision (11) | blurred vision (7) | |
| retrosternal globe (3) | | retrosternal globe (5) |

**Group C (symptoms clearly corresponding with one particular emotion, but which can also become the symptom of another emotion in a specific emotional dynamic):**

| anger-anxiety: | anxiety-anger: | depression-anger: |
|---|---|---|
| headaches (6) | insomnia (11) | loss of appetite (11) |
| hypertension (2) | palpitation (8) | |
| | increased dreaming (5) | |

| anger-depression: |
|---|
| anger (17) |
| belching (16) |
| thoracic depression (9) |
| nausea (6) |

The above data show high correlations with bodily symptoms in some Western diagnostic scales, e.g., the DSM-IIIR version for anxiety and depression.

I present these data more in a qualitative way, just to give an example of the existence of an emotion-bodily symptom correlation in a different medical system. Our task will be to conduct empirical studies among different ethnic groups to demonstrate crosscultural identities and thus biologically based psycho-physiological presentation. One empirical study of this kind has been conducted in Europe by Rimé and Giovanni (1986), who investigated the physiological patterns of happiness, sadness, fear, and anger. The authors argued that "according to these data, physiological changes occurring with emotions can hardly be considered as being predominantly of a non-specific, or undifferentiated kind, as was proposed by cognitive-psychological theories of emotions." Yet, the study revealed some decisive shortcomings. Most importantly, it was a retrospective self-report study that asked the patients to name bodily complaints in previous emotional disorders. Thus only a few symptoms could be remembered.

The Chinese data showed a surprising resemblance to the Western psychosomatic model (Cannon 1922, Wolff 1953; the 'emotional response domain' in the 'Derogatis Stress Profile') that sees emotional action basically in a three-fold way: fight (anger), flight (anxiety, fear), and regression/withdrawal (melancholy, depression). More research is needed to find out to what degree this rather rough emotional pattern has to give way to a more sophisticated one in order to be a helpful tool in psychsomatic diagnosis.

For medical anthropology's renewed interest in the body, the above outlined theoretical approach requires new research tools which focus not only on the cultural uses of the body but on the 'lived-body' as an "existential ground of culture" (Csordas 1990: 5). We must take into consideration that the "discovery of the mind has obscured our experiencing and understanding the body" (Schipperges 1981: 27). There is more to the body than matter and structure."Bodily acts are not merely mechanical, nor are all acts with cognitional and volitional status truly 'mental'" (Leder 1984: 31). A 'thick description' of the rich phenomenology of the body is needed, which makes it necessary that the researcher will have to experience and perceive the 'lived-body'.

## Summary

This paper outlines a critique of current epistemological pathways in psychosomatic medicine: the causal-linear mode of thought in disease causation, the rejection of specificity between emotional and bodily changes, and the wide neglect of patients' subjective bodily complaints. These characteristics are traced back to a long-lasting subject-object dualism in Western thought in which the mind and man's world of culture is valued over and against the body, and nature in general. In scientific discourse this problem arises again and again as the doctrine of intentionless living nature, i.e., only the cognitive mind can convey meaning. It is argued that modern semiotics and phenomenology provide a useful conceptual framework to overcome this problem by redefining the subject-object relation between man and environment, and between body and mind. Traditional Chinese medicine is presented as an example for a semiotic-phenomenological approach towards man that values the patient as subject, and that in diagnosis relies on the patients' subjective symptoms. We might learn from this approach in reconstructing the 'lived-body' of intentional and meaningful experience as contrasted with the 'physical body' of intentionless matter, structure, and form.

Many generations of scholars of different sciences have attempted to overcome the mind-body dichotomy. I argue that it is not this dichotomy that must be viewed as problematic but our cultures' reductionistic misconceptions of the body: the deconstruction of the intelligent 'lived-body' (German: *Leib*) in favor of constructing the intentionless body (German: *Körper*).

## Acknowledgements

The research in China was supported by a grant of the *Deutscher Akademischer Austauschdienst*. Tom Csordas, Susan DiGiacomo, Arthur Kleinman, and Mei Yue gave helpful comments on earlier versions of this paper. I am grateful to the organizers of the inspiring Hamburg conference "Anthroplogies of Medicine" and the editors of this volume. Special thanks to Susan DiGiacomo who - repeatedly - tried to make my badly knitted English more intelligible.

## References

ALEXANDER F. and FRENCH T.M. 1948: *Studies in Psychosomatic Medicine*. New York: The Ronald Press Co.

ALEXANDER F., FRENCH T.M. and POLLOCK G.H. 1968: *Psychosomatic Specificity*. Chicago: Chicago University Press

BERTALANFFY A. 1968: *General System Theory*. New York: Braziller.

BOYCE W.T., KAY M. and UITTI C. 1988: The Taxonomy of Social Support: An Ethnographic Analysis among Adolescent Mothers. *Social Science and Medicine* 26 (11): 1079-1085.

CAIRNS J., OVERBAUGH J. and MILLER S. 1988: The Origin of Mutants. *Nature* 135, 6186: 142-145.

CANNON W.B. 1922: *Bodily Changes in Pain, Hunger, Fear, and Rage*. New York: Appleton.

COHLER B.J. and LIEBERMAN M.A. 1980: Social Relations and Mental Health. *Research on Aging* 2 (4): 445-469.

CSORDAS T. 1990: Embodiment as a Paradigm for Anthropology. *Ethos* 18 (1): 5-47.

DAHMER J. 1988: *Anamnese und Befund*. Stuttgart, New York: Thieme.

DEROGATIS L.R. 1987: The Derogatis Stress Profile (DSP): Quantification of Psychological Stress. In: FAVA G.A. and WISE T.N. (eds.) *Research in Psychosomatic Medicine*. Basel: Karger.

DOUGLAS M. 1980: *Natural Symbols*. New York: Vintage.

DUNBAR F. 1954: *Emotions and Bodily Changes: A Survey on Literature on Psychosomatic Interrelations 1910-1953*. New York: Columbia University Press.

ENGEL G.L., REICHSMAN F. and SEGAL H.L. 1956: Study of an Infant with a Gastric Fistula. *Psychosomatic Medicine* 18: 374-398.

FLECK L. 1980 [1935]: *Entstehung und Entwicklung einer wissenschaftlichen Tatsache*. Frankfurt: Suhrkamp.

FOSS L. and ROTHENBERG K. 1988: *The Second Medical Revolution: From Biomedicine to Infomedicine*. Boston: New Science Library.

FOUCAULT M. 1973: *Madness and Civilization: A History of Insanity in the Age of Reason*. New York: Vintage.

GARMA A. 1958: *Peptic Ulcer and Psychoanalysis*. Baltimore: Williams and Wilkins.

GEERTZ C. 1964: The Transition to Humanity. In: TAX S. (ed.) *Horizons of Anthropology*. (pp. 37-38) London: Allen and Unwin.

-- 1973: *The Interpretations of Cultures*. New York: Basic Books.

GOOD B.J. and DELVECCHIO GOOD M.-J. 1981: The Semantics of Medical Discourse. In: MENDELSOHN E. and ELKANA Y. (eds) *Sciences and Cultures*. (pp. 177-212) Sociology of Sciences Vol.V.

HEGEL G.W.F. 1949: *Encyclopädie der philosophischen Wissenschaften im Grundrisse*. Sämtliche Werke Vol. V, 5. Auflage. Leipzig.

HOLMES T.H. and RAHE R.H. 1967: The Social Readjustment Scale. *Journal for Psychosomatic Research* 11: 213.

HOUSE J.S., LANDIS K.R. and UMBERSON D. 1988: Social Relationships and Health. *Science* 241: 540-545.

KIRMAYER L.J. 1988: The Body's Insistence on Meaning: Metaphor as Presentation and Representation in Illness Experience. Paper presented at the XII International Congress of Anthropological and Ethnological Sciences, Zagreb, Yugoslavia, July 24-31, 1988. (Draft)

KLEINMAN A. 1988: *The Illness Narratives*. New York: Basic Books.

LAKOFF G. and JOHNSON M. 1980: *Metaphors We Live By*. Chicago: University of Chicago Press.

LEDER D. 1984: Medicine and Paradigms of Embodiment. *The Journal of Medicine and Philosophy* 9,1: 29-44.

-- 1990: *The Absent Body*. Chicago: The University of Chicago Press.

MAUSS M. 1985: A Category of the Human Mind: The Notion of the Person, the Notion of the Self. In: CARITHERS M., COLLINS S. and LUKES S. (eds.) *The Category of the Person:Anthropology, Philosophy, History*. (pp. 1-25) Cambridge: Cambridge University Press.

-- 1989: Die Techniken des Körpers. In: *Soziologie und Anthropologie 2*. Frankfurt: Fischer.

MERLEAU-PONTY M. 1962: *Phenomenology of Perception*. London: Routledge and Kegan Paul.

OTS T. 1987: *Medizin und Heilung in China: Annäherungen an die traditionelle chinesische Medizin*. Berlin: Reimer.

-- 1990: The Angry Liver, the Anxious Heart, and the Melancholy Spleen - The Phenomenology of Perceptions in Chinese Culture. *Culture, Medicine and Psychiatry* 14: 21-58.

ROSALDO M.Z. 1981: Toward an Anthropology of Self and Feeling. In: SHWEDER R.A. and LEVINE R.A. (eds.) *Culture Theory*. (pp.137-157). Cambridge: Cambridge University Press.

RIME B. and GIOVANNI D. 1986: The Physiological Patterns of Reported Emotional States. In: SCHERRER K.D., WALLBOTT H.G. and SUMMERFIELD A.B. (eds.) *Experiencing Emotion - A Cross-cultural Study*. Cambridge: Cambridge University Press.

SEEMAN T.E. and BERKMAN L.F. 1988: Structural Characteristics of Social Networks and their Relationship with Social Support in the Elderly: Who Provides Support. *Social Science and Medicine* 26 (7): 737-749.

SCHEPER-HUGHES N. and LOCK M.M. 1987: The Mindful Body: A Prolegomenon to Future Work in Medical Anthropology. *Medical Anthropology* 1(1): 6-41.

SCHIPPERGES H. 1981: *Kosmos Anthropos: Entwürfe zu einer Philosophie des Leibes*. Stuttgart: Klett-Cotta.

STRAUS E.W.M. 1963: *The Primary World of Senses*. New York: Free Press.

TURNER T. 1980: The Social Skin. In: CHERFAS J. and LEWIN R. (eds.) *Not Work Alone*. (pp. 112-140) London: Temple Smith.

UEXKÜLL Th.v. 1963: *Grundlagen der psychosomatischen Medizin*. Reinbek: Rowohlt.

UEXKÜLL Th.v. and WESIACK W. 1988: *Theorie der Humanmedizin: Grundlagen ärztlichen Denkens und Handelns*. München, Wien, Baltimore: Urban und Schwarzenberg.

UNSCHULD P.U. 1985: *Medicine in China: A History of Ideas*. Berkeley: University of California Press.

WEINER H. 1986: Die Geschichte der psychosomatischen Medizin und das Leib-Seele Problem in der Medizin. *Psychother. med. Psychol.* 36: 361-391.

WEIZSÄCKER V. v. 1987: *Gesammelte Werke*, Bd. 5. Frankfurt: Suhrkamp.

WIENER N. 1948: *Cybernetics*. New York: Wiley.

WITTGENSTEIN L. 1979: *Notebooks 1914 - 1916*, 2nd edition. Chicago: The University of Chicago Press.

WOLF S. and WOLFF H.G. 1943: *Human Gastric Function*. New York: Oxford University Press.

WOLFF H.G. 1953: *Stress and Disease*. Springfield, Ill.: Thomas.

YOUNG A. 1976: Internalizing and Externalizing Medical Belief Systems: An Ethiopian Example. *Social Science and Medicine* 10: 147-156.

-- 1983: The Relevance of Traditional Medical Cultures to Modern Primary Health Care. *Social Science and Medicine* 17 (16): 1205-1211.

ZHAO J.D. (ed.) 1984: *Zhongyi zhengzhuang jianbie zhenduan xue* (Differential diagnosis of traditional Chinese medicine). Beijing: Renmin weishang chubanshe.

# Memory within the Body
## Women's Narrative and Identity in a Southern Italian Village

Mariella Pandolfi

*Per affectum intelligo corporis affectiones, quibus ipsius corporis agendi potentia augetur vel minuitur juvatur, vel coercetur, at simul harum affectionum ideas.*

*Spinoza, Etica pars tertia*

### Analogy as Method

Borges' manual of imaginary zoology describes the difficult existence of the A Bao A Qu. An animal without precise shape or color, the A Bao A Qu lives at the foot of the Tower of Victory and waits. It waits for visitors to come and give it color and form. But only if the new arrival is spiritually superior will the A Bao A Qu begin to take on shape and splendor and begin to climb the stairs of the tower. The A Bao A Qu's pursuit of perfection is totally dependent on others. Borges tells us nothing about the 'splendid prince' or, rather, the extraordinary visitor, or whether he really is charismatic. That is not the question. The A Bao A Qu vegetates in lethargy until someone offers him a form that will define him. But as soon as the meeting occurs, the moment the A Bao A Qu achieves form and color and reaches the head of the stairs, he tumbles to the bottom and begins awaiting the next visitor. Identity here is defined in two different ways: one, everyday life consists in waiting; two, identity is delineated through a human relationship perceived as an extraordinary event.

Experience is never an isolated text, and identity is fundamentally defined through relationship. Plato said much the same thing in asserting that man was punished by being divided in half and must find that other half in order to achieve identity. Freud covered the same ground in his writings on narcissism, as did Lacan with his mirror stage.

What caught my interest in Borges' fable was not the moment of the fabulous encounter but the waiting, the lethargy of formless existence. How does the A Bao A Qu perceive himself as a being without form, how does he survive from one day to the next? Of course the encounter with the other is important, but the daily reality of the A Bao A Qu's life is the waiting, his experience of waiting for something out of the ordinary to happen.

There is a reason for mentioning Borges' evocative story in connection with the fieldwork I have been conducting in a Southern Italian village [1] for the past five years. It seems to be a particularly good metaphoric description of daily life and the experience of women, which I came to understand through my ambiguous roles first as a therapist and then as a ethnographer, gradually shifting from 'therapeutic' attention to observing the way suffering is expressed. Elsewhere I have spoken of the space between clinical and ethnographic work as an 'in between' methodological space, a space where emotional experience expressed in a clinical setting meets the emotional experience that emerges in ethnographic work. I realized that the particular way 'feeling bad or ill' is expressed is not just individual but is actually the genre the women use in describing themselves. The grounding of each individual experience was a language shared by all women regardless of age or class, a language that I would call both ancestral and symptomatologic. My initial role as a listening therapist did not allow sufficient room to gather all the valences of a symptomatologic discourse that increasingly proved to be ancestral: the clinical listening stance certainly helped to highlight this particular way of describing one's own 'being', but the psychoanalytic grid alone was too reductive for a diachronic interpretation of the male-female opposition which, in the course of centuries, had organized and periodically reconfirmed a society through life cycle rituals. Listening and therapeutic space could not contain the 'excess meaning', but the particular way the women have of perceiving and describing themselves, their history, and the history of the village would probably not have emerged in an exclusively anthropological approach either.

The 'intertextual' position of my two roles seems to have been a particularly congenial locus for grasping emotional developments that neither the theory of drives nor that of cultural domestication alone could have fully comprehended. There may be a risk of being trapped in the excess meaning and ending up wondering who is the real A Bao A Qu, the ethnographer-clinician or the village women. Is the awaited event, the event that will give form, the meeting of informant and informed, or is it the flow of events of village life that the women experience as something outside the ordinary course of daily life?

## Individual Experience and Extraordinary Event

The ideas of lived experience, of opening to the world, of presence in the world are essentially phenomenological notions (Dilthey 1976, Heidegger 1975, De Martino 1977), and especially Dilthey's notion of *Erlebnis* as 'bringing life close to life' takes on extremely pregnant meaning in my interpretation. More recently a new anthropological perspective focused attention on this notion (Turner and Bruner 1986, Kleinman and Kleinman 1989). Of course as we shall see, the sense of experience changes according to whether the individual approaches life from the center (men) or from the margin (women). Special attention to women informants (women who speak of the darker unvoiced aspects of their lives, of suffering that is sometimes barely whispered) is in line with the anthropological debate that has given new respect and dignity to matters of everyday life and theoretical status to what women have to say (Verdier 1979, Rosaldo and Lamphere 1974, Corin 1985). Following Corin (1985), what turns out in anthropological discourse to be an extremely vital issue is the central role occupied in society by what is seemingly on the outskirts, the borderland, the margin of social exchange. It is along these lines that many writers have been rethinking the centrality of knowledge developed from particular informants (women), who by virtue of the fact that they are not at the center of the social scene can choose special, non normative itineraries to bring out usage, knowledge, and representations established by the group's official tradition. And it is precisely because they perceive life as something felt rather than accomplished that life is anticipated and recounted as an extraordinary event. In this line the notion of the 'extraordinary event' means that everyday existence can only be remembered and anticipated through events that interrupt the ordinary course of events. Existence can only be perceived and narrated to the extent that it breaks the rhythm of daily existence, which is felt to be as an overwhelming wave; past history can only be narrated through events that do not meet the expectations and the hopes of daily life. The notion of the extraordinary event also involves the subjective dimension: the way a woman experiences and expresses her historical context and her position as an historical subject. How are her desires, feelings, and conscious and unconscious emotional dynamics articulated? There is a strong relationship between 'being ill', the sense of daily life as a wave, and the narration of daily life with a series of extraordinary events; there is a strong connection between two forms of expression in the narration: the narration of 'outside' , and the narration of 'inside'. In one sense the narration is always the narration of suffering, in the other the narration establishes a new identity.

## Narratives

For the village women, in fact, identity is established through story-telling. The central point is that the dimension in which the women project themselves is that of expectation or realization of the event, real or metaphoric, past or present. Excluded from the central masculine world of 'doing' and 'making', and excluded from the sense of this making and doing, women wait and often reconstruct their 'own' stories or histories. For example one woman said to me:

"What I would like is another time, another history, another body. Instead, a woman is a gypsy for life. That is her fate. First she is a stranger in her husband's house."

Women rebuild life experience through narrative, through the discourse and the memory of suffering interwoven with the past and the present. But if they want to talk about the village, or about family events, past and present, they begin by narrating the troubles in their own bodies. They describe a constellation of symptoms using terms borrowed from the interpenetration of official and traditional medicine, and delineate a new narrative grammar that transposes the historical extraordinary event into the body's extraordinary event.

"I was sweating and vomiting because my grandson had left me. I raised him like a son. It was the same sweat and the same vomit that I had when my father and my brother went to Australia in 1946."

The narrative of symbolic physiology expands beyond all limits of space and time to take in the whole hostile and uncontrollable world. "My history is written in my body", is how a fifty-year old woman expressed this; and a forty-two year old woman said, "the ailments of the body and soul never let a woman sleep". Generally narration seems to take three forms:

- narration of the historic events upon which the life of San Marco is based;
- narration of personal and family stories, either real or impregnated with traditional beliefs; and narration through one's own body;
- narration through real - or, more often, imagined - illness, feelings of distress, or profound suffering. This third form envelops the other two and delineates a new minimal identity.

Here I am particularly concerned with the third form in which the collective and personal history are transformed into the narration of bodily experience. The symptomatological narratives are expressed in three different dimensions.

The first dimension is the dimension of *severed time* (the history of the village). The memory of foreign domination, times of earthquake, the times of emigrations. The narrative involves the external parts of a parcellized body: legs, back, arms, hands, feet. Legs are paralyzed by fear, the back is bowed as it was in the past; hands and feet are blocked when people emigrate; and eyes go blind in order not to see the ruin of a family and a home. The second dimension, the dimension of *filled space*, refers to social change, to women's roles, involving the internal part of the body. Organs expand, shrink, or change place inside the body:

"When I have to come home from university on the weekend and return to my father's house, my liver shifts places with my heart , and my brain is on fire."

"When my brother asks me about my sexual life my heart gets lost."

The third dimension is the dimension of *blocked movement* and refers to emotions, subjective feelings, the nocturnal plane of sentiments. This dimension involves body fluids, changes in the state of liquids, and changes in the rhythm of the blood.

"The evil of women is envy; the evil of men is strength, it steals your honor. Evil dries the blood, fear of others makes the blood clot."

The narratives that outline the new bodily identity show that it is an emotional event which establishes the operating rhythm of the body's organs and fluids. "Fear kills my veins, it closes my brain." The experience or recollection of a strong emotion immediately shifts the narrative from the normative or common-sense plane to an inner subjective plane that is hidden from the outer world and sealed within every woman's body. This 'minimal' identity, phenomenologically understood, is reawakened and built out of narration in the presence of an emotional event. So the emotional event is the high point of this identity, but it has a dual aspect. The emotion is the event that generates the narrative, but the event is linked to an historical event and the history of the family. An ongoing exploration of one's own body turns up the signs and traces of the history of every woman and the signs and traces of the history of the village. The development traverses a line from emotion to symptom to trace. And it is in the parcelled narrated body that history is petrified, the history of the person and the village. At the same time, a new locus of subjective identity is created, the body, a sore body that hurts and expresses the discomfort of living.

## The Spiral of Suffering

What was fundamental was to give meaning to what I have called "the spiral of suffering" [2], which takes the form of a choice of what is certainly not a linear therapeutic itinerary and the form of narrative. Speaking about oneself by way of the inner parts of one's body, fragmenting one's perception by naming aching, 'ill' parts means that the village women can reconstruct for themselves what might be called a minimal identity, an identity that is not yet social or socialized, nor is it merely idiosyncratic. What in the past was beyond control and too painful to be remembered except in fragments becomes inscribed in the narrative of a body, phenomenologically understood, the only guarantee of actually being in the world. The suffering, metaphorical body is parcelled out and narrated in an idiom that is different than that of psychosomatic manifestations, although it may express symptoms arising from actual physical pathology.

For the women of San Marco, giving a narrative account of themselves by way of the body means narrating symptoms and pathology that may be real or may be the fruit of hypochondriacal tendencies. More importantly, it means transcending the boundary between feeling well and feeling ill, which is to say that the illness is not the central feature. It may exist or it may not exist; it may be something feared or something actually experienced; it may be in an acute phase, or the person may be on the way to recovery. The narrative seems full of symptoms and pathology, but it should rather be understood as the narrative of a body permeated by age-old suffering. Examples of symptomatologic discourse:

"My legs feel heavy, my heart is bursting, I never sleep, my liver has gone to my brain."
"My hands are shrivelling up, my eyes hurt, my heart has seen things that made it shrink so small it no longer beats."
"My blood has dried up, my blood has withdrawn and I no longer breathe as my blood has turned to water and I've lost the strength of my arms."

There is no grammar of illness here, no other order, albeit pathological: it is not a matter of suffering an illness, it is the story of how each woman suffers history, a sense of suffering transmitted from generation to generation and one that is experienced exclusively by women. These traces are, as it were, an inscription of history in the body itself, the idea of history being subsumed in the inscription of a symbolic itinerary in the woman's body, and the body becomes a sort of memorial to one's personal history and the history of the village as well. A woman tells her own story and the history of the village by way of the body's internal parts, parts that give pain, fall ill, or seem out of order. The course of a person's place in history is narrated and recalled by events that break up the flow of everyday history and became inscribed forever in one's body. What might be interpreted as a symptom or a constellation of symptoms, marking the course of an illness, actually turns out to be a trace or a congeries of traces signalling unusual or disturbing events. The area between illness and emotion is not at all clear or well defined in my work, nor is it in the women's narrative. There may at times be actual illness; at other times it may be hidden or expressed together with other bodily metaphors. Being well or ill is not something that depends on physical health, having a good doctor, or taking the right medicine. Feeling well or not depends on whether life spares us or bows us, and you are bowed and brought lower not so much by misfortune or death, which are considered almost natural aspects, as by events that alter and disrupt social relations (emigration in the past, rigid relations between the families of husband and wife, gossip, family quarrels, etc.).

The data that have been collected show that the anthropology of medicine must identify emotional features, passions, and affects that Western medicine often medicalizes (Foucault 1975, Lock 1990). Recent anthropological debate has fostered efforts to remove "suffocating masks" (Boyer 1989) and has led to extremely interesting epistemological revisions. Room has certainly been made for more complex thought even in terms of the anthropology of illness, suffering, and the world of emotional experience via a rhetorical approach (Clifford 1988, Crapanzano 1986, Geertz 1988). Most of all the interpretative, or post interpretative debate in anthropology has helped me move more surely in the area of the women's narrative.

Rather, as I have already remarked, these symptoms are the traces of an interrupted, broken identity, events that are the marks of interrupted personal or shared history. They are aggregated fragments of bodies and events. The body has always represented the first and the ultimate beachhead in defining and outlining a person's subjectivity. Phenomenology from Husserl to Merleau-Ponty and all of phenomenological psychopathology, has always demonstrated this fact. It is hardly worth repeating, but in line with Jaspers, Binswanger, and others, a subject's relationship with a symptom or illness always has a precise valence (Husserl 1931, Binswanger 1947, Galimberti 1987, Jaspers 1963). Hypochondriacal manifestations or obsessive concern with parts of the body are ultimately a person's way of finding anchorage in a world threatening to collapse. The risk of fragmentation is preceded by proliferation of symptoms. A person who has already lost the world tries to use symptoms, traces of the person's body, to find other traces of a history in the world, a relation with other people who are already lost. But in the spiral of suffering narrated by the women of San Marco, symptomatologic discourse does not fragment; rather it reconstructs a view of world and being in the world.

## Knots

I should like to conclude with a story that subsumes the forms of narration I have been describing and the dialogic experience in a global view of world. In the following interpretation social roles, the perception of danger, the sense of suffering and the experience of illness all act simultaneously and in like manner.

Donatella is a witch. At least that's what the villagers call her. She lives in a smaller village near San Marco. After meeting her for the first time, I suddenly suffered a violent attack of stomach cramps. I went to the doctor as soon as I got back to the village. I thought the cramps were due to my own ambiguous attitude. Donatella had frightened me. I had wanted to ask her something very intimate because everybody told me her powers were extraordinary. But I was there to work. And my ambivalent attitude to her probably had its effects. Two years later I went back and, to put to rest my unease about our previous meeting, I asked her point-blank:

"Why did you make me suffer? I got terribly sick right after our last meeting."
"Naturally", Donatella said, "I dissolved all the blood."
"What do you mean?"
"Your stomach was full of knots, knots of blood."
"But why?"
"Why? Because you were afraid of people. You didn't know what was happening to you. And so your stomach, like your whole body, was no longer protected. The evil forces could go in and out through your stomach whenever they wanted to. It's always like that when one is ripped by fear or when pain or mourning rob you of the soul or force to live: all the evil forces get inside you. In your case you suffered a lot. But the others were only able to turn your blood into knots. Another time, you'll not recover."
I wanted to know why the body was the object, and why the stomach.
"Have you had any children?", she asked. "No? Well, then you've got space in there to put everything in ... "
"And men?"
"You know men, men do, they don't feel, and if you make an effort you put it on the outside. You see it's all a question of seeing whether your body receives the external forces suddenly when something happens to you or when you do something. If you do something, if you decide to do something your body closes up and the outside forces can't come in any more. When the body is closed up it gives you all the force you need to move and to act. Men always keep their bodies closed up."

## Notes

[1] S. Marco dei Cavoti. The village is in the Sannio area in the Campania Region 150 kilometers from Naples. The population is now about 3.500. My fieldwork in this area began in 1984 and continues until 1990.

[2] I use this expression in Pandolfi (1990); cf. also Pandolfi (1989) and Pandolfi (in press). This idea was suggested to me by E. De Martino's notion *crisi della presenza* (crisis of presence) which derives from Sartre, Heidegger and Hegel and describes that particular condition when a person faces a situation in life that he cannot handle. See De Martino (1948).

## References

BINSWANGER L. 1947: *Ausgewählte Vorträge und Aufsätze. Bern: Francke.*

BOYER P. 1989: Pourquoi les Pygmées n'ont pas de culture. *Gradiva* 7: 3-17.

CLIFFORD J. 1988: *The Predicament of Culture.* Cambridge: Harvard University Press.

CRAPANZANO V. 1986: Hermes' Dilemma. The Masking of Subversion in Ethnographic Description. In: CLIFFORD J. and MARCUS G.E. *Writing Culture: The Poetics and Politics of Ethnography* (pp. 51-76). Berkeley: University of California Press.

CORIN E. 1985: Centralité des marges et dynamiques des centres. *Anthropologie et Société* 10, 2: 1-23.

DE MARTINO E. 1948: *Il mondo magico*. Torino: Einaudi.

-- 1977: *La fine del mondo*. Torino: Einaudi.

DILTHEY W. 1976: *Selected Writings*. Cambridge: Cambridge University Press.

FOUCAULT M. 1975: *The Birth of the Clinic: An Archaeology of Medical Perception*. New York: Vintage.

GALIMBERTI E. 1987: *Psichiatria fenomenologica*. Milano: Feltrinelli.

GEERTZ C. 1988: *Works and Lives: The Anthropologist as Author*. Stanford: Stanford University Press.

HEIDEGGER M. 1975: *Sein und Zeit*. Frankfurt: Klostermann.

HUSSERL E. 1931: Cartesianen meditationen. In: *Husserliana Gesammelte Werke*. Den Haag: Nijhoff.

JASPERS K. 1963: *Psychopathology*. Manchester: Manchester University Press.

KLEINMAN A. and KLEINMAN J. 1989: Suffering and its Professional Transformation: Toward an Ethnography of Experience. Paper presented at the first conference of the Society for Psychological Anthropology. San Diego, 6-8 October 1989.

LOCK M. 1990: On Being Ethnic: The Politics of Identity Breaking and Making in Canada, or, *Nevra* on Sunday. *Culture, Medicine, and Psychiatry* 14, 2: 237-254.

PANDOLFI M. 1989: La femme est une gitane à vie. Emotions et corps feminin dans l'Italie du sud. *Culture, Santé, Health* 6, 3: 4-24.

-- 1990: Boundaries Inside The Body: Women's Suffering In Southern Peasant Italy. *Culture, Medicine, and Psychiatry* 14, 2: 255-274.

-- (in press) L'altro corpo. identità femminile fra tradizione e neotradizione. In: INGHILLERI P. and TERRANOVA CECCHINI R. (eds.). *Nuove frontiere della psicologia transculturale*. Milano: Franco Angeli.

RISSO M.A. and CASTELNUOVO F. 1982: *A mezza parete*. Torino: Einaudi.

ROSALDO M. and LAMPHERE L. (eds.) 1974: *Woman, Culture and Society*. Stanford: Stanford University Press.

TURNER V.W. and BRUNER E.M. (eds.) 1986: *The Anthropology of Experience*. Urbana, Chicago: Illinois University Press.

VERDIER Y. 1979: *Façon de dire, façon de faire*. Paris: Gallimard.

# Section 2

# Sociosomatics and Ethnicity

# The Symbolic and the Physiological
## Epigastric Patients in Family Medicine in Flanders

René Devisch

### Introduction

In Flanders, the northern part of Belgium, family physicians estimate that half of the epigastric complaints they face cannot be given a clear biomedical diagnosis. Heyrman (1986) reports that in 75% of the cases complaints are very vague and do not permit an unambiguous definition of their origins [1]. The current psychosomatic approaches do not seem to be able to cope with these problems. Both diagnostically and therapeutically, they offer little to go by (Verdee 1986). I believe that a broader anthropological research may help here. The experience of studying a foreign culture yields new concepts, perspectives and techniques that may prove fruitful to the problem at hand.

I have conducted anthropological research among the Yaka living in rural southwestern Zaire and in the capital Kinshasa [2]. I have focused on the healing of the physically, socially and mentally disabled (1981, 1985), or on the ritual transformation of the commoner into a paramount chief (1988). My prime concern has been the *khita* ritual meant to treat gynaecological problems (1984). Furthermore, I have studied the systems of divination (1990b) and misfortune management (1990a). These investigations helped me very much in studying the epigastric complaints of Flemish, Turkish and Moroccan patients. At the University of Antwerp, a special unit was formed [3] to find out what lies behind such complaints. Attention was directed toward the *cultural* and the *symbolic* meaning of the symptoms.

Since it takes in food, the stomach is in a very direct and physical contact with the outside world. Particularly through what people call "the stomach's mouth and exit", the stomach can take in and pass through or hold up. Epigastric complaints say something about the patient's body in relation to its environment. That is, they reflect something about maintaining and crossing boundaries. They may be related to the domestic or conjugal realm involving intimate relationships between partners, parents, children, and close kin (including also deceased relatives). The complaints often refer to the family's values regarding the 'daily bread'. The role of the bread-winner is essential in this respect. The complaints are further dependent on the family's habits regarding cooking, sharing meals, purchasing and exchanging food, recycling rests and handling waste (Châtelet 1977, Giard 1980, Verdier 1979).

Attention for the relationship between culture and health is shared by both family physicians and anthropologists. It must be admitted that medical research has made spectacular progress precisely by isolating man from his cultural life-context and by studying him in an invariable and experimental setting. However, the family doctor is confronted with the complex role played by the family and the cultural context in the evolution of the disease and cure. In Belgium, the family doctor has the privilege to visit the patient at home. This everyday context is the traditional research domain of the anthropologist. When one considers this environment, the disease is no longer purely physiological. A patient and those concerned look at the illness in a certain way, they experience and act upon it. As a result, the sickness is both an individual, subjective problem and a social, moral one. Ill health is a part of the person's life-style, life-cycle, and family history ('le roman familial'). It is therefore not efficient to reduce the patient to a physiological function. I propose to consider the patient primarily as a *homo semanticus*, a person who creates meaning.

## Body, Society, And Culture - A Semantic Interplay

I focus on the relation between the bodily experience and the particular culture. A subculture (say of Moroccan families in Brussels) influences the way in which its members experience their bodies. In turn, the subculture is shaped by these bodily experiences. The interplay centres round rhythm and movement, touch, hygiene, smell, breathing, ingestion, listening, speech, gaze, posture, dress etc. I suggest that a particular (sub-)culture imposes a particular form of rhythm, movement, touch etc. Consequently it incites its bearers to choose a particular form of somatization. Among less urbanized people, for example, somatization seems to be more common than psychologization (Kirmayer 1984) [4].

### *Theoretical horizon*

As suggested, the semiotic approach adopted here finds its origin in the study of Yaka culture, most particularly in the scrutiny of healing cults and transition rituals (Devisch 1984, 1988, 1990a). I have been influenced by Lévi-Strauss' structuralist approach to sensuous logic (1962), and by Pierre Bourdieu's praxiological study (1980) of the human body as an 'analogic operator'. For Bourdieu, the body is the mediator between the social order of objective conditions and the cosmological or cultural order of practices and representations. The objective conditions are inscribed in the body, that is, embodied in the habitus and exteriorized in ways of thinking and doing.

Yaka healers and healing cults do not develop much verbal discourse or elucidating comments. I consider rituals and culture as creative activities or dramas that make their own scripts while performing. The prime role in the drama is performed by the body as the analogic operator. This goes against the study of ritual as a text or a discoursive process understandable by exegesis and by the study of its grammar and syntactic rules. I compare my task as anthropologist to the one of the art-specialist who tries to gain insight in the esthetic production and creativity. My account aims at a description of the creative interplay rather than at an analysis of the architectonic or syntactical structure. It is my task to grasp the autogenesis of the illness and the therapy as a multi-layered form of drama. The significance of such a drama appears only with the appearance of its forms. Hence the crucial importance of participation, observation, and description. The participants' subjective interpretation are of secondary importance. Unlike Turner (1968), I do not use popular exegesis as an explanation; these comments are as much to be interpreted as any other data. To reach my aims, I developed the so-called three-field method to analyze the spatio-temporal congruence between the body, the social order and the cosmological domain (that is, the world-view). This congruence I call symbolization. I learned from Yaka that the human body provides the subject with a reservoir of symbols that order the relationships between person, society and world. The physical body (with its properties of marking and crossing boundaries) is man's first and most natural source of symbols. The skin and the body's orifices are perceived in Yaka culture as marking the boundaries and passages between inside and outside, between self and others. Bodily functions such as breathing and touching involve the skin and the various orifices. As such, they immediately involve the marking of a boundary and its transgression. The boundedness and openness of the body provide a symbolic model which becomes a metaphor for the boundedness and openness of units on the social and cosmological levels: the domestic unit, the house, the village. Corporeal activities involved in alimentary traffic (eating, digestion, excretion), olfactory exchange, sexual communion, child-birth, speaking and listening always imply transitions of body boundaries. These transitions display a spatiotemporal order that most often involves three terms, for example before - simultaneous - after, inner - surface - outer, or high - middle - low. This order is transferred to the social and cultural forms of such exchanges. The transference is double. On the one hand, the corporeal activities exteriorize the order and thus project it upon the social and cosmological contexts in which these actions take place. On the other hand, the context imposes itself upon the activities. The meeting of the two influences is the drama. For example, in Western society, the space of the partners is reshaped into a conjugal home and domestic space via the sharing of table, bathroom and bed. Man and woman give form to their being husband and wife when they allot conjugal functions to one another in specific spaces in the home. The shaping of the space depends upon such corporeal activities as ingestion, digestion, excretion, fecundation, pregnancy and child-birth. These, however, only become meaningful within this socio-cultural space: they shape it only to the extent that they are shaped by it.

The same goes for healing rituals. Corporeal acts such as dancing, gesticulating and eating shape and express a particular social order and a particular world-view. By means of incantations, dancing, miming, body decoration, massage, fumigations, ingestion, orificial transactions, and trance-possession, the patient *gives body to* and *embodies* the social and cosmological orders. He not only incorporates these orders but also transforms them in their exteriorisation. This is again a double activity.

In this way, a metaphoric correspondence is brought forth between bodily, social and cosmic processes. When the therapist manipulates one aspect, all the others are affected. He induces the patient to adapt a new bodily experience, posture and activity. In doing so, the patient receives a new place in the social and cosmological domains. He is reborn in a new body, a new identity and a new world-view.

### *Symbol and symptom in bodily space-time*

I argue that good health has to do with the social and cultural definition of the proper balance between boundaries and transitions on the bodily, social and cosmological domains. Ill-health is linked with the social and cultural definition of the improper interactions between the three domains. A boundary (a body orifice or a door for instance) involves a subtle balance between connecting and separating, hiding and uncovering. To express it in quite normative terms: proper corporeal activities have to re-define the body boundary within the transition itself, and have to evolve according to the socio-culturally sanctioned body order.

When boundaries act as both mediators and differentiators, they connect and separate at the same time. They intertwine elements while retaining the uniqueness of each single and separate element. Such connections I call ternary: they re-define the limit while crossing it. They always imply a three-term relationship.

A *symptom* is a crack in the ternary relationships. It implies binary relationships which may be intrusive versus extrusive, or dualistic.

*Intrusion* versus *extrusion* is the first form of symptom. It weakens the body boundaries and is unable to distinguish properly between the three domains. The boundaries act as pure conjoiners, they destroy the separateness of each element. Another person's presence may be experienced as an intrusion of a form of bewitchment. The vital force of the victim consequently dwindles and disappears or may spill over uncontrollably. The envious and deprecatory gaze of another may penetrate one's bodily space and obstruct one's vital flow. These anomalies result in symptoms such as depression, respiratory difficulties, disorientation of the senses, anxiety and nervousness, suffocation, syncope and anaesthesia. Current concomitants are insomnia, nightmares, words and acts of aggression, intoxication, obscene speech and sexual misdemeanor. It is usually through agnatic bonds that intrusive afflictions are effected. Victims living close to offended agnatic relatives may be attacked by means of envious looks, backbiting, by "ghosts" sent at night, or even by an aggressor who enters the victim's body.

The body boundaries may be transgressed in other unacceptable manners as well. It may be witnessed in extrusive behavior that inverts or counteracts the individual's functions of sociability. In this case the afflicted individual suffers from a severe temper. He or she is said to have an "impatient heart", and one which "bursts out" in rages; the individual is no longer able to "keep cool" and control his temper. Such a disorder is manifest in hypocritical speech, inconsistent thought and discourse, inability to get along with others. Concomitant anomalies are the failure to agree with the elders, habitual irascibility, sexual aggressivity, and disrespect of family ties. These forms of asocial behavior may eventually lead to "a loss of the senses" because they undermine the vital flow by subverting the social norms regarding personal boundaries.

*Duality* is the second form of symptom. It involves withdrawal: The individual closes up, "turns inward" or no longer responds when spoken to. The boundaries now act as pure differentiatiors, as mere walls or fences. There are many interrelated expressions of closure: grief, anger, pent-up rage, refusal to speak or to share one's income, senseless speech, being stiffened by sadness or despair, melancholy, withdrawal from social contact and apathy. These symptoms are said to go hand in hand with disorders such as dumbness, deafness, blindness, madness, and sterility. The prime means of infliction is bewitchment: it turns the victim's body and vital flow inward "like a fermenting casssava paste that is indissolubly bound in a bushel".

In conclusion, symbol stands to symptom as ternary relationship stands to a binary relationship. The source of both symbol and symptom is to be located in body acts in their relationship with the social world and the cosmology. Body acts are symbolic if the transgressions they involve link and separate the corporeal (cf. *orectic* pole, in Turner 1968) to the socio-cultural (cf. Turner's *eidetic* pole). They are symptomatic if they merely link or merely separate. The multidimensional interaction either becomes an uncontrollable flow or withers. Healing aims at reinstauring the multidimensional interaction of the metaphoric linking and separating.

## Epigastric Complaints - A Traditional Approach

Nutrition plays a fundamental role in the development of the body image. It is also vital in shaping the relationships between partners, parents, children, siblings, and grandparents. In short, it is central in shaping one's identity and one's socio-cultural belonging.

Most psycho-medical approaches develop a specificity-hypothesis: they install a relationship between a psychophysiological complaint (dyspepsia) and a specific phenomenon that is con-sidered to be the cause. This cause may be sought in an emotional tension, a characteristic attitude or behavioral pattern, or in any stressful environmental events.

Behavioral therapists relate many forms of dyspepsia to the subject's insufficient capacity (called "deficient coping mechanism") to deal with stress and fear. They are concerned with the influence that behavioural stimuli may exert (via endocrinal reaction) on the acidity and motility of the stomach.

From a psychodynamic perspective, Alexander (1950) and Bonfils et al. (1982) characterize the patient suffering from duodenal ulcers in terms of "hungry for love" and "oral-aggressive". The patient is said to have the attitude of one who "has been deprived of what he is due and wants revenge". The oral hunger for tenderness would incite a chronic hypersecretion of gastric juices.

Such a reduction of epigastric pathology to orality and dependency ignores the fact that chronic dyspepsia-patients not only have complaints about food-intake, but very often also about breathing, headaches, itching, and impotency. This is not surprising since food-intake has been connected with tactile, thermotactile, and acoustic experiences from the very start of one's life (Anzieu 1974, Jeddi 1977). Food-intake is just one form and phase in the development of internalization and externalization. Other parts of this development include crying, playing, anal retention, and demeanour discipline. One must look at the whole network. Only then can one understand why and how gastric complaints may be linked with other problems regarding the body boundaries such as skin rashes and breathing difficulties.

## Epigastric Complaints - A Semantic Approach

### *Living, cooking, eating and digesting*

For Flemish culture, I should distinguish between two forms of eating and the types of social relationships that go with it. First, there is "our daily bread". This brings about a horizontal relationship between mother and child, and between the individual and his body. Bread is the basic food symbolizing man's minimal and absolute right to earning his livelihood. On the cosmological level, bread symbolizes the dependency of child on mother, of family on bread-winner, and of inner on outer. On the social level, the importance of bread varies with the income-class and the sort of work that is necessary to obtain "the *daily bread*".

Second, there is the elaborate and festive *meal*. This brings about mediation and differentiation between the participants. These are said to become "one big family". They are together and yet apart from each other. In the more traditional family, the father cuts and distributes the meat, he serves the wine and thus induces the participants to start the meal as commensals. Through eating, the participants are integrated.

Ingestion, having a good meal, and sexual union are bodily practices that relate inner to outer, self to other (Châtelet 1977). In terms of the cultural body order, ingestion is a horizontal movement from out to in. Its inversion - as in vomiting - is in many families experienced as polluting. Ingestion not only entails a horizontal movement, but also a specific space and time. In numerous families, cooking and eating take place at fixed periods during the daytime. They are restricted to the kitchen and the dining room. Sexual communion is reserved for the night and the bedroom. But food intake and sexual union are not only separated, they are also linked. Having a good meal may stimulate the sexual desire "to eat (swallow, devour) the other" (Giard 1980). Festive meals may celebrate the transition from one phase in the sexual life-cycle to another. Thus, Verdier (1979) mentions that the soup and wine at wedding-meals in rural Burgundy symbolically refer to the sexual act.

### *Epigastric complaints*

The complaints are seen as part of a life-style. They need not deteriorate into major symptoms but may offer opportunities for emancipation and (self-)healing.

In the Flemish population under consideration, I have distinguished a continuum of four major *profiles* or *subcultures* of chronic epigastric complaints. The distinction draws on the logical combinatory possibilities offered by the spatio-temporal logic of marking and crossing boundaries at the level of the human body and the socio-cultural group (commensals, family, domestic space).

The distinction also takes into consideration the ways in which the patients give form to this logic in their complaints, their behavior, their daily life-style and their organization of the domestic space. The profiles may overlap, and I stress that they are idealized pictures. They show quite different spatio-temporal logics underlying the marking and crossing of boundaries. As such, they have a *heuristic* value: they lay bare semantic congruences, but they do not pretend to offer causal explanations. Neither do they offer a typology based on statistical evidence. They are meant to be eye-openers and grids which may help the interpretation of the patients' complaints. They focus on the relation between the body language, the cultural body symbolism and the social family context. That is their most important advantage over other typologies.

(I) Patients with the *hyperactive* profile seem to shut themselves off. At the same time, however, they have an irresistible tendency to penetrate in the other's space. As for their bodily experience and interaction, they seem to be running ahead of themselves: always busy and on the move, they hardly find time for the essential corporeal activities. They dislike corporeal contact and hate being in a receptive position. In their conversations, one detects their scrupulous commitment to their jobs, and their dislike for compromises. They avoid the ambiguity of transitional zones and in-between states. They seem to concentrate their attention on the socialized borders of their corporality.

Observation A concerns a man of 55 years old who has been married twice; his second wife has a daughter from her first marriage. Five years ago, he had an ulcer. Since then, he has been complaining of stomach-aches, heartburn and insomnia. He always seems very tense. Three or four times a year, he comes to consult the doctor. Dressed sharply, smelling of perfume, he walks into the consultation room and asks the doctor if he has carefully scrutinised the latest blood-tests and discussed matters with the specialist. "I lead an extremely busy life, which is unavoidable because of my job" (he is a representative). There is competition for the promotion he himself seeks. He is overparticular in the choice of his food. He enjoys a good meal, preferably in a restaurant "because my wife, who goes to work, is then freed from all the bother and problems of cooking". His appetite has dwindled, and he has had severe retrosternal pains combined with uncontrollable belching. He then feels out of balance and reluctant to eat, he withdraws from company and feels at ease in bed or in comfortable chair. His wife says he smokes too much and drinks too much whisky. He thinks nothing can be done about his high blood pressure. Moreover "the doctor knows that my state is not so worrying. If only I obtained my job-promotion, I would surely take things easier."

On the *bodily* domain, these patients regard the stomach as a hindrance rather than as a mediator. They are reluctant to take in: they are able to appropriate but not to assimilate food. Complaints are concerned primarily with the intake of food, losing weight, difficult swallowing, belching and vomiting, heartburn, particularly after having had a formal dinner. In other words, these patients consider their stomach as a subordinate whose messages from the body interior they do not acknowledge. They only report facts and factors of stress outside of themselves, in their job and way of living. They point to the painful zone vaguely and quickly. The patients never complain about the lower abdominal zone, nor about defecation (outward movements); this is a tabooed zone. Genital problems are only alluded to, and most of these patients wear exceptionally clean clothes.

Regarding the *social* relationships and the *cosmology* of these patients, we find a similar model. The patients often exhibit a formal and extremely distinguished conduct. In reference to the norms of their income group, they are usually dressed colourfully, perfectly. Their exuberant way of communicating is restricted almost completely to verbal contact. This must be started by themselves: they take the initiative in conversations, often raising direct questions about empirical and instrumental situations and relationships. They usually succeed in structuring the interaction from their perspective. They do not allow the conversation to be reciprocal, the hearer may not become an equal speaker. Patients often want to stress hierarchical and even sexual division of roles: "In my job (as a bank employee), I dislike people who call me by my first name: I am acting there in a professional capacity". "I (a butcher) don't like clients who ask too insistently for my choicest meats, I throw them out of my shop."

This means that the other is forced to become a passive recipient. The spatio-temporal scheme in these relationships is hierarchical and dualistic: upper/lower, before/after, etc. The intrusion in the other's space implies a shutting off on one's own space. Patients strongly insist on their autonomy and continuously try to be at the beginning of all sorts of initiatives. Paradoxically enough, they seem to be acting out of compulsion. They look for a well-known physician who confirms their self-perception. They are anxious that the doctor should arrive at a competent and careful diagnosis but they try to divert his attention from their problems, they undermine his findings. Some patients practice self-medication. The logic is always the same: patients will enter the other's private sphere, but no one is to enter theirs.

Complaints of this type occur especially after a serious loss (divorce or marriage), a failure, or an intrusion in their private sphere (for example, an ailing parent having joined them at home). Progressively, these complaints may become chronic: a few days of rest provides sufficient relief. This indicates that the expansive corporality (giving, intruding, and controlling the other) is underlaid by a contracting corporality (taking, being intruded upon and dependent).

The disturbance of the dependency relation goes together with a contradictory view in parental roles. Often, these patients had maternal expectancies of their fathers. Significantly, they will marry passive women who are not able to fulfil their maternal roles in the family. The female patients' husbands do not appear virile and do not share very much in the domestic tasks. Other patients with this profile have an all-too-good, almost maternal father and/or an ever intrusive mother. In all instances, the culturally prescribed roles of mothers and fathers are undermined.

(II) With patients of the *implosive* profile [5], boundaries are too open. Such patients are not intrusive themselves. Rather, they are victims of intrusion. As a result, they have lost all sense of a private space. They are 'doubly alienated' in their body. On the one hand they live 'beside themselves' and do not relate inside to outside, upper to lower. They stand to their body in an observer's role, looking at it from the perspective of the physician: "Doctor, I took too long coming. Isn't there something wrong with my stomach?". On the other hand, their body is lived in as a peel around an empty space, a tube hiding a parasite core, the ulcer. In 1964, patient B was worried: "Doctor, I have been eating like a wolf and have not put on weight: food just passes straight through"; and from 1974 onwards, he had been expecting an ulcer: "Doctor, couldn't there be an ulcer?".

Observation B concerns a woman of 40. She is thin, well-dressed, and usually very tense. Her husband is a cook and caterer. He is 48. He runs the business with his 20 year old daughter and two female shop-assistants. His wife seems superfluous in this space. There are no contacts with other members of the family, who all live in their natal region far removed from the present home of the patient. In 1976, the woman's facial muscles are paralyzed. In april 1977 she complains about abdominal pains, coughing, hoarseness etc. In 1979, the pains in the right half of the face recur. (Later, it turned out that at that time she had had herself sterilized.) Two months later, she complains of pains in the thorax and the right arm. She feels tired and fears she will develop a gynaecological tumor. In 1979, she discovers her husband's adultery; she starts complaining about insomnia and contractions in the right half of the face. At the beginning of 1980, she feels severe pains in the back and a swollen feeling in the stomach. There are no medical explanations for these pains: tests only indicate a spastic stomach. Usually the patient herself comes for consultation. In periods of crisis (end of autumn and winter) the doctor is called to her home. All day she lies on a couch in the darkened living room. Motionless, with half-open eyes, she speaks with a hoarse voice: "There is something wrong with my abdomen". She fears she will develop a gynaecological tumour.

On the *bodily* level, one notices that the patient's complaints shift from the head (for example, ear-aches, paralysis of the facial muscles, etc.) to the respiratory system (hoarse voice, shortness of breath, etc.), the epigastrium and finally to the back and the limbs. Some patients may even present most of these complaints simultaneously. A male patient: "For several weeks I have been suffering from low-back pain. I am eaten up with pain in my stomach. My stomach is now paralyzed, and in the morning, it is filled with wind: isn't it all a problem of nerves? Someone gave me a medicine that would make my hair grow again."

The emotional and communicative capacities of the body (breath, voice, eyes, sexuality etc.) no longer form any reciprocal link with the outer world (mediation), nor are they supported by any specific feeling of inner strength (differentiation). The abdominal zone becomes the centre of the bodily experience and the relations with the outer world. Body parts are knotted up in themselves and function almost autonomously: "I feel as though my chest is knotted up". Any traffic in and out of the body is disturbed. Speaking is blocked or hindered, sometimes giving way to stammering. The sexual desire decreases, intimate contact becomes rare. Thus, patient B complains about insomnia and about pains in the thorax and in the upper abdomen.

The corporality of these patients becomes dualistic. Sometimes they live outside their bodies, inattentive to pain. Sometimes they shrink away in their stomach. In patient B's terms: "Just before going to bed, I eat half a loaf so as to fall asleep more easily". This can be called an *implosion* of the body: the body collapses, becomes internally empty and thereby creates a strict (peel-like) boundary between the hollow inner space and the outer world. Inside, all powers seem to converge on one point that attracts the whole interior: "Doctor, could there be a tumour?" This tumour is like a foetal life-form that the patient generates by himself and that he keeps hidden in his inside. It is not surprising that female patients fear a gynaecological tumour. It is like the fruit of incest and it turns the vital flow on itself, stopping any outward flowing. Without this outward movement, the internal fruit becomes rampant and polluting.

Since the interplay between in- and out-going movements is disturbed, the patient no longer sees the stomach as a transformative mediating centre between eating (inward movement) and defecating (outward movement). There is, in his view, no interaction between in and out, upper and lower; there is only a void with an uncontrollable core that has taken the place of the transformative space between inner and outer. This inner void is progressively enveloped, starting from the upper parts (head, chest) and going to the lower parts (abdomen, back and legs).

As far as their *social* relationships and their *cosmology* are concerned, these patients do not seem to locate the centre of their acting and being in their body. They are strangers to themselves. In relation to their partner and children, they fail to assume their age and sex role. Quite exceptionally for Flemish conjugal partners, a considerable disparity in age has been noticed between the partners. They seem to relate this to a parental model: "My mother was up to twelve years the junior of my father". In other words, the differences between generations are becoming blurred. This hinders well-balanced sexual, marital and parental relationships: one partner (viz., the patient) is at the same time partner and child, standing in an inferior, vertical position towards the elder. There is a case of a male patient who lives in with the parents of his wife. The two women compete to appropriate the domestic space. The alienated marital relationship may also be evidenced by the division of labour. A few female patients are married to men who fulfil a feeding or caring role according to the

Flemish cultural symbolism: they may be cooks, nurses, bakers or hair-cutters. In all instances, the contribution of the patient to the family's earnings is thought to be completely redundant. "I feel useless and marginal in the house." At the margin the patient takes up very few functional tasks: he seems to have "no room" in this space.

Some patients seem to be alienated from their family of origin and estranged from their roots. The father figure and the patrifiliation are absent from the spontaneous discourse of such patients. Asked for information about their father, they remember a feeding and loving father who was all too aften away from home. It is striking that usually these patients have left the natal region where their parents live. This migration may be a search for differentiation to guard against the complete unison but, as we have seen, this often leads to a complete implosion.

These patients talk of themselves and of others in an instrumental and distanced way. Superfluous in the household, they live in the shadow of their partner or even of their children. They are thus alienated from themselves. This is reflected in three essential ways. First: having no clear social space of their own, these patients live beside themselves but also beside the others in extreme self-confinement and dependence. They seem to live in a void that is anxiously protected. They usually stay home, do not enjoy eating at anybody else's place, and are afraid to "show off". Second: living beside themselves, they find some identity in what the other is saying about and to them (the physician's words give them a feeling of "That's what I am"). In this sense they are extremely dependent. Third: they do not constitute their own centre of perception and communication. As a result, that may in good faith present gastric complaints that do not tally with their good appetite and energy. They speak in clichés, and will provide contradictory information to the family doctor and the specialist. They hardly seem interested in any medical intervention that may be suggested. It is as if those patients are shadows of themselves.

In the illness period, some patients allow themselves to be carried away by political convictions, fashions and cultural models that induce them to subordination. In a vertical hierarchy, they nestle themselves in the subordinate role, groping for some sense of security and boundedness. In this period, they "shrink" without expanding: they show little interest, do not make a point of selecting their own sort of car, clothing, or interior decoration. Like their body boundaries, their house-walls cover an empty space in which these patients cannot recognize themselves.

(III) Patients with the *amoeboid* profile oscillate between the expansive appropriation of profile I and the contractive surrender of profile II. This oscillation may involve subsequent phases in time: moments of expansion are then followed by moments of contraction. It may however combine the two aspects at the same time. Thus, expansion in the public sphere may be combined with contraction in the private sphere. The term *amoeboid* suggests this double nature: The amoeba's cell-boundary can flow out into pseudo-tentacles that can enclose food in an inward movement. Patients with this profile usually relate with people whom they either assimilate to their own space or put on a pedestal so that they become inaccessible. In the first case, there is absolute mediation driven by an internal dynamic; in the second case there is absolute differentiation and an external dynamic. Because of this constant oscillation, all sorts of boundaries fade away: boundaries between generations (for example, the child is sleeping constantly in the bed or bedroom of the parents), between café owner and café frequenter, man and domestic animal, private and public zone.

On the bodily *level*, the patients show asexual, vague and 'undervalued' bodily boundaries. They usually dress in an old-fashioned way and blur sexual characteristics. In their contacts, they are intrusive. They tend to point to the aching stomach with one finger, or with the fingers joined together in one point: they then press with a horizontal and inward moment on the painful zone, sometimes using the doctor's hand or finger. This clearly suggests the intrusion, the fading of boundaries.

The domestic arrangements concord with this bodily intrusion. If cooking is at all done, it is not as the preparation of an elaborate meal presided over by the father. One feeds oneself primarily with 'maternal' basic food (such as bread), either alone or as a common body eating out of a common pot. Eating is done at irregular moments; great amounts of sweet things are eaten; small children are fed in the baby-chair. As a result, the domestic space is not clearly marked of by the symbolization of eating. Eating is not even strictly separated (in complementarity) from defecating; boundaries become diffuse.

Independently of the income-group, the house decoration is dull and full of kitsch. The various living areas and activities overlap, for example, clothes will be left in the living-room or the kitchen, provisions may be stored in the bedroom. There is no specific arrangement for dining. The bed of the bedridden patient can be placed in the living-room. Alternatively, it may mark the scene from the open door of the bedroom. Often one is confronted with a masterless dog (or some other domesticated animal) that intrusively approaches visitors, that sleeps on the bed of the patient, and that gives a somewhat promiscuous character to the setting. The world of these patients hesitates between nature and culture, order and disorder.

This brings us to the *cosmology* and the *social* world of these patients. Social relations of this type are characterized by continuous transition and ambiguity. Often the patient is the eldest or youngest or only child of parents whose descent, age and life-style were widely different. The patient reproduces this asymmetry in his own marriage and descent.

Thus, patient C was born in 1928. He is very thin. He is the youngest: his two corpulent bachelor brothers are still living in the parental house where the mother controlled the whole space, up to her recent death. He married a woman, twelve years his junior. Of this marriage, a daughter was born. When entering the office, patient C holds the doctor's hand a few moments, and looks at him with a mixture of tenderness, familiarity and shyness. The patient mixes French and Dutch vocabulary. In September 1978 he consults his family doctor: "I feel weak and dizzy"; in October he complains about stomach-aches; in November he again feels weak and dizzy, and vomits blood: a duodenal ulcer is diagnosed. In January, May and September 1980 he comes to consultation: "I am depressed, I feel heavy in the stomach, I sweat at night; I have no power, no appetite, I feel dizzy and often vomit:" He says he has bad dreams but he cannot remember them. In September 1980 he complains about eye-pressure and shortly after about loss of urine though there are no signs of prostatis-pathology. His wife and adolescent daughter often drink in a local pub: there is no communal cooking and eating in the family. C then becomes much concerned with his daughter and her marriage. Early 1983, patient C suffers again from stomach-aches, pressure on the thorax, and lack of energy. There are traces of a recent duodenal ulcer with bleeding. When visiting the patient's home, we notice that he is interrupted by his wife as soon as he alludes to the daughter's addiction to alcohol and her bad housing.

All the complaints of patient C show that boundaries fade away, thereby disturbing the inner-outer order. We find an analogous situation with his daughter. She is obese and suffers from chronic amenorrhoea. In 1981, she married a man six years her senior; they live with the maternal grandparents of her husband with whom he has lived since his mother left him as a baby. She does not cook, stays in bed for days, and does not take care of herself. Both in the family and in that of her father (patient C), there is no bounded domestic space: the private merges with the public, the sleeping arrangements with the kitchen and living room.

Several patients have parents with a great disparity of age; some were "forced to marry" or live in a culturally or ethnically mixed marriage. Their complaints arise and become intense shortly after events that involve a breach or a change of the domestic sphere: divorce, death, children leaving home, menopause. Patient and partner appear to be intrusive and competitive: the husband will seek to align himself with his daughter against the collusion between mother and son. All this destroys the ternary relationship between parent and child, man and woman, inner and outer.

In terms of Mary Douglas (1978), the families of these patients tend to be of the strong group-strong grid type: they are built upon horizontal relationships of intrusion and role-inversion. Thus, the husband will do the house-keeping while his wife has her job elsewhere. The family develops a collective and hardly differentiated bodiliness that may be extended to friends, visitors, and clients. In such an expansive phase, these people are extremely friendly and genial: distinctions between in and out, self and other tend to be obscured. The family doctor is often called for a visit. The medical consultation is seen as an opportunity to ask for a prescription for an absent family member. Sometimes, husband and wife, parent and child come together to see the doctor at the office. In such cases, one family member will seek in the doctor an ally against the other. I noticed some cases where one partner no longer showed medical needs when the other was ailing.

(IV) Patients with the *passive-receptive, "greedy mouth"* profile take in without bringing out. There is no expansion coming from them. Such patients surrender to a dominating figure who cares for them in an intrusive and expansive manner. This figure may be a parent, the partner, the physician, or the life-community. The patient undergoes the protecting embrace. To the outer world and in their professional activities, they are the antagonists rather than the protagonists of their actions.

The bodily posture often reflects this weak and dependent position. Patients with this profile are often very fat and extremely soft: they walk with a stoop, as if they were running behind their bellies. They show no erect, vertical posture. On the whole, their bodily praxis is dominated by a continuous inward movement: they bow their heads, cast down their eyes, hide themselves in thick clothes, speak in a hesitant voice. With hand stretched out and rubbing in a broad manner over the painful zone, these patients indicate the "pressure on the stomach", the "swollen, extended belly". Many similar complaints make up the picture. Patients suffer from a cold, from shivering or sweating, from pins and needles in the legs. They get vertigo when standing up and they often have potency difficulties. For these complaints, a placebo turns out to be as effective as the actual medicine.

In their domestic environment, these patients are equally weak and self-effacing. They take up very little room. Some feel the need to lie down after eating; at the table they are silent and inconspicuous participants. They long for a "warm and cosy" house and bed. Only rarely do they eat alone. They prefer 'maternal' food and lots of sweets. They make little or no personal contribution to the house-decoration. They seem to be sinking away in taking in: they show no expansive dynamics.

This holds true for the *social* and the *cosmological* domains. In their relationships, these patients are very dependent, anxious, and depressed. They are easily upset. Their sorrows and fears concern their inability to fulfill all sorts of obligations and desires. This goes together with a strong feeling of self-depreciation. The complaint is unending: "I have the impression that I always have to prove something to the other".

Patients with this profile have difficulty in actively and creatively giving form to their marital and parental role. Patient G remembers his father as "a man without backbone", a man who was himself very dependent upon his mother. For patient H, the intrusive presence of his father's mother in the family inhibited conjugal and parental relationships. He remained a son and was unable to become a husband and a father. He thereby made it impossible for the married woman to become a wife and a mother. Patient I remembers her mother as "very good, caring, but cool". This female patient stands in a competitive relation to her mother. She married a much younger man whose activities are still under his mother's control. To their children, these patients are soft-hearted and over-protective. They desperately want to be the giving side. To their parents they are dependent and cannot escape being the taking side. There is no integration of this giving and taking: both poles are unconnected and thus become extreme. The stress is on intake. The patient passively surrenders in a defenseless opening-up. In other words: the differentiation is given up for a complete mediation. Boundary-transitions go from outer to inner. The patients regard the others as intruders who threaten their and their children's existence. In this case, the abdominal complaints recur periodically. Marital problems, difficulties with the children, or an impending dismissal often form the background of the complaint.

## Clinical Relevance

How can this anthropological perspective lead to clinical applications? I think anthropology may enrich the doctor's attitudes and skills. As to the attitudes, anthropology broadens the view doctors hold on illness. Ill health can be considered, partially at least, as a knot to be untied or unraveled on the bodily, social and cosmological levels. This may encourage doctors to take into account the larger context of the patient. Concomitantly, doctors will get a clearer view on the self-healing capacities of the patient. If illness is regarded as a significant part of a person's biography and development, the role of the doctor becomes one of mediation and guidance in that particular development. He shows possible roads and outcomes from which the patient can gradually learn to make a choice. In this way, ill health can lead to growth and emancipation of the patient and of the relevant others (De Vries 1985, Sheldrake 1981, Weiss 1977). As to the skills, I will focus on communication and observation. These two capacities may be enhanced thanks to our anthropological perspective.

### *Communicational skills*

The most common form of communication in the medical setting concerns the so-called anamnesis. Like any classical biomedical discourse, *anamnesis* abounds in logico-causal reasoning. Anamnestic questions explicitly deal with: "What brings you here? What's the problem? Since when and how? What makes it better or worse? Do you smoke? Any alcohol abuse? Do you feel stressed? Does the week-end bring any amelioration?" Although this type of communication is well-known to doctors, it should be stressed that it may be alien to, and hence alienating for the patient. At the onset of the consultation, the patient often offers vague complaints. To facilitate the consultation, the patient readily adopts the medical vocabulary. Some patients may say something like: "It is again my gastritis that hurts me"; "it gives me the sensation of perforation". This is called the proto-professionalisation of the patient. Some of them read medical books in order to present their symptoms "in a proper way".

Semantic anthropology may help the family doctor to understand the original, uncensured language of the patient. When the patient uses his own words, the doctor comes closest to the patient's personal experience on the bodily, the social and the cosmological level. Instead of using technical terms like gastritis, hypersecretion, and gastralgia, the doctor could rely on common or lay expressions like trouble, discomfort, digestion, and restlessness. *Empathy* is the art of enliving the patient's experience and life-world. It is the skill to set up the contact in such a way that the patient feels fully understood and accepted. It is also the capacity to mediate between different types of lay and specialist discourses. The gap in cultural meanings and values of patient, family and health-care providers may be bridged in this way (Kleinman 1980).

To arrive at this empathy, a new sort of communication is needed. *Metaphoric* speech is an indirect way of communication that discloses or unravels much of the latent semantics and the uncensored experience of the patient. Sticking to the patient's figurative language and using body metaphors, the doctor can produce insight and change on one level (say, regarding eating), thereby favouring a congruent change on related levels (regarding rhythm of activities, commensality, body hygiene, relaxation). Here the doctor appears in a mediating role analogous to that of the maternal uncle or midwife in Yaka culture: the patient-doctor relationship becomes maieutic and whole-making. It evokes resonances within and across different fields, and in that way it helps to re-integrate the bodily, the social and the cosmological domains.

To reach the subjective body experience, the doctor could pay special attention to the patient's spontaneous speech and body metaphors. As we have seen in the profile discussion, the patient's spontaneous complaints provide insight in his corporeal identity, his social relations, and his cosmology. "I'm wound up inside, there is a heavy feeling in my stomach, a tight, knotted feeling in my stomach" - expressions like these must not be discarded as 'unscientific' but accepted as useful indicators of the patient's life-world.

The doctor could incite the patient to speak in his or her colloquial language. Usually, this language contains a vast number of metaphors about crossing and marking boundaries, about intake, digestion, and excretion. Moreover, these figures of speech show the feelings and emotions that go with the discussed physical phenomenon. Bodily states and changes are invariably linked up with emotional states and changes. Both aspects influence and reinforce one another. It is no good trying to look at the body while discarding the emotional background. Metaphors have the advantage of showing this intricate relation. The following questions may prove successful in eliciting such metaphors from the patient:

"What sits heavy on your stomach?"
"What turns your stomach?"
"What does your stomach try to say?"
"Is there something/someone you cannot stomach?"
"What is there you cannot swallow?"
"Is there a heavy/tight knotted feeling in your stomach?"
"What makes your blood pressure go up?"

A semantic network analysis (Good 1977, Bibeau 1978) may help to discover the concatenations of central concepts and root metaphors about the body and the place in the social world.
Metaphoric communication is able to uncover and arouse endogenous *self-healing* capacities. The following type of questions may provide a helpful start:

"What will happen when you wouldn't do anything?"
"How would you evaluate your health?"
"How would others evaluate your health?"
"In your present condition what is the spontaneous evolution?"
"In your present condition what are the possibilities for a spontaneous remission?"
"How would you feel without diet, prescription, rest, therapy, medical contacts?"

Or the doctor may ask questions which are even more projective:

"Whom do you resemble physically, or in your stomach troubles?"
"What will your stomach, body, health be in five years' time?"
"How do you think your life-style and life-conditions will evolve over the next five years?"
"How will the evolution be if you don't do anything?"
"How do your relatives look upon your problem?"
"Are you considered as weak or strong?"
"Draw your line (course) of life"
"Which book did you read in the last months?"
"What is your most favourite character in film or fairy-tales?"
"In what country do you feel best?"
"If the kinsman who usually advises you shold now be invited for advice, what would you expect from him or her?"

After such contact via metaphoric communication, a Moroccan gastric patient witnessed: "He (that is, the anthropologist) asked me questions that people had never before asked me: I think he loved me". And a few weeks later, the patient regained his job after six months of disability due to gastric complaints.

Metaphoric communication allows for subtle and indirect *intervention*. This may be done most successfully through eliciting information with reference to a *third pole*. This method of course goes back to our theory of triadic relationships. It entails that the doctors appeals to some third party in order to bring the patient to insight and change. One way to install a third instance is to actually confront the patient with such an instance. For instance, the doctor may ask the patient to write a letter to the specialist or to a relevant other. Whether this letter will be sent or kept will very much depend on the patient's pofile. A mirror can be used as a third pole: "What do you (like to) see in the mirror? What do you think you could develop in yourself?" The doctor can bring in a third person: for example the partner or a family member. This person may act as co-therapist in discussions regarding diet or other projects. Imaginary third instances may be invoked by asking the following questions:

"Whom do you resemble; is there any attribute of him/her you would like to develop in yourself?"
"What do the others think of your problems?"
"From whom would you like (expect) support?"
"What book are you reading now, and what character in the book do you resemble?"
"What film or fairy-tale do you prefer now?"
"What animals do you like most now?"
"What do you admire now, why?"

Drawings (maybe with closed eyes) may also bring in this third instance. The doctor could give the patients some tasks: "Draw your body, your house and locate the family members, the different rooms, boundaries, mirrors, the transition zones". Or: "Make a diagram of the successive periods of good/bad feeling during a day and describe the activities going on during those periods". Transitional spaces such as the waiting room and the consultation room offer further facilities that may help to build up these ternary relationships.

### *Observational skills*

Consonance and dissonance between the different levels and modes of communication offer important hints for therapeutic work. Therefore the doctor should not only listen to the patient's words, but also watch the bodily motions accompanying the patient's story. Tics, gestures, posture, facial expressions - they form more or less uncensored expressions that may be consonant or dissonant with the patient's verbal communication. Video-taped doctor-patient encounter may be extremely revealing when looked at from this point of view. The dissonances often point to a central problem that the patient is trying to hide.

Visual contact in the medical encounter should however not be limited to observing without participating. The doctor's look should be diffident to prevent intrusiveness. Doctor and patient direct their view to a common third instance: the patient's healing. To combine observation and participation, the medical visit at home presents excellent opportunities. This visit is the family doctor's very privilege. In non-urban Flanders, family doctors pay home-visits in a third of the contacts. In the joint venture with anthropology, doctors renewed their enthusiasm for visiting patients at home. The doctor may come in at lunch-time when the family is cooking. He may try to visit different places in the house; with gastric patients, special attention would be paid to the kitchen to see the cooking, the food, the meals, the handling of foodrests. The doctor may observe the behavior of the family members, the role of domestic animals. He may ascertain the importance of transitional spaces and elements and may look at the pictures and photographs on the wall. After an investigation of the material thus collected, the doctor may seek with the patient to rearrange the domestic space, doors and thresholds.

An important aspect in this family observation is the role of the third generation. In several cases, the patient's problems became understandable only after we had discovered the significane of a grandparent or a maternal uncle. In many cases we have found useful information in inquiries concerning the transmission of first names, the role of the grandfather, or the inheritance of treasured objects. In this way, we have continually broadened our scope and our field of research. This may be the most valid contribution anthropology has to offer to medicine: to broaden the scope until finally the patient emerges as a person in a multilayered network that may tie him down or lift him up. Therapy tries to make the most of this network. It realizes that changes on any single domain (say, on eating or dressing) will produce changes on the other levels (say, in social relations and world-view). It pays special attention to those activities, moments and places that interrelate and differentiate the three domains, that is, the bodily, the social and the cosmological levels. Transitional activities such as getting up and going to sleep would be reoriented in therapy to reinstall ternary relationships. Other aspects that call for the doctor's special attention are the patient's bodily hygiene, manner of dress, table manners, interior decoration, spatial division of the house and so on. This enumeration can do no more than scratch the surface. A detailed elaboration of new therapeutic approaches and techniques deserves a study on its own. The findings in our research group have shown that this is a new and promising field waiting to be explored by anthropologists and doctors in close co-operation.

## Acknowledgements

The research among rural and urban Yaka has been financed by the Belgian National Fund for Scientific Research. I am indebted to the members of the "Medical Anthropological Research Unit" at Brussels and the "Research Unit on Symptom and Symbol" in Antwerp. Many arguments in the text have grown out of the bimonthly seminars in both research units, in 1981-86. The text has further benefited from comments on former partial presentations at the GIRAME in Montreal (autumn 1986) and in seminar sessions at the medical faculties of the following universities: British Columbia at Vancouver (convener Prof. A. Marcus), Harvard (Prof. B. Good), McGill (Prof. M. Lock), Washington at Seattle (Prof. R. Walker). The sections dealing with the clinical aspects in family medicine owe very much to Dr. H. Stuer. I thank Dr. B. Vervaeck for his editorial help.

## Notes

[1] According to Heyrman (1986), a family physician in Flanders has in average some 5300 consultations a year for about 1500 patients. Heyrman further reports that in family medicine, in Flanders, epigastric complaints make up anything to 20 and even 30% of the presenting symptoms; 75% are vague complaints; 12% of these patients produce an ulcer, only 2 or 3 of these are new cases, the others are relapses. Spontaneous healing seems to occur in 55% of the cases. From general medical literature, we learn that peptic ulcers are rare: the acute duodenal ulcer correlates positively with personality type and family factors; its incidence in all occupational classes both in town and country exceeds 5 to 10 times that of gastric ulcers and it occurs up to 5 times more among men than among women. Chronic gastric ulcers do not show a significant correlation with personality type, or with genetic or family factors. In the menopausal period, women seem to form the risk group and cases are more common in the lowest occupational classes. Prescribed medical care in Flanders is similar to other European countries: during the first week (except for alarm symptoms), antacida are prescribed beside some advice regarding food, tobacco and alcohol; if no change occurs after a few weeks, acidity secretion inhibitors are further being prescribed.

[2] The Yaka live in western Bandundu, a region bordering Angola and lying between the Kwaango and Waamba rivers. Half of the present Yaka population, numbering approximately 600,000, currently resides in the capital Kinshasa. The research took place in the Taanda territory of thirteen villages which average a populace of 120 inhabitants. The territory is situated about 450 km to the southeast of Kinshasa. In the field, I complemented particpant observation with lengthy conversations with elders responsible for the preservation of the essential traditions. I acquired insight into the people's life and learned their language at the same time by sharing their way of living. My annual stays of some seven weeks, since 1986, among the Yaka in Kinshasa, allow me to say that the Yaka healing protocol tends to become simpler in the city and also to many non-Yaka clients for problems that are related to cash economy, urban jobs, school education.

[3] At the Institute for Family Medicine in the Department of Medicine at the University of Antwerp, Hugo Stuer MD and R. Devisch have coordinated the Research Group on Symptom and Symbol since 1982. Approximately 10 staff members (Gerard Beretta, R. Devisch, Louis Ferrant, Remy Maes, Christine Rouneau, Leo Schillemans, Paul Smits, Hugo Stuer, Eddy Springael, Bart Van Den Eynden, Hans Verrept), nearly all of them with a part-time practice, have developed in their fortnightly seminars an anthropological approach to epigastric complaints (Devisch 1986, Devisch, Stuer and Verrept 1989). The research is based on some 50 extensive case-studies, complemented by a biographical anamnesis and an analysis of the family history ('roman familial') over at least three generations. All patients consulted their family physician with chronic complaints regarding the alimentary canal. Some 25 Sicilian, Moroccan and Turkish patients in Brussels, and some 25 Flemish patients, have been documented. This has been done through observing the patients's life-world, through studying the medical records, and through investigating the family history. We have also collected data via a biographical anamnesis recorded during semi-structured interviews (The information constitutes some 1000 type-written pages of data and some 50 hrs of video-taped interviews).

The medical doctors involved in this project have testified how deeply this approach has modified their relationship with their patients and therapeutic options. The anthropological approach makes it possible for the family doctor to make better use of his or her unique knowledge of the concrete, daily lives of his or her patients and their family histories. The results of this project have shown how important it may be to know the unique development of the biography of a certain patient if one wants to be able to help him or her.

Applying a similar semantic perspective, Dr. B. Huyghe has studied anorexia nervosa (1985), and Hans Verrept (anthropologist) and Leo Schillmans (MD) have done research on incest patients, whereas Devisch and Gailly (1985) have set up in Brussels a self-help group among middle-aged emigrated Turkish women who suffer from depression and repeated hysteriform crises. A similar perspective was adopted for the study of psychiatric art-therapy in Tunis (Devisch and Vervaeck 1986).

[4] In family medicine in Antwerp and Brussels, Sicilian men at the age of 30 to 45 present 4 to 6 times more gastric complaints than their Belgian peers. In the matrazza syndrome the corporal identity seems to contract into the stomach. This can be seen as a form of implosion (cf. infra). Almost 10% of Turkish immigrant women at that same age develop epileptiform crises, up to several times a week. The problems seem to be concerned with the growing up of the eldest son. Turkish men at that age complain of pains between the shoulders or in the loins. Abdominal pains, insomnia, and apathy are regular troubles in that period. Some seem to overcome this implosive surrender momentarily through some expansive (and hence explosive) behavior like gambling or physical violence, while others complain of disability and withdraw from the labour-scene after a minor labour accident. Middle-aged Moroccan men may similarly withdraw from social contact while complaining of "pain in the right knee that hinders them to stand up and walk".

[5] Implosion offers a more active and spatial image than the psychological notion of depresssion.

## References


ALEXANDER F. 1950: *Psychosomatic Medicine; its Principles and Applications.* New York: Norton.

ANZIEU D. 1974: Le moi-peau. *Nouvelle Revue de Psychanalyse* 9: 195-208.

BIBEAU G. 1978: L'organisation Ngbandi des noms de maladies. *Anthropologie et Sociétés* 2, 3: 83-116.

BOURDIEU P. 1980: *Le sens pratique.* Paris: Minuit.

BONFILS S., HACHETTE J.-C. and DAMNE O. 1982: *L'abord psychosomatique en gastroentérologie.* Paris: Masson.

CHÂTELET N. 1977: *Le corps culinaire.* Paris: Seuil.

DEGRYSE J. (ed.) 1986: *Maagklachten opnieuw bekeken.* Leuven: Acco.

DEVISCH R. 1981: La mort et la dialectique des limites dans une société d'Afrique Centrale. *Archivio di Filosofia* (Roma) 2-3: 503-527.

-- 1984: *Se recréer femme. Manipulation sémantique d'une situation d'infécondité chez les Yaka du Zaire.* Berlin: Dietrich Reimer Verlag.

-- 1985: Symbol and Psychosomatic Symptom in Bodily Space-Time: The case of the Yaka of Zaire. *International Journal of Psychology* 20: 589-616.

-- 1986: "Ik blijf me raar voelen in de buik": Een semantisch-antropologische verkenning van bovenbuikklachten in Vlaamse huisartspraktijken. In: DEGRYSE 1986: 61-91.

-- 1988: From equal to better: investing the chief among the Northern Yaka of Zaire. *Africa* 58, 3: 261-290.

-- 1990a: The Therapist and the Source of Healing among the Yaka of Zaire. *Culture, Medicine, and Psychiatry* 14,2: 213-236.

-- 1990b: Mediumistic Divination among the Northern Yaka of Zaire: Etiology and Ways of Knowing. In: PEEK Ph.M. (ed.), *African Divination Systems: Ways of Knowing (pp. 104-123).* Bloomington: Indiana University Press.

DEVISCH R. and GAILLY A. 1985: Dertlesmek: the Sharing of sorrow - A therapeutic self-help group among Turkish women. *Psichiatrica e psicoterapia analitica* 4, 2: 133-152.

DEVISCH R., STUER H. and VERREPT H. 1989: La signification psychoculturelle des plaintes gastriques. *Santé, Culture, Health* 6, 1: 39-52.

DEVISCH R. and VERVAECK B. 1986: Doors and thresholds: Jeddi's approach to psychiatric disorders. *Social Science and Medicine* 22: 541-551.

DE VRIES M. 1985: *Het behoud van leven. Ziekte en heling in nieuw perspectief.* Utrecht/ Antwerpen: Bohn, Scheltma & Holkema.

DOUGLAS M. 1978: *Cultural bias.* London: Royal Anthropological Institute of Great Britain and Ireland.

GIARD L. 1980: Faire-la-cuisine. In: GIARD L. and MAYOL P. (eds.), *L'invention du quotidien. Vol. 2: Habiter, cuisiner (pp. 149-234).* Paris: Union Générale d'Editions 10/18.

GOOD B. 1977: "The heart of what's the matter": The semantics of illness in Iran. *Culture, Medicine and Psychiatry* 1: 25-58.

HEYRMAN J. 1986: Huisartsen en patienten met epigastrische klachten. In: DEGRYSE J. 1986: 9-18.

HUYGHE B. 1985: Structuration métaphorique du corps et identité dans l'anorexie mentale. *The International Journal of Psychology* 20: 501-520.

JEDDI E. 1977: Ontogenèse du confort thermique et hypothèse théoretique sur sa place dans le développment affectif. In: DURAND J. and RAYNAUD J. (eds.) *Thermal comfort - confort thermique*. INSERM 75 (Dec.): 101-144.

KLEINMAN A. 1980: *Patients and Healers in the Context of Culture.* Berkeley: University of California Press.

KIRMAYER L.J. 1984: Culture, affect and somatization. *Transcultural Psychiatry Research Review* 21: 159-188, 237-262.

LÉVI-STRAUSS C. 1966: *La pensée sauvage*. Paris: Plon.

SHELDRAKE R. 1981: *A New Science of Life*. London: Blond and Briggs.

TURNER V. 1968: *The Drums of Affliction*. Oxford: Clarendon.

VERDEE G. 1986: Specificiteit bij psychofysiologische klachten. In: DEGRYSE J. 1986: 55-60.

VERDIER Y. 1979: *Façons de dire, façons de faire: la laveuse, la couturière, la cuisinière.* Paris: Gallimard.

WEISS P.A. 1977: The system of nature and the nature of systems: Empirical holism and practical reductionism harmonized. In: SCHAEFER K.E., HENSEK H. and BRADY L.(eds.) *Toward a man-centered medical science.* New York: Futura.

curare Anthropologies of Medicine Special Issue 7/91, 87–103

# Nerves and Nostalgia
## Greek-Canadian Immigrants and Medical Care in Québec

Margaret Lock

### Introduction

The British North America Act of 1867, the founding document of Canada, in contrast to the American Constitution, proclaims no majestic vision of a new society with equal rights and freedoms guaranteed for all citizens. From its inception, Canada recognized collectivities as fundamental units and emphasized group rights over those of individual citizens. Whereas in America "life, liberty, and the pursuit of happiness" were enshrined as fundamental ideals, in Canada "peace, order, and good government" were laid down as overarching goals. In the Canada of today, despite its conservative past, perhaps because of an historical sensitivity to group rights, cultural survival is actively supported hand in hand with equal economic opportunity. To this end a minister of multiculturalism was recently appointed by the federal government (albeit in all probability as a piece of political opportunism in election year).

During August and September alone of this year (1988) more than 2,500 refugees applied for immigration into Canada, and during the first six months more than 11,000 immigrants and refugees were accepted into Québec. One Montreal school reports that students come from 69 ethnic groups, and that they speak 32 first languages between them, this same article notes that racial hostilities are becoming a problem, and describes an incident between rival gangs of Vietnamese and Salvadorian youths (Gazette 1988). The Canadian mosaic is becoming increasingly fragmented, but at the same time, items relating to ethnicity and the status of immigrants and refugees figure prominantly in the media.

Health care policies at both the federal and provincial levels are explicitly concerned with ethnic rights. Children's hospital in Montreal is the site of the first federally funded program in multiculturalism in Canada, and the health ministry in Québec has made statements in connection with the rights of people to receive health care and social services in their own language. At first glance these changes appear to be unequivocally for the better, but brief reflection reveals glaring contradictions inherent in both federal and provincial policies. For example, the present Canadian constitution has ambiguity built into its 1985 Charter of Rights and Freedoms, which states on the one hand that discrimination on the basis of ascribed characteristics such as ethnicity or race is prohibited, while on the other hand it recognized ethnic groups as having rights, and actively promotes multicultural policies. At the provincial level, the official language is French, all official transactions should take place in French, and immigrants must have their children educated in French, but at the same time, provincial policy now states that people have a right to have someone available who can speak their own language when they seek out medical help.

Given the current influx of refugees and immigrants, the question of ethnicity in relation to health care is of some urgency in Canada. The need for debate on the subject was recently brought home to me in connection with the research I have been doing with Greek immigrant women in Montreal. Not only are there contradictions in government policy, but this is compounded by contradictions within the health care system itself. Several nurses and social workers whom I interviewed brought up the same objections that had been raised with Zborowski in the 1960's when he set out to do his now classic comparative study on pain (1969), namely, that medicine is a scientific enterprise, in which one deals with the neutral terrain of biology. Hence attention to ethnic differences is not only not necessary, but actually racist. The two most vociferous nurses were themselves immigrants, one from Honduras and the other from Hong Kong. When I talked about traditional Greek ideas in connection with illness to a second generation Greek origin social worker, she implied that a graduate student and myself were reactionaries for being interested in 'superstitions', and stated emphatically that immigrants must get rid of their old wives' tales and make their way into the 20th century. A physician who sat through a talk I gave on the Greek concept of *nevra* (nerves) and its meanings, stated that it all sounded very poetic, and demanded to know if I was trying to make literary critics out of the medical profession. Those opposed to the recognition of ethnic differences in the health care system apparently accept with little or no questioning the ideology of mainstream biomedicine. Their reasons for opposition to multi-culturalism vary, I believe, from a personal desire to become part of what they perceive to be the dominant Canadian culture wedded to rationalism and progress, to a need to preserve the superiority of medical authority. The end result is the same, however, and while they may acknowledge variation in the basic etiquette or even nutritional needs of other peoples and perhaps attend to these differences in caring for them, anything more is 'unscientific' and, in any case, beyond the medical domain.

What contribution, if any, can a medical anthropologist make to this debate? Should one be committed to individual or to ethnic rights, or some fluid mixture of the two? Is the fostering of ethnic rights merely a ploy to either conquer and divide or to cull minority votes, and therefore where does one stand as an anthropologist? Is it possible to maintain anthropological integrity in the presentation of a group of people in health care settings, where time is of the essence and where rationalization of procedures and behavior is the order of the day? Does one slink away, when it becomes patently obvious that most ethnic groups do not like their cultural baggage being displayed by one outsider for the edification of other outsiders (a refined version of Orientalism in one's own backyard?) and when, in any case, many immigrants clearly want to sever all ties with the past?

In this paper I will use the experiences of Greek immigrant women to Montreal in order to explore these dilemmas. I will start with the concept of *nevra* / *nervios* / nerves, that polysemic term which occurs in every geographical setting where traces of Galenic medicine (in some locations reinforced by 19th century European physiology and early psychiatric beliefs (Figlio 1975)) survive today. The focus is on the discourse(s) of *nevra* in order to explore the relationship of identity, the "cunning of the weak" (Fischer 1988), ethnicity, and political consciousness. It will be shown that many Greek women (and perhaps the majority of immigrant women coming into Canada at present) have experienced numerous losses, both tangible and cultural, prior to and as a consequence of immigration and that their sense of self, of identity, is severely damaged. Furthermore, their traditional means of expressing belonging and dissent and of manipulating the system, are no longer available, leaving them psychologically and physically vulnerable, and often full of despair. Disenfranchisement is extreme, but there is little in either Canadian federal or provincial policies, the health care system, or the Greek community which functions to relieve the situation.

The project is applied in intent. My purpose is not merely to describe, but to make recommendations to those directly involved in multiculturalism and health care. This means that I must deal with one of the current concerns in anthropology, namely, "can ... interpretations be valid if they imply meanings of which the actors are unaware" (Bibeau 1988)?. Does one accept at face value what many of the women themselves say, that they endure the unendurable for the sake of their children, and leave it at that? Does one sit by and watch the medicalization of their bodies which many of the women themselves seek out, when one is convinced that the source of their pain is lack of self-esteem and lack of political power? Or does one actively chip away at dominant ideologies to instill a modicum of political consciousness, and an awareness of that most modern of concepts, individual rights? Such an approach would be in clear violation of the expressed wishes of some members of the Greek community who want to retain a sense of ethnicity and also control over recent Greek immigrants to Montreal.

I will start by situating the present study of *nervios* / *nevra* / nerves in relation to previous work on this topic, and then will move on to juxtapose the lives of immigrant women in Montreal with the lives, values, and aspirations which they had been raised to accept in Greece. It will be shown that *nevra* functions as a metaphor around which narratives about distressfull and painful events can be elaborated. The latter part of the paper will be devoted to the medicalization of distress, through an examination of the discourse(s) produced in health care settings in connection with *nevra*. In conclusion I will discuss the applied dimension of the project by setting the findings in the context of contemporary Canadian policy towards multiculturalism and health care.

## The Anthropology of Nerves

A survey of research on nerves over the past thirty years provides an illustrative history of medical anthropology in North America. In the early years efforts were largely directed towards achieving a clinical definition of nerves (Weisel and Arny 1952). This concern is still very evident today, and nerves is regularly used as an illustrative example in the ongoing debate about psychiatric nosologies and culture bound syndromes (Low 1985). Research into nerves from a meaning-centered perspective took root about a decade ago (Good 1977) and also continues to the present day. It is evident from these earlier studies that the cultural construction of nerves cuts across several conventional psychiatric diagnostic categories so that one cannot have a simple dialogue between a psychiatrically oriented viewpoint and an anthropological one.

In the latest collection of papers, the approach taken by all of the authors is one in which nerves are interpreted as an expression of distress resulting from familial, social and cultural disruption (Davis 1988, Dunk 1988, Guarnaccia et al. 1988, van Schaik 1988). Both macro and micro-contexts of power imbalances are examined in these papers. It is argued that the weak and dispossed (usually women), experience and describe debility and distress in the muted language of nerves; that nerves can therefore be thought of as metaphorical for the structural inequalities of gender, class, and ethnicity inherent in the societies under study. In her commentary on this particular set of papers Low praises them, but is also critical because the authors have not specified to what exactly they are referring when they use the term nerves. Are nerves, asks Low, a syndrome or a symptom? She believes that for purposes of comparison and for communication with medical practitioners, there must be clarity as to what, in physical terms, is being discussed (1988). Although Low states that diagnosis of nerves in biomedical categories is not the work of anthropologists, she nevertheless, in effect, accepts a culture as epiphenomena position. Not that she wishes to ignore culture entirely, but in using biological markers as the starting point for comparison, she imposes a straightjacket on the culturally constructed meaning of nerves which, while it may facilitate some sort of communication with the medical world, severely limits what is to be taken as relevant ethnographic data, and hence the possibility of producing a sound interpretive approach to the problem.

In contrast to Low, I would not criticize the papers but on the contrary praise them for retaining a fluidity and fuzzyness in terms of the relationship of linguistic categories to physical sensations. However, I think we can push the arguments made in this set of papers further, beyond an analysis which essentially describes nerves as distress, and how this is articulated with structures of power. It is important, I believe, to regard nerves as just one cultural genre among a large, perhaps infinite, number of possibilities whereby the powerless, sometimes consciously, sometimes unconsciously, flex their muscles. In order to situate nerves thus, one must be willing to classify them alongside, not only similar culturally constructed forms of natural symbols (that is, embodied emotions such as *solidao* (in Portugal), *pena* (in Ecuador), *ki* (in Japan), and so on), but also spirit possession, gossip, witchcraft accusations, carnival, and certain forms of theatre and poetry, and even what Scott describes as "everyday forms of resistance" such as "food dragging, dissimulation, false compliance, pilfering, feigned ignorance, slander, arson, sabotage, and so forth" (1985). Secondly, an awareness of the *longue durée* is essential in order to recognize the way in which nerves and similar forms of dissembling "get in the way of history, hinder its flow, and in hindering shape it" (Braudel 1980: 31).

Once this broad sweep of the canvas is in place, then *nevra* can be thought of not simply as 'somatization' or a 'disvalued body state', or even merely as a form of distress, but in a much more positive and political light: as a form of resistance. The various discourses of *nevra* can be interpreted *by an outside observer*, as products of and reactions to the dominant ideology of the Circum-Mediterranean honour and shame complex, at other times to the deployment of labor in capitalistic society, at yet other times as a product of Galenic medical thinking, or as various combinations of all of these influences. Very few of the individuals who experience *nevra* articulate a clear relationship between dominant ideologies and *nevra*, however they often pick out oppressive working conditions and social relations as causal agents and to this extent they are alert to the social and political origins of their problem.

Devisch, following Bourdieu, has stated that natural symbols "produce and are part of ideological systems, they disguise and justify the social artifice as a natural part of the unchangeable 'order of things'" (Devisch 1985: 409). However, I would agree with Lila Abu-Lughod when she suggests that "Bourdieu ... sometimes takes a position granting official ideology a totalitarian role in structuring experience" (1986: 256). Certain natural symbols, particularly those such as *nevra* which are used to express individual pain and mental anguish, have the potential of indicating all too clearly to the individual that there is something amiss with the 'natural' order of social life. Those in power may seek to reduce the symptom to biology, to convert it into an amoral 'it'. Alternatively symptoms may be construed to be due to a lack of personal responsibility, or to a lapse in morality. But none of these explanations questions the prevailing social order. The sufferer, on the other hand, while she may for a time accept the neutrality of her symptoms, or, alternatively, accept individual responsibility, may well come to believe eventually that her suffering in all probability has something to do with the 'sickening' social order.

If we accept the notion of sickness as cultural performance (Frankenberg 1986) and include in this the tiny dramas of daily life where individuals express their "inchoateness" through the use of culturally acceptable metaphors (Kapferer 1979), then we can see how natural symbols such as *nevra* can function to promote a self-reflexive critique, and potentially puncture a few holes in dominant ideologies. Individuals must constantly account for the contradictions between what their body is feeling and 'saying' with what other people say their body is feeling, and with what other people say their bodies *should* feel (Lock and Scheper-Hughes 1990). The study of nerves to date has dealt with the contradictions between what people say they feel and how, in the process of medicalization this language is appropriated and reconstructed into 'neutral' medical categories, often to the detriment of the well-being of the patient. We now need to consider further the dialectic between individual symptoms and dominant ideologies in general, not only to show how symptoms are culturally constructed and sustained, but also to show how, if healing does not occur, symptoms return again and again, forcing a search for causal explanations beyond the physical body. Sometimes the end result is despair, but sometimes symptoms are slowly recognized as repressed rage which can then become a source of political awareness - not that this awareness necessarily leads to dramatic social change.

## In Search of Security: Greek Migration to Canada

The Greek population in Montreal numbers over 70,000 and is one of the largest outside of Greece. It is composed primarily of immigrants of rural origin who emigrated between 1941 and 1975. Greece experienced a Nazi occupation during the second world war and thirty changes of government between 1944 and 1975 culminating in a right wing dictatorship between 1967 and 1975. This political instability coupled with regional struggles inside the country caused great hardship involving the loss of land for large portions of the rural population many of whom immigrated to Canada, Australia, and Germany where the majority of immigrants are employed in semiskilled or unskilled jobs in the service and manufacturing sector. The women who participated in the present study are for the most part employed as sewing machine operators in factories or alternatively are occupied with piece work in their homes. Others are salesclerks, cleaners, or waitresses. Thirty per cent of the sample migrated from villages of under 1,000 people. In all 59% of the sample is of rural origin, 26% from small towns, and only 17% from metropolitan areas with populations of over 50,000.

Data for this study were obtained initially from eighty-three first generation Greek immigrant women living in Montreal who were given semi-structured interviews in Greek in their homes. It became apparent in the course of carrying out these interviews that the concept of *nevra* was central in the narratives given by virtually all of the respondents. This term was inevitably used to link environmental and psychosocial variables to distress and painful physical states. In the second phase of the research, 19 of the original women were selected for in-depth open-ended interviews during which the concept of *nevra* was discussed in detail. A further data set was obtained from 25 women who were interviewed in the office of a Greek Canadian general practitioner, and 15 more attending a poly-clinic in a Montreal teaching hospital. These interviews were carried out in order to establish if *nevra* is used as a presenting complaint in clinical settings, and if so, how it is diagnosed and managed by health care professionals. Several internists, psychiatrists, family and general practitioners, and social workers who have daily contact with Greek patients were also interviewed.

## Attacks of Nerves

*Nevra* is most usually described as a feeling of loss of control, of having ones nerves "burst out" or "break out" or "boil over". An attack is often accompanied by shouting or screaming and throwing things about. Headaches, dizzyness, pain, and feelings of melancholy are associated with *nevra.* Women have the following to say on the subject:

"When you have *nevra* you are very, very different - you are not yourself. Your head hurts, you are tense all over. You scream and shout."

"When I have *nevra* I feel like killing a person, in a manner of speaking. It helps if I shout or scream. To be truthful, when I have *nevra* I cry a lot."

"I feel as though I am boiling inside. I want to hit the small ones so I must leave the children when I have *nevra.*"

Getting nerves is regarded as normal, although an unpleasant and disvalued affect. Some people are said to be more constitutionally or temperamentally vulnerable than others to attacks. Women are believed to suffer from it more than men, and virtually all of the respondants stated that it occurs more in Canada than Greece; its increased incidence is explicitly linked with the immigrant experience. It is only when *nevra* becomes chronic and disabling that women take on the sick role and seek out medical help. In clinical settings, not surprisingly, their complaints focus more on physical symptomatology which often includes sharp or stabbing pains in the chest or the base of the skull in addition to other non-specific symptomatology.

## Narratives of Nevra

Panagiota recounted her story in the waiting room of a Greek-Canadian general practitioner:

"I was born just outside Sparta and came to Canada when I was 16. I've worked ever since cleaning people's houses. I didn't have much education - just until I was eight or nine. My three children are all in college now and I'm really happy about that. My husband is a house painter. He has a regular eight hour a day job.

I came to the doctor about my nerves. I'm tired and really nervous. I get headaches and pain down the back and recently I've been screaming around the house a lot and fighting with my husband and the kids. I have a fear inside me all the time. I have this dream where I'm fighting for my life by myself. There's nobody to help me out. I'm in a vast desert with no way back and I have this dream several times a night. I see it more and more these days.

Whenever I try to talk with my husband we fight - maybe it's me. My husband jumps up and down and insists that I have to agree with him. I can't say anything for myself. He's tight with money - If I just ask him about money he's going to scream all day. And now I have these nightmares - I want to go far away. I fight with the kids too. I try to tell them how to behave but they think I'm too fussy. It hurts. My husband is always mad and screaming. I want to go somewhere by myself. Just me. Not with other people at all. My husband doesn't believe I'm sick but I've come to the end.

I have a sister in Greece. My brother died last year and I have nobody else. I have a few friends here but I can't open up with them. It's shameful to tell them family things. Something is wrong, I know. I tried to send my brother's ashes back to Greece like he wanted but my mother didn't want to deal with the papers and so on at that end. I kept seeing my brother in my kitchen after he died. He wanted to go to Greece, but my mother wouldn't accept him and my father is too sick to do anything.

I don't want to live in Greece. I've been back three times. It's 27 years now I've been in Canada. I think that Greece is more claustrophobic than ever. My family doesn't care about me anymore. I have to live with that inside me. My husband is nervous too, a very nervous guy. I told him to go to the doctor but he says he's alright. It's like a prison, my house. I can't go out without his permission and I have to call home as soon as I arrive at my friend's house. I can't go to the movies. I have to be home for him otherwise he screams. He doesn't hit me, just shouts so as everyone can hear. I'm 44 and I don't have a life of my own. My family are strangers; my friends mean more to me than my family. I miss my father most. He's 91 now. Life isn't easy but I try to laugh.

I've seen many doctors and taken lots of medicine. I've had a lot of tests too but they can't find anything wrong with me. I don't understand why. I am sick. I feel sick. I like this doctor, he's kind but I feel sorry for him. He is trying so hard to help me and I come to see him so often. Yes, I take valium right now, but it doesn't do any good, except it helps to impress my husband a bit."

Maria was born in a Peloponnesian village:

"I'm 55 now and I have three children all over 20. I came to Canada in 1952 and started out by cleaning houses but I soon switched and started to work in a garment factory. They fired me a year and half ago for being too old, but I do a good job, I'm a good operator, it's just my age that is no good. My unemployment insurance has just run out and I'm sure no one else will hire me now.

I'm a nervous type. I've been getting nose bleeds recently and this is because of my *nevra.* I get *nevra* because things aren't too good at home. My husband and I don't get on well and we are so poor. My husband gambles and I was always left at home with the kids - that's the Greek way - men leave the family alone and the wife is at home like a slave with the kids. He stopped gambling a year or two ago but it's worse - now he stays at home and makes me nervous. He only likes to go out by himself and be free - that's how Greek men are."

These accounts with their emphasis in marital discord and isolation are similar to those which immigrant women from many different backgrounds recount, but when they are contextualized in light of ethnographies about rural Greek life they take on new significance.

## Honour, Shame, and Gender Roles

In common with many other societies of the world (Griaule 1965, Hugh-Jones 1979) an explicit relationship is made in Greece between a healthy and 'correct' human body, a clean and orderly house, and moral order in society at large. The house is the focus of family life in Greece, not only in terms of furnishing all the physical and social needs of family members, but because it is also a spiritual center, replete with icons and regular ritual activity where family members seek to emulate the Holy Family (DuBoulay 1986). Management of the house is the special responsibility of the woman who is both functionally and symbolically associated with it (Dubisch 1986). Cleanliness and order in the house are said to reflect the moral character of the woman, and a discussion of private, 'inside' family matters should not cross the threshold into the threatening domain of the outside world. Ideally, a woman should never leave the house for frivolous or idle reasons and venture outside where dirt and immorality abound. A woman who spends too much time out of the house can be accused not only of neglecting her household duties, but is open to suspicion of illicit sexual activity or gossip, either of which activities can damage the all important social reputation of her family.

Just as a distinction is made between inside and outside the house, so too is a distinction made between the inner and outer body (Dubisch 1986). Contact must be avoided between what goes into the body and what comes out of it. Dirty clothes and polluting waste human products must be strictly segregated from food preparation. While the fulfillment of male sexual needs are considered imperative, a woman's life is hedged with taboos in connection with menstruation, marriage, the sexual act, and childbirth, designed to combat her supposed innate negative characteristics, mute her sexuality, and contain the polluting nature of her body products. Dubisch postulates a parallel between the kitchen in the traditional Greek house and the vagina:

"The kitchen, the point of entry into the house, is protected by the porch, an area for collecting or deflecting dirt, just as entry into the woman is protected both by modest clothing and the propriety of her own deportment, which deflects improper sexual advances or gossip. Both kitchen and sexual entryway are subject to cultural rules regarding the passage of substances, rules that serve to turn a natural product or impulse into a culturally approved one." (Dubisch 1986: 211).

A woman who conducts herself appropriately, therefore, is responsible for the maintenance of cleanliness, order, and harmony in both her personal conduct and the household. Her task is to bind the family unit together, keep it ritually pure, and protect it from the potentially destructive outside world. Together with the raising of children, this role is the prime source of a Greek woman's pride and self-esteem.

Relationships among Circum-Mediterranean families have been described as competitive in terms not only of economic advancement but also in connection with the honour of the family. Everyday behavior of Greek men and women is characteristically structured around the concepts of honour and shame as organizing principles (Campbell 1964). There is regional variation in the elaboration of these beliefs, but the assumption that men should protect the family honour through their *filotimo* (manliness), while women must show *dropi* (sexual modesty) is universally accepted. Men are responsible for the behavior of the women in their family and a woman's body "becomes the symbol of family integrity and purity and, more generally, of society as a whole" (Dubisch 1986: 211).

The oppositions described above between inside and outside set up unresolvable contradictions for Greek women. As part of the maintenance of honour and prestige, it is expected that a family should keep up an active social life and be 'open', an auspicious state which is actively sought after (Hirschon 1978), but for a woman 'openness' is a potentially dangerous state equated with a lapse of sexual modesty. Ideally, women should be 'closed' and controlled, but paradoxically, for many this becomes associated with feelings of ill health, with withdrawal, melancholy, and depression, the Greek word for which is *stenahoria* (literally, a 'narrow place'). A woman who becomes too 'closed' and withdrawn is thought of as antisocial and 'cold'. Women, therefore, spend considerable energy trying to preserve a suitable equilibrium between inside and out, open and closed.

Classical Galenic concepts of health and illness causation promote a belief in balance and moderation in connection with health maintenance: excess of worry, work, food and alcohol, and sudden changes in environmental conditions should be avoided. Emotional stability is also highly valued so that loss of control, and most particularly feelings of bursting out of one's body boundaries, of blurring inside and outside, are feared and thought to contribute to illness causation. *Nevra* is the concept used to mediate between distressing social events, disvalued emotional states (of either a sensation of loss of control, or of confinement and withdrawal, or both) and somatic symptomatology.

### *Yiannoula the Athenian*

"I came to Canada 28 years ago and I'm 60 years old this year. My husband was a mechanic on the railroad but now he's retired. We still have our son at home. I've been a sewing machine operator ever since I came to Canada, but I've recently been laid off.

I feel tired all the time these days and I've been having *nevra* and headaches for several years now. I get upset really easily and feel as though I'm losing control. It's the factory that's the problem. That's the big reason why everyone gets sick and has pressures. You work like a horse for seven hours. I say, like a horse, because if I speak to you at work I lose money because it's paid by the pieces. Every minute the packet [the bundle of clothes to be sewn] is no good [in disarray], you lose money. You understand the pressure? ... I used to make good money. I made $500 each week ... because I am fast in the factory. I made 1,400 pair of pants and skirts each day ... but the bosses decided the workers were making too much money. They broke up our section ... and they lowered the price paid for the pieces. ... This sort of thing makes you nervous. You don't sleep at nights. I see the mountains of pants. I finish this color in half an hour and then they bring me the next. I finish the black and then they bring me the red mountain. I never stop."

I used to leave the house at 5 a.m. and you're not hungry then, you just drink a fast coffee and go. At 8 o'clock you're hungry but you're working so hard that you can't eat. At 10 o'clock you have a fast coffee break. The organism is not an electric button. The organism knows when its night and when you should be sleeping and when you should eat lunch. That work schedule left me all out of control.

I've been taking pills for my *nevra* for a few years now and seeing doctors. But I think it's too late for me. I can't relax at home after all those years in the factory, you give your life for the pieces. My dreams are broken.

I've never got used to the Canadian winters. I'd like to go back to live in the Greek countryside but my husband likes Montreal and my son is Canadian. I feel as though I've lived my life half in two countries."

### Immigrant Women and the Garment Industry

The garment industry in Montreal is notorious for exploitation of the labor force. There is little job security, no guarantees of full shifts and the work is seasonal with periods of excessive work followed by long periods of unemployment. The average number of work weeks per employee per year was reported in a 1984 study as 28 (Goldenberg 1984). When workers are employed, the hours are long, pay is often below minimum wage, noise and dust levels are high, ventilation is poor and there is constant pressure to work quickly. The scene is very reminiscent of the Manchester cotton mills at the end of last century since managers parade up and down in front of the rows of women reprimanding them for slovenly behavior.

The effect on health of poor working conditions and poverty are not limited to women, of course, but gender distinctions are manifest on both the shop floor and in the union structure. The few skilled jobs - those of cutters - are held mostly by men. These jobs are paid considerably more than are sewing machine operators, who are principally women, usually recent migrants to Canada. Union membership is composed primarily of women, but the executive is made up of men, and a two-tiered system of bargaining often takes place whereby good contracts are secured for male employees at the expense of any increase in wages for sewing machine operators. Many immigrant women choose or are forced to do homework, that is, they purchase their own machine, pick up 'pieces' at the factory, sew them up at home and then deliver the completed article to the factory door. There are estimated to be about 20,000 such workers in Montreal at present who have no security of any kind, and recent surveys indicate that this kind of labor is on the increase.

Reasons for the existence of homework today are the same as they have always been: for the workers it is a way to combine household commitments with paid employment and for the employer it reduces overheads, acts as a hedge against a recession, and keeps wages down. Many of the women work in the factory until they have children and then take up homework because they cannot afford childcare. Since many men and women of Greek origin are fearful when a married woman has to work outside of the home because of a concern with honour and sexual propriety, this is often an added incentive to do homework. When asked to identify the disadvantages of homework, the most frequently cited problem was that the workers never felt able to relax. "Because there are no fixed hours and no breaks, homeworkers feel that they are under constant pressure to produce" (Johnson 1982: 79). Women cite the "double work day" as an extra burden since they retain full responsibility for housework and the care of children in addition to working long hours.

For Greek women whose identity is invested in a clean and orderly house homework presents a great burden. Some Montreal apartments reveal more clearly than words the contradictions which the women feel. The largest room is usually filled with the 'best' furniture which is often covered with plastic coats; delicate embroidery adorns the coffee table, and everything visible is scrubbed to a shining perfection including the interior walls, which are washed as though they were part of a traditional Greek house. The sewing machine and piles of fabric are seconded away, as is every sign of daily life. The children are clean and well behaved, husbands are usually absent; outward signs are of order and control, cleanliness and serenity.

### Immigration and the Loss of Traditional Values

Traditional values are not abandoned with immigration, on the contrary, the uncertainties produced by a new life may promote them. Greek women in Montreal often complain that they are very rarely allowed out of the house unaccompanied by their husbands except to go to and from work and the supermarket. A Greek-Canadian physician described many of his patients as suffering from the 'hostage syndrome'. In traditional life the subordination to and dependence upon men which women must abide by is curtailed by numerous social supports. For example, a man who is too harsh with his wife will be chastised by the community. Moreover, women derive strength and friendship from female relatives. Cramped apartment life in Montreal, often with no relatives in the same city, leads to a greater dependence than in Greece on the part of the women upon their husbands, especially because they have virtually no opportunities to learn either French or English. In addition to unwanted feelings of dependence, themes of isolation and nostalgia run through the stories given by women to account for *nevra*. The combination of a small apartment and the Montreal climate adds to a sense of loss and confinement when compared with the warmth of an airy Greek house. Many people had their house and land forcibly taken from them before they decided to come to Canada, others simply chose to emigrate, usually in search of financial security, a dream which is rarely materialized in one generation. The women we interviewed spoke with great longing for their homeland, and mourned their homes and gardens where they had maintained a self-sufficient life-style and from which they had derived much pride. Their other traditional source of self-esteem is also irrepairably damaged with immigration: women are not able to control to the extent they would like the socialization of their Canadian children who grow up with a very different value system from their parents.

## The Causes of *Nevra*

Narratives which women give to account for either their own or the *nevra* of other women indicate very clearly that, although they believe some women to be more constitutionally vulnerable than others, they attribute ultimate causality to social and political variables. Moreover, they link *nevra* explicitly to the immigration experience and to a nostalgia for their lost heritage. Immigrants, like Yiannoula, often think of themselves in exile:

"The Greek comes to a foreign land and they have this *anxos* (anxiety) to return to the *patridha* (homeland). With the family, they can't manage (to return home) and this *anxos* stays within them and they always suffer."

"I believe that all Greeks suffer from this illness because they attempt to get money and return to Greece. This creates *nevra* in the family situation and they have *anxos* from the work in the factory."

I think that *nevra* is caused by the type of lifestyle we live. Life is too fast, too quick, we are always rushing about. Even if I don't work and I stay home I always have this stress. We all do. In Greece it is different because we relax in the afternoons, we nap."

"We were born in Greece and that is where we grew up. We came here as adults. Naturally we had gotten used to a different way of life and we were used to freedom. All of us Europeans suffer more here. It must be the way people live their life here, even the doctor tells you this."

People get *nevra* more in Canada because they come with nothing and they are trying to get ahead."

"They have *nevra* in Greece and they have them here too, but here in Montreal life is much narrower [more constrained]. I hear that a lot of women in Montreal are using pills for *nevra*, more than in Greece."

"There is more pressure here ... I come from a city of about 40,000 in Greece and they don't have the pressure we have here. My aunt works in a factory in my city and she says the work is hard, but not like here. You know here you have someone on top of your head all the time pressuring you to work more."

"Women are much more nervous because they have many more responsibilities: the house; if they work they have that plus the boss; if they are home there are household responsibilities and the children and for the Greeks the woman does everything. The man doesn't wipe the dishes, use the broom or anything."

"In the afternoon, when the woman comes home from work there is the house, the need to cook, and the children in the middle. All these things will cause her *stenoxoria* (to feel confined, depressed) and *nevriasi* (to become nervous)."

"I think that *nevra* is an illness that everyone is accustomed to today. Without a doubt, when you don't lead a normal life, when you are constantly tired, in the house and have *stenoxoria* and melancholy ... All these things cause *nevra*. And monotony ... when you stay inside the house constantly, and you lead a life of the same things all the time, you get *nevra* ... Or when you have to work all of the time, and you have to be there on time, or you have trouble at your job ... all these things. Life today is just *nevra*."

This everpresent background of contributory factors is punctuated with precipitating events, crises of various kinds ranging from being fired or laid off from work, to quarrels with family members: "*Nevra* is usually set off by your husband or your children. When you tell them something and they don't listen to you it makes you nervous." Feelings of a lack or loss of control are cited as causal in acute episodes of *nevra* which are considered normal. It is only when the episodes become very frequent, or osscilate rapidly back and forth between *stenohoria* in which one feels confined, and agorophobia in which one feels terrified at the thought of going out even with the family on Sundays, that gradually a woman takes on the sick role.

## A Lament for the Homeland

Greek women suffer a massive onslaught to their traditional sense of identity as the result of migration. The losses, cultural and material, which they sustain before and as a result of immigration are multiple, and the economic rewards which they seek in Canada are only very rarely realized in one generation. Immigrants from rural Europe have, of course, formed the backbone of the east coast North American labor force for more than one hundred years, but whereas in the early days people came from rather extreme rural poverty to a continent full of hope and opportunity, today the situation is rather different. Rural Greece is no longer the back of beyond, and North America is no longer the land of promise it once was. Moreover, there is fierce competition from refugees and immigrants from all over the world for even the most menial of jobs.

The stereotype of the life of a Greek woman is one of subservience and exploitation. Recent ethnographic work demonstrates, however, that analysis of gender roles in terms of simple oppositions between men with power and women with none is very inadequate. Although descriptions given by women about women are unambiguously negative: "Men are intelligent but women are 'gossips', ... man in his work is said 'to have a blessing' but 'the woman ... is cursed' ... men are reliable while women are credulous and fearful ... men carry responsibility, but women are 'the weak link'" (Du Boulay 1986: 139), there is also a contrary, more symbolically embedded set of values in which women are associated with religious power.

The word for house in Greek *(spiti)* is synonymous with the family itself, and belonging to a family defines all of a villager's categorical rights and obligations. It is only through identification with the family that one can attain an adult identity. A man is defined all his life as a 'guest in the house', whereas a woman, from her pivotal position within the house is responsible for not only all the basic nurturing in connection with the family, but also as the mediator between the family and the realm of the sacred. The roles of men and women are more appropriately considered as complementary rather than hierarchical, with possibilities for pride and self-esteem in the fulfillment of either role. Women do complain at times of beatings and of sexual abuse, and they openly criticize public male behavior when it is regarded as either foolish or too aggressive, but Du Boulay has pointed out that women do not set out to overturn the ultimate authority of men (1986: 153). She describes the close association of women with the sacred by quoting a village woman: "the woman is the Mother of God in the church. The woman is the Mother of God, she is the church" and goes on to conclude:

"Thus while it is true that in the imperfect world of humanity the men of rural Greece tend to overstep the bounds of their proper authority and power, and the women to suffer for this, it is true also that the women wish only to modify this behavior, not to reverse the order of reality on which it is symbolically based, for ... they would, in consequence, be forced to overthrow also their association with the myth of the birth of Christ and the honor given to women through the figure of the Mother of God (Du Boulay 1986: 167)."

Since the family is the primary focus for one's loyalty and since outsiders cannot be trusted, women do not band together to create a feminine subculture which might serve as a focus for social change. Moreover, as is the case in many virilocal systems of residence there are inbuilt tensions between mothers-in-law and daughters-in-law so that even within a household women do not naturally unite against male oppression. Women, nevertheless make life-long friendships with other women which they often value highly and from which they derive great pleasure (Kennedy 1986).

Although apparently no obvious efforts are made to change the restricted lifestyle assigned to them, there are, however, certain traditional cultural forms available to Greek women for the expression of dissent, the most notable of which is the ritual lament. Most laments are performed in cemeteries, but they can also be performed in the home, the fields, alone, or in the company of other women. The tradition is oral formulaic, but improvisation is admired. Caraveli shows how ritual laments are used not only to mediate between the living and the dead, but also as a form of protest in which disallowed emotions can be expressed in muted register; the lament is an expressive genre very close to that of the poetry of Bedouin women described by Abu-Lughod (1986) and the 'boundary-dissolving' songs sung by outcaste Tamil women (Trawick 1988).

In contrast to women in the village, immigrant women find themselves in a twilight zone; one where traditional sources of prestige are not longer available, where traditional codes of conduct in which one could take pride in the homeland can actually become crippling, where there is no access to alternative sources of satisfaction, and where there is no time to learn the languages of their new country despite free language programs. Their husbands, themselves often victims of exploitation and massive culture shock, may demonstrate their anxiety by exercising tighter controls than usual over their wives, or by resort to physical abuse.

Family ties are ruptured with migration, so too are valuable friendships. Traditional idioms for the expression of distress, such as the lament, are no longer available. This tradition is on the decline in Greece itself, as is unquestioned religious belief, but in Montreal, the genre is virtually finished, and the orthodox church is no longer viewed as a source of support (Dunk 1988). Losses due to migration are massive, gains from immigration apparently very few, except one - the reason that most of the migrants stay and do not flee back to the sunshine, olive trees, and oregano covered hillsides: the children. One endures everything; the factory, an abusive husband, a cold climate, a tiny apartment, and isolation, for the sake of the children.

### *Nevra* as Sickness

Anna is 29 years old. She is married and has no children. She was educated to grade 6 in Greece and came to Canada eleven years ago with her brother who has since returned to Greece. She is employed as a sewing machine operator in a garment factory and her husband is a short order cook in a restaurant. Anna works a 9 hour day and does the housework when she comes home. Her husband works the 5 p.m. - 2 a.m. shift at his restaurant so that she rarely sees him on workdays.

Anna has been having problems with *nevra* off and on for a few years. Lately she feels that her nerves have been getting worse. She says, "I have *nevra* all the time. I get headaches and pain in the back of my head, and sometimes I feel like I am going crazy. I seem to spend a lot of time crying." Anna's husband gets frustrated with her when she is upset which causes her more worry.

Anna describes work as something which she has to do but doesn't really enjoy, although she likes to meet her friends at the factory. She says, "the machine's are no good, the work is no good, the boss is telling you 'vite, vite, vite'. I have too much nerves. Too much nerves. I go to work and I come back. I stay at home, I can't relax. Sometimes I just feel like crying and I don't know why."

Although she has worked at the same factory for 11 years, Anna has recently begun to have trouble on her job. Organizational changes have been made but without any worker consultation. There is more pressure to work quickly and any protest brings a threat of firing. Anna attributes this to high levels of unemployment and the awareness of her employer that there are plenty of desperate people who will soon replace a grumbling employee. "Before they didn't push you. Now they push all the time and say if you don't like it, go."

Anna would like to move back to Greece, but her husband enjoys life in Canada. She believes that life is more relaxed in Greece, work is different there and more time can be spent outside and with friends. She describes her life in Canada as monotonous - "You go to work, come home at 7 p.m., cook dinner, clean the house, do the laundry and go to bed. The next day you get up and start the same thing over again." She continues: "I can't really talk to my husband, but if I talk to someone it feels better. You see, I am talking to you now and it feels better. It's crazy like that. My sister doesn't like to talk, and my husband doesn't like to talk."

Anna has recently begun taking medication (ativan) but believes it to be ineffective. She likes to talk to her doctor about her problems: "The doctor helps a lot. We have a talk about what my trouble is, why I am nervous. Afterwards I feel a little bit better."

## The Medicalization of *Nevra*

Very few women start out by talking about *nevra* when they go for medical help, but because many of the physicians they see in Montreal are Greek-Canadians, the term is recognized and actually appears on the diagnostic charts in one of the hospitals which holds a 'Greek clinic'. From the biomedical point of view, *nevra* is a condition which is difficult to diagnose. The condition is usually described as 'psychogenic pain of unknown origin' and many of the women are diagnosed in addition as either depressed or suffering from anxiety. Several medical professionals, including those of Greek origin, stressed that, in their opinion, Greek patients make a great fuss and become upset easily, shout and scream and tend to disrupt the waiting room. Physicians agreed that is is important to allow patients to ventilate their feelings, but added that there is a tendency on the part of female patients to focus on somatic complaints *because* it is not acceptable for women to express hostile feelings about their family life. Physicians talked about 'covering-up' on the part of patients, because they believe that many women have in fact come to see them for comfort and support, but are unable to describe the reasons for their distress.

None of the physicians interviewed denied the existence of disabling symptoms, moreover all of them considered psycho/social and sometimes political factors to be of causal significance, but also agreed that intervention at this level was extremely difficult, especially since it is not acceptable to intrude into the sacred realm of Greek family life. Physicians, like patients, freely acknowledge the limitations of medication and agree that medicine often merely acts as a source of legitimation for the woman's sick role with her family. At the same time they often feel frustrated and unable to be of much help. Several of the Greek Canadian doctors stressed that women should give up their futile dreams about returning to Greece, but only one doctor appeared blatantly unrealistic about the lives of his patients. He suggested that they should learn how to relax more, and not worry so much about making money.

Whereas physicians are, on the whole, sympathetic, social workers exhibit much less tolerance. In particular, those of Greek origin expressed their feelings very strongly. It was stated emphatically that the women should break out of their shackles, 'adapt' to Canadian life, and learn to put up some resistance vis-à-vis their husbands and employers. The social workers appeared to be insensitive and unrealistic to me, and I thought that their hard line was in part a result of the fact that Greek women, while they respect physicians on the whole, do not like to be seen by social workers and are explicit in their dislike of them. Social workers, after all, pry into family matters. But there is more to the story than this. The statistics are no doubt somewhat inaccurate, but indicate that wife beating and child abuse are high in the Greek immigrant community. Compared to other immigrant groups, Greek women, for example, rather frequently appear in shelters for battered women. The overworked and harrassed social workers who must deal with the horrifying consequenses of these problems are impatient with the women whom they see as allowing themselves to be victimized. Some of the social workers have grown up in Greek families themselves, they have been encouraged by their own mothers to create a style of family life different from that of their childhood, and they have little patience with traditional values.

Health care for the majority of Greek Canadians in Montreal is centered around one particular community clinic, one of many clinics distributed around Québec known as CLSCs *(centres local de services communautaires)* which have served as primary care centers since their inception in 1972. One of the purposes of such clinics is to organize local citizen groups around specific health related issues such as preventive medicine, maternal and infant care, the problems of the elderly, and so on. The social workers at the CLSC in the heart of the Greek community, with physician and nurse support, have tried several times to organize a women's group at the clinic, but with little success. Clearly women who labor in factories all day are exhausted in the evenings, furthermore, many of their husbands will not give permission for them to go out of the house. In addition, for those who come to organizational meetings, the experience is often a disappointment because factional arguments, so evident in Greek society, often emerge to dominate the evening. Factions are formed around perceived ethnic divisions within Greece, and favoured political affiliations. These divisions are so strong that women, given their culturally nourished suspicion of strangers of any kind, soon become disinclined to take part in any group activities.

## Ethnically Sensitive Health Care: Myth, Mirage, or Realistic Dream?

The reactions, verbal, physical, and behavioral, of Greek women to the losses associated with immigration reflect their socialization into Greek culture. If one specifies these losses in detail then they can be thought of as 'typically' Greek, but at a more general level of abstraction the losses reflect a situation which all of us in this rapidly changing world, but most particularly immigrants, and even more so, refugees have to face. Berman has described the experience as the "contradictory forces and needs that inspire and torment us: our desire to be rooted in a stable and coherent personal and social past, and our insatiable desire for growth" (1982: 35). The difficulties faced by first generation immigrants are, of course, severe and it is a truism that they often think of themselves as a lost generation where all hope for the future is invested in the children. The exploitation which takes place in the labor force is disgraceful, but it is naive to assume that this is simply exploitation of minorities by those in the powerful majority. In the case of the garment industry in Montreal most factories are managed by immigrants who in turn exploit more recent arrivals.

It is clear from talking to Greek women that they are 'individuals' in that they have attained a sense of self which is to some degree independent of what is culturally expected of them. Unlike the Hofriyati women of Sudan whose selfhood is "culturally overdetermined" in that they appear to attain only a minimal capacity for criticism of the social order (Boddy 1988: 16), the Greek women in this study demonstrate, for the most part, a keen awareness of oppression, and link their situation to gender inequalities which are built into Greek traditional culture, and class exploitation which is built into capitalist society. The majority of women would support the position of the Cretan woman who stated: "Of course we want equality. How could we not want to be equal? My dreams for my daughters are for them to try and educate themselves ... I think that when she is educated, my daughter will be able to decide by herself what to do" (Kennedy 1986: 127).

Women reveal in the interviews the extent to which they are aware of the resistance which can be expressed through *nevra*. They are often explicit about the way in which attacks, whether infrequent or chronic, can be manipulatory and produce small victories for change. This is not to say that the pain is not 'real', or that *nevra* is some kind of culturally sanctioned charade, but there is often a degree of conscious control and manipulation involved. It is rather rare, of course, for a woman to be able to articulate her criticisms in the broadest of terms, although many are very devastating in their comments about traditional Greek society and about factory conditions, only one or two were critical of the 'modern' dominant concepts of economic progress and development.

What do these perigrenations tell us, if anything, about ethnically sensitive health care in Québec? Clearly concepts such as *nevra* should be taken seriously for what they represent and not reduced to psychiatric nosology. At the level of health care provision, some very practical suggestions can be made such as, for example, that isolated women can be introduced to one or two other women from the same part of Greece in order to form friendships (not support groups). Or that doctors and social workers should try to understand the antipathy of Greek patients towards social workers and why. If physicians take time to talk to the women this is apparently greatly appreciated, much more so than being sent off to see a social worker. Since some women are already highly critical of traditional Greek values perhaps there is room at the CLSC to focus on this subject.

At the level of government policy, my conclusions are that efforts so far, federal and provincial, are indeed mostly window dressing, although they include some very fundamental rights such as health care in one's native language. I believe that focussing on ethnic needs as a policy *may* be well intentioned but is nevertheless a red herring, in that the implicit values embedded in modern medicine and accepted by most health professionals remain unexamined. So too do the unequal power distributions and exploitation of immigrant labor which is the source of so many of the health problems of recent arrivals in Montreal. Canada, after all, wants immigrants to do the work that no self-respecting Canadian will do. The irony of this situation is that Greek immigrants who came to

Canada early this century, like so many generations of established immigrants from many places before them, revel in the dominant value system which I as an anthropologist criticize in connection with the treatment of their newly arrived compatriots. Perhaps the Greek women are quite right - the dream really lies with the children, perhaps that is the most realistic hope for an uneducated first generation immigrant woman until such time as the fabric of capitalist society with its associated values is transformed.

## References


ABU-LUGHOD L. 1986: *Veiled Sentiments: Honor and Poetry in a Bedouin Society.* Berkeley: University of California Press.

BERMAN M. 1982: *All that is Solid Melts into Air: The Experience of Modernity.* New York: Simon and Schuster.

BIBEAU G. 1988: *Meaning, Context and Experience in Trialogue*. Opening remarks to the Symposium "Between Semantics and Rationality". 12th International Congress of Anthropological and Ethnological Sciences. Zagreb.

BODDY J. 1988: Spirits and Selves in Northern Sudan: The Cultural Therapeutics of Possession and Trance. *American Ethnologist* 15: 4-27.

BRAUDEL F. 1980: *On History.* Chicago: University of Chicago Press.

CAMPBELL J.K. 1964: *Honour, Family and Patronage: A Study of Institutions and Moral Values in a Greek Mountain Community*. Oxford: Clarendon Press.

DAVIS D. 1988. The variable character of nerves in a Newfoundland fishing village. *Medical Anthropology* 11: 63-78.

DEVISCH R. 1985: Approaches to Symbol and Symptom in Bodily Space-Time. *International Journal of Psychology* 20: 389-415.

DUBISCH J. 1986: Culture Enters through the Kitchen: Women, Food, and Social Boundaries in Rural Greece. In: DUBISCH J. (ed.) *Gender & Power in Rural Greece*. (pp.195-214). Princeton: Priceton University Press.

DuBOULAY J. 1986: Women - Images of their Nature and Destiny in Rural Greece. In: DUBISCH J. (ed.). *Gender & Power in Rural Greece*. (pp. 139-168). Princeton: Princeton University Press.

DUNK P. 1988: My Nerves are Broken: *The Social Relations of Illness in a Greek-Canadian Community*. M.A. thesis, Montréal: Department of Anthropology, McGill University.

— 1988: Greek women and broken nerves in Montréal. *Medical Anthropology* 11: 29-46.

FIGLIO K.M. 1975: Theories of Perception and the Physiology of Mind in the Late Eighteenth Century. *History of Science* 12: 177-212.

FISCHER M.M.J. 1988. Aestheticized Emotions and Critical Hermeneutics. *Culture, Medicine and Psychiatry* 12: 31-42.

FRANKENBERG R. 1986: Sickness as Cultural Performance: Drama, Trajectory, and Pilgrimage. Root Metaphors and the Making Social of Disease. *International Journal of Health Services* 16: 603-626.

GAZETTE 1988: School steeped in Racial Tension, Students Say. Friday, October 14.

GOLDENBERG S. 1984: Chained to machines. *McGill Daily*, March 8: 10-11.

GOOD G. 1977: The Heart of What's the Matter: The Semantics of Illness in Iran. *Culture and Psychiatry* 1: 25-58.

GRIAULE M. 1965: *Conversations with Ogotemmeli.* Oxford: Oxford University Press.

GUARNACCIA P.J., de la CANCELA V. and CARRILLO E. 1988: The multiple meanings of atagues de nervios. *Medical Anthropology* 11: 47-62..

HIRSCHON R. 1978: Open body / closed space: The transformation of female sexuality. In: ARDENER S. (ed.) *Defining Females*. London: Croom Helm.

HUGH-JONES C. 1979: From the Milk to the River: *Spatial and Temporal Process in Northwest Amazonia.* Cambridge: Cambridge University Press.

JOHNSON L.C. 1982: *The Seam Allowance: Industrial Home Sewing in Canada.* Toronto: The Women's Press.

KAPFERER B. 1986: Performance and the Structure of Meaning and Experience. In: TURNER V.W. and BRUNER E.M. (eds.) *The Anthropology of Experience.* Chicago: University of Illinois Press.

KENNEDY R. 1986: Women's Friendships on Crete: A Psychological Perspective. In: DUBISCH J. (ed.). *Gender & Power in Rural Greece*. Princeton: PrincetonUniversity Press.

LOW S.M. 1985: Culturally Interpreted Symptoms or Culture-Bound Syndromes: A Cross-Cultural Review of Nerves. *Social Science and Medicine* 21: 187-196.

— 1988: Health, Culture and the Nature of Nerves: A Critique. *Medical Anthropology* 11: 2-14.

SCOTT J. 1985: *Weapons of the Weak: Everyday Forms of Peasant Resistance*. New Haven: Yale University Press.

TRAWICK M.1988: Spirits and Voices in Tamil Songs. *American Ethnologist* 15: 193-215.

VAN SCHAIK E. 1988: Paradigms Underlying the Study of Nerves as a Popular Illness Term in Eastern Kentucky. *Medical Anthropology 11: 15-28.*

WEISEL and ARNEY 1952: Psychiatric Study of Coal Miners in Eastern Kentucky. *American Journal of Psychiatry* 198: 617-624.

ZBOROWSKI M. 1969: *People in Pain.* San Francisco: Jossey-Bass Inc.

# The Development and Change of Health Research among Migrant Workers in West-Germany
## Ideologies and Practice Between 1956 and 1986

Beatrix Pfleiderer

### The Beginnings of Research Interest

The onset of large-scale worker immigration to West-Germany in the 1950s opened a new chapter in the history of medical care, a chapter that today, more than thirty years later, has not yet come to a close. Because the migrant workers came from other cultures and accordingly brought with them different concepts and customs concerning therapeutic measures, this wave of immigrants - like the later medical challenges in the countries encompassed under the term "third world" - had an impact upon both the forms of comunication and the forms of knowledge in the medical world.

In his contribution to the handbook on the "Psychiatry of the Migrant Worker", Böker adresses the resulting questioning of conventional medical self-understanding by taking this position with regard to the symptomatology and etiopathogenesis of psychically ill migrant workers:

"Are these *typical* psychiatric conditions whose pathoplastic is *characteristically* formed *merely* by the culture-specific behavior of the foreign patients and which appear in the same manner and same frequency as in their country of origin - or are we faced with phenomena which are first elicited or perhaps even caused by the migratory process and which may possibly be describable as a *typical* 'migrant worker syndrome"' (Böker 1975: 430, emphasis mine).

Böker's choice of words provides us with a glimpse of the helplessness of the expert. The complexity of otherness and the changes in disease manifestations surprised the support agencies to such an extent, that a multitude of studies and research projects were triggered as a result.

At first, one looked at similar experiences from other countries with a history of immigration, e.g., the United States and Canada. The beginnings of health research (Malzberg and Lee 1956) into the culture of migration is to be found in psychiatry, where it was possible to draw upon the methods and the goals of transcultural psychiatry (Kraepelin, Bastide, Kleinman, Pfeiffer, Ødegard) as well as reflecting - on a more fundamental level - upon the relationship between nature and culture in pathogenesis. This displacement in psychiatry is also reflected in the fact that the *Handbuch für Psychiatrie der Gegenwart* contains two articles on transcultural psychiatry and the psychiatry of migration, while the Handbuch der Sozialmedizin, which also appeared in the 1970s (1975, 1976, 1977; edited by M. Blohmke et al.), made no mention of the problems. This interest on the phenomenology of the forms of psychic disturbances among migrant workers initially shifted the psychology of migration into the domain of ethnopsychiatry or "cultural psychiatry". This closeness, however, was merely apparent, for the methods and intellectual interests differed, since the "identity of the basis" as well as the "equivalence of meanings" (Bastide 1973: 58) was not guaranteed. The question of comparability of "emotional worlds" will be taken up again later when we examine the work of the American anthropologist Kleinman on the question of the cultural influences upon somatization and psychologisation (e.g., depression) (cf. Lutz 1985, Kleinman and Good 1985, Kleinman 1986).

## Early Literature on Migration: The Interest on Psychiatric and Sociological Questions

The early psychiatric literature on migration which appeared in America and Europe examined and classified hospital admissions. As a result, the evaluation of first admissions of immigrations in psychiatric institutions in the USA, Australia, and Canada provided the first evidence of a culture-specific (or, more appropriately, nation-specific) patterns of morbidity.

The fact that patients born in their native country outnumbered at admission patients born in the country of choice appeared to suggest that the social stress associated with the readjustment could have a disease-inducing effect. Social stress and selection were the contrasting mechanisms which concerned the psychiatric epidemiologists in the migration situation. During this initial phase of research, national characteristics were not so apparent, unless one considers the frequent emphasis upon the high rates of alcoholism-related psychoses among Irish migrants to the USA (Ødegard 1972: 248). Her observation of this fact was what led the American anthropologist Nancy Scheper-Hughes to investigate the question of selection in the country of origin. This resulted in a year-long fieldwork study of the predominant personality structures and the associated ways of life in the rural Irish areas from which many of the immigrants came.

'Deviant behavior' or psychic disturbances were found to be integrated into life-style in such a manner that they were no longer considered noticeable in the face of the culture-specific concepts of normality (Scheper-Hughes 1979). When applied to the ideology of the hospital admission studies, this makes it easier to provide an answer to questions concerning the selection of emigrants, in this case from Ireland.

The problem can not be seen in any psychiatric presdisposition of the migrant worker who becomes psychically ill during a certain measurable period after his immigration. Instead, it becomes necessary to examine the particular ways of life (isolation vs. integration, etc.). These will either reveal and make manifest the psychic illness, or they will have an integrative effect and stylizes this integration to a way of life.

This early literature on migration, characterized as it was by its interest in causality, related disease terminology to class, sex, origin, and country. What was lacking at this time was an interest on the course, interpretation, and methods which the afflicted person utilizes to deal with his illness. In his above-cited article, Böker (1975) wrote that the difficulties in working on the incidence and prevalence of psychiatric diseases lies in the fact that epidemiological studies into these questions on foreign populations were not able to clearly distinguish between permanent emigration and more short-term residency abroad.

The early relevant studies on immigrant workers in the Federal Republic and the diseases they manifest include those of Frießem (1972) and Häfner, Moschel, and Özek (1977). Frießem counted the number of psychiatric first admissions of foreign patients in a psychiatric hospital in Stuttgart between the years of 1963 and 1970 and compared these with a group of German psychiatric patients. He found that schizophrenia and affective psychoses were distributed uniformly, while neuroses, abnormal, and psychosomatic reactions were significant higher among immigrants. Suicide attempts and alcoholism were lower than among the group of German patients. Frießem also attempted to statistically measure national characteristics and found that Spanish immigrants had a higher rate of schizophrenia and Turks and Greeks a lower rate of alcoholism. From the perspective of the present study, it is interesting to note that Frießem was able to recognize at this relatively early period of migration that Turkish immigrants exhibited high rates for psychosomatic, neurotic, and abnormal-reactive symptoms.

The findings of Häfner, Moschel, and Özek (1977), which are primarily based upon the investigations of the Turkish psychiatrist Özek, are of similar interest. Following the early studies of Ødegard and Malzberg from the 1930' s, Özek interviewd a group of 200 migrant workers from central Anatolia directly after their arrival in the Federal Republic. Three months after their arrival, a quarter of this group had been reported to be psychically ill; a rate which rose to one-third after eighteen months. Özek found that while depressive symptoms predominated during the first three months, the emphasis had shifted after eighteen months almost exclusively to psychosomatic complaints without any substantial depressive components (Häfner, Moschel and Özek 1977: 268). Özek, who conducted his hospital admission studies along the lines of the American approach and obtained his sample through the employment office, interpreted his findings in the following manner:

"There are two possible ways to explain this shift of symptoms which we observed between the first and the second study. Either the separation from home and family had its greatest impact during the initial phase, and the reactions of grief lessened with their deactualization, while the psychosomatic syndrome remained as an expression of the reaction to continuing stress. Or the psychosomatic syndrome is an expression of an internal partial accomodation to a separation which has never been overcome, and which it may lead to rejection or desensitization of the grief or the memories which forms its basis (Özek 1977: 274).

I shall discuss the problem of partial adaptation and the forms of psychosomatic changes below.

In his article on the incidence and prevalence of psychiatric illnesses, Böker discusses a number of studies, including one by Bingemer et al (1968), who studied the perception of symptoms among Spanish, Italian, Turkish, and Greek migrant workers. They reported that complaints about "grief and depression" came almost exclusively from Turks (Böker 1975: 442).

What is striking about these (quantitatively oriented) studies from the early phase of migration research is the persistence in comparing non-comparable data. The concluding remarks of the article we have just looked at should suffice to arouse our suspicion. There, Blöker speaks of a relative "uniformity" of the observed manfestations, of the "lack of cultural variability" (Böker 1975: 274). For the purpose of simplicity, this is often referred to as the so-called "depression of uprooting". Here, the cultural moment makes itself apparent through the fact that it is not seen. And if it is seen, e.g., in a "description of characteristic manifestations, such as psychoses with hysterical behavior, magical forms of experience, and culturally-conditioned meanings", then this becomes a deviance within deviance. Here, it is the product of "subjective attention" which is, however, "objectively apparently small" (Böker 1975: 275).

Summing up, it can be stated that voices critical of the evaluation of admission data were soon heard, as too little knowledge of the type, form, reasons, and aims of migrations were taken into consideration. What is also remarkable is the ignoring of data about patterns of socialization and personally structures, which could provide the most important indicators of premorbid dispositions (Danna 1980). At least part of the reasons for this lie in a cross-cultural and international divergence of research interests. Whereas in the United States and in other countries there was a wide-spread interest in cross-cultural studies of socialization, an interest which also found an echo in transcultural psychiatry, interests in Germany were initially limited to more epidemiological matters. This represents something of a paradox, as the early literature on migration was concentrated upon an *explicit search for causes*.

In contrast, the next class of studies we shall examine reveals a *distinct tendency to what is 'characteristic'* by the medical support agencies. Now, attempts were made not just to 'understand' how the observed 'characteristics' which the patient himself spoke of belonged to the context of the patient, but to even integrate where appropriate the culture-specific conceptions of the patient into the therapeutic process. I shall refer to this shift to the conceptual environment of the migrant worker as the anthropological interest.

## The Anthropological Interest and Disease Study in the Context of Migration

In this section, we shall be examining studies which report about the necessity of linking the clinical eye with the anthropological. We may follow the lead of Bastide, who wrote that the sociology of the mentally ill does not permit him to leave his cultural background. Instead, he asserted that it would be a good idea to make use of the 'methodological spirit of the anthropologist'. "For the anthropologist is attracted by *what is different*" (1973: 57-58, emphasis Bastide's). Accordingly, he suggested that the first rule for psychiatric support agencies and experts be that they 'combat their ethnocentrism and *not utilize the categories* of Western psychiatry to evaluate the neuroses of other peoples" (Bastide 1973: 58, emphasis mine).

The study by Risso and Böker (1964), "Verhexungswahn - Ein Beitrag zum Verständis von Wahnerkrankungen süditalienischer Arbeiter in der Schweiz" (Hexomania - A Contribution to the Understanding of Delusionary Afflictions among South-Italians in Switzerland), was a most important German-language study which both paved for the way and helped to effect such a change. Following a detailed presentation of their case histories, the authors reported the following insight. Although the "symptomatology of our patients revealed a surprising uniformity (emphasis mine), a diagnostic ordering (emphasis Risso and Böker) of the disease manifestations met with substantial, almost insurmountable difficulties ... Thus, our experience showed that attempts to encompass South-Italian patients in the customary psychiatric schema proved to be quite generally difficult" (Risso and Böker 1964: 61). Those patients who were admitted into the clinic with complaints of having been hexed had falsely interpreted the culture-specific symbols used by young (Swiss) women with whom they had "fallen in love".

This cognitive dissonance and the distress resulting from these false interpretations eventually led the young men to 'delusionary' conceptions and to 'disturbances of the body experience'. The acceptance of these patients' concepts and the therapist's consideration of their causality is an importatnt step in the development of a therapeutic culture in an age of massive culture-contact. The works of E. Zimmermann report similar steps. Based upon his observation of clinical reality from an anthropological perspective over a period of several years, he arrived at the following interpretation of mistakes in the therapeutic process. The (in this case South Italian) concepts of disease, which differ from the discourse in the guest country, impede the understanding between the doctor (who may consider magical explanations to be 'irrational if not absurd') and the patient (for whom disease units lie outside of the body, so that they must be 'incurred'). Zimmermann's works demonstrate how incorrect diagnoses and therapies inevitably result. He explains why patients are often deeply disappointed when, in spite of their 'dogged insistence' on the questions concerning the cause of their disease, they are not told about this in a manner they can comprehend, but rather in a complex and symptom-oriented manner. As Zimmermann reported on a pediatric hospital in Freiburg, the work of the doctor becomes more difficult when the parents of patients present the doctor with 'signs of disease' specific to their interpretation of the disease. And he accepts these as genuine symptoms and bases his diagnosis and therapy on this (Zimmermann 1981, Zimmermann et al. 1983). Time and again, Zimmermann experienced how important it is that the cognitive and symbolic environment be taken into consideration by someone familiar with the area when a patient is describing symptoms and that this information should also be made available to the support agencies. This view, in which an anthropological perspective is joined to the clinical, was only slowly beginning to spread. In another study which tended in the same direction, it was reported how patients with predominantly psychogenic disturbances consulted both traditional healing possibilities such as the Hodsch and medical agencies such as hospitals. Using four case histories, Röder and Opalic (1987) have reported on the role of the traditional healer in the consolidation of four psychosocial cases of conflict within Turkish families. The traditional healing agency advised that the conflicts be solved in a manner appropriate to the country of origin.

The authors assume that the utilization of 'reliable symptoms' had a favorable effect upon the abatement of the disturbances. They explicitly argued that the doctor tolerate this possibility for the foreign patient and, what is more, that they be encouraged to incorporate their experiences with the Hodscha into the therapeutic process. The acceptance of the patient in his cultural totality is, the authors add, conducive to the healing process (Röder and Opalic 1987).

In American medical anthropology (German: Medizinethnologie), there are a number of studies which deal precisely with questions concerning how symbols presented and utilized by traditional healers effect the healing process. Two authros who have discussed this problem in a number of articles and have each written a monography on the subject are K. Finkler (1985) and A. Kleinman (1980). The power of symbols in the healing process has long been known to anthropologists. Healers mediate between culture and nature; as Moerman (1979) writes in his well-received survey of the anthropology of symbolic healing, they initiate what may be thought of as a 'cultural physiological' reaction. In this article, Moerman shows that the metaphorical structure and the system of meaning of a therapeutic unit is decisive for its effectiveness, and is just as important as 'real physical' factors. He advocated viewing symbols from the perspective of their effects upon the hormones produced by the hypothalamus. A sympathetic reaction based upon symbolism may result in just as much stress as a 'real fight or flight' situation. Stress which arises through participation in healing rituals arises through the meanings attributed to the central metaphors of the healing ritual and not from actual physical danger, although the nature of the stress does remain the same (incist hands, dizziness, etc.). It is this context of stress, as Moerman writes, which we must take into consideration when thinking about the effects of healing metaphors which undoubtedly change and 'move' us. The *construction* of healing symbols *is* itself healing (Moerman 1979: 66, emphasis Moerman's).

Kleinman, in his study in Taiwan, described how patients reacted to the therapy of traditional healers and doctors and made the following distinction: those patients who had a biophysical disturbance and visited traditional healers and doctors had more chance of a successful healing than one afflicted with more psychosocial problems (Kleinman 1982).

## Recent Clinical Studies of Illness in the Migration Culture

During the 1980's, it is possible to observe a change in the Federal Republic with respect to the therapeutic *handling* of foreign patients. The literature which shall be considered in this section is representative of this change. It is striking that the view which was concentrated so long upon the causality of 'such and such' a syndrome (the reader may recall the 'depression of uprooting') shifted to the meanings of the complaints. Similarly, the sequences of disease episodes as well as their meaning with reference to the social environment of the patient should be observed and induced within the therapy.

In an analytic understanding of disease, E. Leyer describes forms of intervention that were developed in the context of the Giessen Model Project for the psychosomatic provision of Turkish immigrant workers. The patients, who were distinguished by three features - membership in the lowest social class, the presence of physical and/ or psychosomatic illnesses, and their experience of problems specific to the migration culture - were offered a *gesprächstherapeutisch* treatment. The patients, most of whom had a long history of illness with a number of contacts with doctors, almost always complained of pain. Psychosomatic symptoms are symptoms of speechlessness, Leyer writes, and refers to the anthropologist and psychoanalyst Devereux (1978), who sees these symptoms as an overadaptation to the cultural patterns of illness in industrial cultures. With Balint, Leyer considers body pain to be an unbearable psychic pain which has become a physically perceptible affliction. This proposed symptom is usually received in the sense of an 'institutionalized practice of conflict defense' between the doctor and the patient (Volmerg 1984). The life crisis which may be seen in association with unattained migration aims is *medicalized*: "sociostructurally produced conflicts are translated by the patients into somatic disturbances and responsibility for correction is given over to the health care system" (Leyer 1987: 309). Many patients have *precise conceptions about the causes of their afflictions*. These concern the conditions related to their migration. Leyer reported from her therapeutic experience that Turkish patients who were offered a "solid emotional relationship were able to perceive the conflicts expressed through their physical symptoms surprisingly quickly" (Leyer 1987: 311). At the same time, her presentation makes it clear that psychotherapeutic intervention, often in conjunction with help from social workers, will be accepted.

In 1983, W. Lehmkuhl published a study entitled "I can't keep this up" ("So kann' s nicht weitergehen"). Once again, a case history was seen in the context of migration and the socio-cultural environment and the interpretation of the "exceedingly clamorous symptomatology" was taken into consideration in the attempt to understand. The case concerns a conversion reaction elicited as a result of a family conflict. The adolescent female patient who was being clinically treated for a paralysis in the legs, was forced by her family, and in particular her mother, to agree to her engagement with a young Turkish man she had never met. The author demonstrates how the therapeutic discussions led the patient to realize that she "herself had created the connection between her inability to walk and the marriage plans of her mother and was able to comment on her situation "I can't keep this up", which led to the alteration with her mother. Her inability to walk sabotaged her mother's marriage plans. Under the influence of (therapeutically) agreeable surroundings which offered her alternative solutions, the patient was able to abandon her symptoms (Lehmkuhl 1983).

Another - almost extraordinary - form of intervention is reported in a very interesting article by Devisch and Gailly (1985). They discuss the foundation and methods of a therapeutic 'self-help' group called Dertlesmek, whose structure, course, and forms of communication were designed to conform to the traditional meetings of Turkish women (e.g., at the Turkish henna festival). Like the authors previously mentioned, they also describe how Turkish patients are able to manifest their somatically perceived complaints in a very selective and 'theatrical' manner and that they adhere in this to the pattern of suffering in traditional Turkish culture. Devisch and Gailly show how the physical pattern of suffering expresses, and in fact reflects, the predicament in the life-style of the migration context. The disease is often transferred to the epigastrium. The pain resides in the place where reception and transmission occurs in the body. The center, which is characterized as yürek, is the seat of vitality, sentiment and temperament, while the stomach symbolizes the exchange of substances between the body and the outer world. Thus, according to Devisch and Gailly, when the patients speak of an 'inner distress', this refers to a kind of retreat from the outer world. To interrupt the circle of somatization and medicalization, patients who express their life crises in a certain physical manner were invited to join the Dertlesmek group therapy. In this way, the external isolation in which they found themselves could be overcome and the inner self could be accepted as meaningful for both the other and the patient herself. This group offers women in menopause a new social opening under changed, but nevertheless culture-specific signs. This was not a symptom oriented therapy, for the continued presence of symptoms was not considered as a failure, but a 'rite of transformation' (Devisch and Gailly 1985).

Studies of this type make it clear that the study of disease among immigrants is involved in a 'paradigm shift' (I may be allowed to make use of this overworked concept). A clear distance has become apparent to those interests manifested as catalogization and the search for causes. The fact that this latest phase of dealing with ill immigrant workers has assigned problems associated with communication a more central role makes it clear that an interest has arisen in the representation of the disease and its semantic function in the social context of the patient.

In his sociolinguistic analysis of medical consultations with Turkish parents, Rehbein (1985) pointed out that the often fruitless results were frequently the result of more than just linguistic barriers, and also reflect differences in structures of knowledge which are culture-specific and institutionalley conditioned. Rehbein demonstrates how the 'institute of consultations' aids in shaping different kinds of interests because two interactants from different professional groups and (in this case) cultures confront one another. This 'basic paradox' of the consultation, Rehbein points out, is in itself sufficient to make communication between Turkish parents and German doctors more difficult, so that misunderstandings should not always simply be attributed to any 'intercultural conflict'. In consultations between such a combination of interactants (German doctor - Turkish patient}, the professional often fails to hear the patient's own evaluatory processes or his or her acquired and traditional knowledge (which is expressed in the effects of self-help within the family context) during the 'listening' and 'eliciting' of the consultation and for this reason does not 'tune into' this during the process of consulting. In a detailed analysis of selected consultations based upon his linguistic-pragmatic approach, Rehbein used the speech ductus to show that the clients clearly referred to knowledge previously acquired in the collective process of consulting. Generally, this reference was not linguistically or contextually perceived by the German partner and, accordingly, could not be realized. For this reason, Rehbein suggests in his 'conclusions' that "neither in Turkey nor the Federal Republic should the structures of culture-specific self-help and self-consultation be destroyed. Rather the structure of familial practical knowledge should be held intact in the face of urbanization and as an effective communicative processing of disease, thereby making modern medicine accessible to migrant groups" (Rehbein 1985: 414). To achieve this, it is necessary that the doctor linguistically and semantically discuss with his non-professional clients in order to avoid "refocussing, dethematisizing, and to bring his patients' knowledge into line with his own (Rehbein 1985: 415, cf. Pfleiderer 1984, Pfleiderer and Bichmann 1985).

Collatz also reports on some failed attempts at consultation and communication. Drawing upon an empirical study (complete survey of all childbirths) as part of a pilot project in the German states of Bremen and Lower Saxony to improve health care both prior to and following delivery, it was possible to compare the consultation and delivery behavior of Turkish women with other foreigners and Germans. It was found that Turkish women tended to carry a greater burden of medical risk, that they had more need of consultation, and that they were less informed than German women about the meaning and frequency of preventative examinations and periodic check-up of young children. Qualitative analyses revealed that Turkish women often had such a slight understanding of preventative examinations during pregnancy or were even repulsed by their treatment, and that the advice given and the forms of communication used were so deficient that they no longer went to the examinations after two or three visits because they saw no *sense* in them (Collatz 1986: 80; emphasis in original). If we recall Rehbein's findings, then Collatz' report provides additional confirmation from a different quantitatively investigated domain. In another quantitative-qualitative study in the Rhein-Neckar district, the disease behavior of Turkish patients in Turkey and in the Federal Republic were compared for the first time in a large-scale project (Kroeger et al. 1986). The report stated that whereas traditional concepts of 'folk medicine' still influence how the sick are treated in Turkey (and in larger ghettos of foreigners), in the migration context it was the more 'readily available Western medicine' which, as the only really accepted healing agency in the industrial society ... had become the standard for interpreting diseases (Kroeger et al. 1986: 279). Culture-specific concepts of disease produce, in the eyes of those investigated, no reasons to delay in resorting to modern medical services (1986: 279).

In a 1985 anthology from Collatz, Kürsath-Ahlers, and Korporal entitled 'Gesundheit für Alle' ('Health for Everyone'), we find other important evidence that forms of prevention specific to a particular immigrant culture were being related to curative and rehabilitative care. Thus, Korporal (1985) showed that during the preventative care of pregnant women, foreigners received less extensive care. He pointed out that, in general, there is still little specialist (and instead more general) care of foreign patients. One result of this is that in comparison with German women foreign pregnant women are treated less frequently with vitamin and calcium preparations and more with such ill-advised (during pregnancy) medicines as tetratcycline, androgens, etc. Korporal suggests that similar research findings must be taken into consideration in order to be able to achieve a social and cultural opening of medical care agencies for the needs of foreign patients (Korporal 1985).

In another contribution dealing with qualitative social research on female Turkish immigrants, Grottian showed how fundamental ethnomedical structures of knowledge influenced the understanding of health and the body among the women who were interviewed. The relativity of the concept of health which was utilized was demonstrated in a very clear manner and elucidated with original statements. For example, a thirty-eight year old Turkish woman whose menstruation had ceased after the birth of her fourth child thirteen years previously, stated that 'my body is rotting'. For us, this means that this woman adheres to a concept of the body in which menstrual blood is in any case polluting. In her case, this pollution had attained the state of illness because it had accumulated through lack of bleeding. She traced a number of symptoms she had noticed back to the fact that her body was in an impure state per se (Grottian 1985: 282).

In her contribution, Theilen discussed the metaphorical shape in which this perception was cast. This pointed out the relationship between a culture-specific scale of value and a metaphor of the organs. Such metaphors for all that is terrible as 'my liver is decaying, ... is cut into pieces' or 'my heart and my blood hurt', find their ways to the doctor, where they are intended as statements about the patient's general condition. 'Unfortunately', as Theilen (who has had a long history of experiences in clinical intervention among Turkish patients) writes, "the condition which lies at the basis of this description, namely a holistic disturbance of a person's well-being, is ... not understood by the representatives of our culture and our medicine in its holism and it thus frequently misinterpreted“ (Theilen 1985: 296). In the further course of the article, Theilen provided an interpretation of the conceptual environment of such utterances as well as the implications they have for disease behavior. Thus, the liver and the lungs (white liver) are seen to be the seat of sorrow, disease, and pain. Yet liver also stands for the 'inside of a thing' as well as for 'person and human'. Since the liver is mentioned metaphorically when describing one's condition (e.g., as the seat of sorrow, Turkish uzüntü ), numerous Turkish medical journals make efforts to rule out any supposed disease of the liver itself.

In the same volume, Dietzel-Papakyriakon (1985) reports in a similarly graphic manner about the relativity of symptoms among female Turkish workers who were compared in an interval study to a group of female German workers. Brucks et al (1985) made an attempt to differentiate among earlier questions in health and disease research. Theirs is an attempt to overcome the mere relativity of symptoms and to systematically remove some of the clichés (e.g., 'culture shock') which arose during what has become known as 'guest worker research'. They point out that the earlier methods of cultural comparison did not understand that the migration culture triggers learning processes and therewith changes which cannot be studied using the earlier models. One of the interesting perspectives opened up through this study is its establishment of the relationship 'human-body-work', which is elucidated on the basis of exhaustive interviews. In general, this volume represents a pivotal turning point in migration research. It distinguishes among many of the previous attempts and suggests many ways for new research.

## Prospects for the Future: Somatization versus Psychologization. New Comments about an Old Controversy

At this point in our survey of the literature into the problems associated with cross-cultural health research, I would like to call attention once again to the thoughts of one of the leading medical anthropologists and psychiatrists in America, Arthur Kleinman. As was shown at the beginning of the paper, the comparative approach to culture-specific patient behavior had its beginnings in comparative psychiatry. I attempted to show the reasons why this was so. Here, it may be added that as (bio-) medicine once again began to spread through the world in the 1950's after the colonial structures began to be dissolved, psychiatry was assigned a pioneering role with respect to comparative questions during this process of expansion. It was precisely the 'psychic disturbances', the metaphorical moment involved in being ill, which helped to express the role of culture. For example, it was comparative psychiatry which discovered the so-called 'culture-bound syndromes', and it was through the use of psychiatry and anthropology that attempts were made to recognize the cultural relativity of the expressions and perception of emotions. In contrast, comparative anthropological research is still in its infancy. Another point to consider is that comparative psychiatry has been succeeded by a cross-cultural approach to health research. This is a result of the fact that both physicians and anthropologists who work in other cultures have been involved in systematic efforts at documenting how purely physical perceptions and presentations of symptoms can be related to cultural conditioning (cfr. Pfleiderer and Bichmann 1985). While the export of medicine out of the West and into countries like India and China has been taking our understanding of disease (in such 'fashionable' forms as neurasthenia versus depression) into other countries for hundreds of years, there are still many regions (and not just on the periphery) which have not yet been affected by the spread of Western medicine. This is true for many parts of one of the countries which has given the Federal Republic so many migrant workers: Turkey.

While 'modern' psychiatry was being propagated in Istanbul at the turn of the century by Rasit Tahsin (a student of Kraepelin, the father of comparative psychiatry) and established by the actual father of Turkish psychiatry, Mazhar Osman Uzman (Volkan 1975: 391), away from the centers of learning, disease and psychic disturbances were being dealt with by centuries-old traditional or family healing agencies (Öztürk and Volkan 1971: 259-260). What this means for a person caught up within the context of migration is that they were forced to turn to their own, ancient empirical knowledge which had developed in independence of the developments of Western science. With a doctor-patient ratio running from one to several thousand or more in the peripheral areas, it seems unlikely that Western medical knowledge constituted a component in the lifes of migrants before they left their country (at least to the extent that they did not arrive in Europe as part of a slow process of internal migration). This aspect of the migration context is frequently reflected in the courses which diseases take as well as the manners in which those afflicted express their concept of the body (Theilen 1986). The perceived symptoms (e.g., "everything hurts; cf. Ete 1986: 143), the frequent lack of detailed accounts about the causes and perception of diseases, and the models of disease progress, is capable of being understood if we take the above-mentioned hierarchization of knowledge into consideration. In the migration context, models for disease forms have been composed of folk medical units (1), Islamic and Galenic remnants of the doctrines of the humors (2), and the processed knowledge of biomedicine (3) as practiced in their host nations. The statements of Turkish patients often surprise Western doctors because the products of these different traditions are difficult to predict. Such a patient's statement as "it is the eye", which is referring to a knowledge of magic (1) and speaks of the evil eye which he considers responsible for the disease, is, when uttered as a phrase, as confusing for the doctor as the statement "my blood is bad", which is based on the doctrine of humors (2).

The literature is more or less unanimous in supporting the view that the intolerability of psychic or social life is expressed among Turkish patients through somatization and the fact that they describe their afflictions in terms of organs. The clinical experience with immigrant workers from a relatively simply structured society corresponds to the ideas which have been taken over by comparative health research, and thus appears to provide evidence that "simple people" from "simple societies" are "only" able to express their afflictions in a physical manner. That is, individuals with such backgrounds experience personal and social problems which they interpret and articulate by means of their bodies (Kleinman 1986: 51). Yet there is also another variation which may lead to somatization. In a cross-cultural survey, Marsella (1979) showed how the somatizing expression of depression exhibits a high prevalence in the clinical samples of non-Western cultures.

Other studies have shown that Nigerians, Turkish groups in Iran, rural Mexican women, and Chinese in Taiwan all develop their own specific somatic forms for their afflictions (Kleinman 1986: 52). As a clinical study demonstrated, Chinese suffering from depressions tend especially to exhibit such physical complaints as tiredness (90%), pain (89%), and gastro/ cardiological symptoms (87%) to go with their statements (sorrow, despondency, et.c.) about their condition (Cheung 1981; cited in Kleinman 1986: 53).

In his important study into the social backgrounds of disease and suffering in China, Kleinman notes that many of the above-mentioned studies overlook the fact that (1) somatization is a frequent phenomena within the Western industrial context and that (2) somatization is not merely a translation of depressive disturbances (Kleinman 1986: 55). Medical sociological studies in England and America have shown that three-quarters of all of the patient visits to general practitioners involve cases of somatization (ibid). In Western societies, Kleinman writes, somatization appears among low-income groups and groups with little education, in rural settings, among patients who are religiously active, and with those who consider themselves part of a tradition with respect to their ethnic membership. Somatization, according to Kleinman, was prevalent in the West up to the period when the Victorian middle class (and its Wilhelmian equivalent here in Germany) had admitted and produced psychologizing idioms for suffering. Kleinman considers the psychologizing idiom to be analogous on the personal level to the social process of rationalization as described by Max Weber. What this means is that the terminology and the experience of 'affect' has been transformed among the Western middle class to a 'deep' psychological experience with specific terms for the specific emotions of depression, fear, anger, etc., emotions which were only a short time previously felt and categorized as physical experiences. Because of the fact that bodily experiences are seen and understood by means of social categories and live in our thinking as so-called natural symbols, it becomes possible to understand both psychologization and somatization as cultural constructs for psychological processes (ibid). The psychologization of suffering is thus the result of the Western form of modernization, and also affects to elite of the non-Western world. Somatization, on the other hand, remains restricted more to the rural, traditionally-oriented side and to the poorer, less educated classes in the West. In contrast to the view generally held by researchers into comparative health, however, it is not the somatization in China, India, Turkey, or the West which should surprise us or appear unusual and demand explanation; instead, it is the psychologization of suffering in modern Western societies which is so unusual and in need of explanation (Kleinman 1986: 56).

Kleinman also discusses the fact that the dichotomy of psychologization here and somatization there should not be further pursued because it is based solely upon clinical observation. He cites a number of (longitudinal) studies of somatization which show that somatization is not related to (suppressed) psychic disturbances alone, but is in fact based upon a dialectical model. This model, which is primarily the product of Mechanic, a medical sociologist (Kleinman 1986: 59), maintains that a certain degree of introspection and processing of physical and psychological experiences based upon an expansion of the somatic and cognitive-affective complaint does take place. Thus, this model views emotions within an integrated and non-dualist framework as embodied experiences (ibid). Under this assumption, somatization should also be seen as a question of the cognitive style of the patient, who is led by his experiences with a system of medicine and his familial experiences to a certain style of disease.

Somatization may thus be considered as adaptive behavior, as the effect of cultural and personal features, as the result of 'labeling' processes, and as a transformation of emotional states (Kleinman 1986: 60).

In an anthology co-edited by Kleinman which appeared in 1985 under the title "Culture and Depression", a group of anthropologists, psychiatrists, and psychologists concerned themselves with the role which culture plays in the disease of depression, which has been known to medicine and psychiatry for at least 2000 years. We are told of the cultural plasticity of dysphoric feeling in association with a culture's concept of normality and from the cognitive explanatory models which a culture uses to interpret its culture-specific scales of feelings. Thus, for example, a practicing Buddhist considers it self-evident that he voluntarily turn towards dysphoria as the first step on his path to salvation. This deliberate acceptance of a 'depressive' or 'dyshoric' mood has no place in Western normative scale, and it is thus removed from any therapeutic efforts with a Western origin. For it is implicitly assumed that the 'normal' condition is neither depressive or melancholic, but rather its opposite, i.e., happiness.

Western researchers have been greatly surprised by, e.g., the manners in which the Balinese deal with feelings. In a number of anthropological and psychiatric documents, we are told of the emotional performances of the Balinese people, who exist in a great life crisis. Existential convulsions are hidden behind a smile, or are they rather expressed through a smile. These are questions which a comparative anthropology and psychiatry must face if they are to apply a psychiatry which is based upon folk theories of emotional expression.

If we think further about culture-specific distinctions, then we see that among us, depression is an emotional annoyance and schizophrenia (in contrast) a disturbance of thought. Here, we see again the role which 'feeling' and 'thinking' play in our folk theory of the person as a feeling and thinking unity. If we wish to use language as a means of making the 'emotional work' (Lutz 1985: 89, cf. Hochschild 1979) of different peoples accesible to one another, then we should remember that our language is hardly able to escape from our folk theories of the concept of the person and the concept of emotion. For this reason, Catherine Lutz (1985) has argued in her article on the "Translation of

Emotional Worlds" that the translation of emotional states must pay more attention to the social and cultural contexts and begin to make use of analyses of semantic networks as well (Lutz 1985: 89, Good 1977). When this occurs, we may be sure that the processual nature of the attribution of meaning and changes that occur in the meaning of perceived symptoms and complaints will begin to become understandable. For those of us in the field of medical anthropology, this means that we will be able to learn to see case histories as texts whose understanding require that their cultural uniqueness be taken into consideration (Pfleiderer 1981).

## References

BASTIDE R. 1973: *Soziologie der Geisteskrankheiten*. Köln: Kiepenheuer und Witsch.

BINGEMER K., MEISTERMANN-SEEGER E. and NEUBERT E. (eds) 1972: *Leben als Gastarbeiter - geglückte und mißglückte Integration*. Opladen: Westdeutscher Verlag.

BLOHMKE M. 1975ff.: *Handbuch der Sozialmedizin.* Heidelberg: Springer.

BÖKER W.1975: Psychiatrie der Gastarbeiter. In: *Psychiatrie der Gegenwart* (2. ed., Vol. III: Soziale und angewandte Psychiatrie, pp. 430-466).

COLLATZ J. 1986: Probleme der gesundheitlichen Versorgung ausländischer Mitbürger am Beispiel schwangerer türkischer Frauen. In: HESSE G.; MEYER R. and PASERO U. (eds.) *Krank in der Fremde*, Evangelische Akademie Nordelbien (Dokumentationen II)

DANNA J.J. 1980: Migration and Mental Illness: What Role Do Traditional Childhood Socialization Practices Play? *Culture, Medicine, and Psychiatry* 4: 25-42.

DEVEREUX G. 1978: *Ethnopsychoanalyse*. Frankfurt: Suhrkamp.

DEVISCH R. and GAILLY A. 1985: A Therapeutic Self-help Group among Turkish Women: Dertlesmek: The Sharing of Sorrow" *Psichiatria e Psicoterapeutica Analitica* 4 (2): 133-152.

DIETZEL-PAPAKYRIAKOU M. 1985: Krankheitsverhalten türkischer Arbeiterfrauen. Ergebnisse einer Intervallbefragung. In: COLLATZ J. und KORPORAL J. (eds.) *Gesundheit für alle* (pp. 323-338). Rissen. E.B. Verlag.

ETE E. 1986: Besondere psychische Krankheitsbilder türkischer Patienten: Möglichkeiten und Grenzen der Behandlung. In: HESSE G.; MEYER R. and PASERO U. (eds.) *Krank in der Fremde*. Evangelische Akademie Nordelbien (Dokumentationen II).

FINKLER K. 1985: *Spiritualist Healers in Mexico*. South Hadley, Mass.: Bergin and Garvey.

FRIEßEM D.H. 1972: Psychiatric Disorders of Foreign Workers in a Metropolis of South-Western Germany. *4th International Congress of Social Psychiatry, Jerusalem* 25-26.

GOOD B. 1977: The Heart of What's the Matter: The Semantics of Illness in Iran. *Culture, Medicine, and Psychiatry* I: 25-58.

HÄFNER H, MOSCHEL G. and ÖZEK M. 1977: Psychische Störungen bei türkischen Gastarbeitern. Eine prospektiv-epidemiologische Studie zur Untersuchung der Reaktion auf Einwanderung und partielle Anpassung. *Der Nervenarzt* 48: 268-275.

HESSE G., MEYER R. and PASERO U. (eds) 1986: *Krank in der Fremde.* Tagung der evangelischen Akademie Nordelbien. Dokumentationen II.

HOCHSCHILD A.R. 1979: Emotion Work, Feeling Rules, and Social Structure. *AJS* 85 (3): 551-575.

KLEINMAN A. 1980: *Patients and Healers in the Context of Culture*. Berkeley: University of California Press.

KLEINMAN A. and GALE J.C. 1982: Patients treated by physicians and Folk Healers: A Comparative Outcome Study in Taiwan. *Culture, Medicine, and Psychiatry* 6 (4): 405-423.

KLEINMAN A. and GOOD B. (eds) 1985: *Culture and Depression. Studies in the Anthropology and Cross-Cultural Psychiatry of Affect and Disorder*. Berkeley: University of California Press.

KLEINMAN A. 1986: *Social Origins of Distress and Disease*. New Haven: Yale University Press.

KORPORAL J. 1985: Probleme der kurativen, präventiven und rehabilitativen Versorgung von Arbeitnehmern aus der Türkei und ihren Familien. In: COLLATZ J. und KORPORAL J. (eds.) *Gesundheit für alle*. Rissen: E.B. Verlag.

KRAEPELIN E. 1980: Vergleichende Psychiatrie. *Curare* 3 (4): 245-250.

KROEGER A.; KOEN E.; WEBER W., STREICH K. 1986: *Der Umgang mit Krankheit in türkischen und deutschen Arbeiterfamilien. Heidelberg*. Universität Heidelberg, Institut für Tropenhygiene und öffentliches Gesundheitswesen.

LUTZ C. 1985: Depression and the Translation of Emotional Worlds. In: KLEINMAN A. and GOOD B. (eds) *Culture and Depression* (pp. 63-100). Berkeley: University of California Press.

MALZBERG L.; EVERETT S. 1956: *Migration and Mental Disease*. New York: Social Science Research Council.

MARSELLA A. 1979: Depressive experience and disorder across cultures. In: TRIANDIS H. and DRAGUHUS J. (eds) *Handbook of Cross-Cultural Psychology* 6 (pp. 237-290). Boston: Allyn & Bacon.

MOERMAN D.E. 1979: Anthropology of Symbolic Healing. *Current Anthropology* 20 (1): 59-80.

ØDEGARD, Ø. 1972: Epidemiology of the Psychosis. In: *Psychiatrie der Gegenwart* (2. ed., Vol. II/ 1) Klinische Psychiatrie I (pp. 214-258). Berlin, Heidelberg, New York: Springer.

ÖZTÜRK O. und VOLKAN V. 1971: The Theory and Practice of Psychiatry in Turkey. *American Journal of Psychotherapy* 25: 240-271.

PFEIFFER W. 1971: *Transkulturelle Psychiatrie*. Stuttgart: Thieme.

PFLEIDERER B. 1981: Der semiotische Ansatz in der Medizinethnologie dargestellt an einem Heilritual in Nordindien. *Curare* 4: 223-229.

— 1984: Der traditionelle Patient: Wie kann man ihn verstehen und ihm helfen? In: DIESFELD H.J. and WOLTER S. (eds.) *Medizin in Entwicklungsländern* (pp. 72-78). Universität Heidelberg: Institut für Tropenhygiene und Öffentliches Gesundheitswesen.

PFLEIDERER B. and BICHMANN W. 1985: *Krankheit und Kultur*. Berlin: Reimer.

REHBEIN J. 1985: Medizinische Beratung türkischer Eltern. In: REHBEIN J. (ed.) *Interkulturelle Kommunikation* (pp. 349-419). Tübingen: Narr.

RISSO M. and BÖKER W. 1964: Verhexungswahn - Ein Beitrag zum Verständnis von Wahnerkrankungen süditalienischer Arbeiter in der Schweiz. *Bibliotheca Psychiatrica et Neurologica*, Fasc. 124. Basel.

RÖDER F. and OPALIC R. 1987: Der Einfluß des Hodschas (magischer Heiler) auf türkische psychiatrische Patienten in der Bundesrepublik - Eine Auswertung klinischer Fallbeispiele. *Psychia und Prax*. 14: 157-162.

SCHEPER-HUGHES N. 1979: *Saints, Scholars, and Schizophrenics. Mental Illness in Rural Ireland*. Berkeley: University of California Press.

THEILEN I. 1986: Der Einfluß des Islam auf Leiden und Krankheitsverhalten türkischer Menschen in der Bundesrepublik. In: HESSE G.; MEYER R. and PASERO U. (eds.) *Krank in der Fremde* (pp. 106-131). Evangelische Akademie Nordelbien (Dokumentationen II).

VOLKAN V. 1975: Turkey. In: HOWELLS J.G. (ed.). World History of Psychiatry (pp. 383-399). New York: Brauner Matzel.

VOLMERG B. 1984: Zur Sozialisation psycho-somatischer Symptome in der Arbeitssituation von Bandarbeiterinnen. *Psychosozial* 23: 91-119.

ZIMMERMANN E. 1981: Kulturspezifische Deutungsmuster psychischer und somatischer Erkrankungen bei süditalienischen Migranten in der Bundesrepublik. *Ausländerkinder* 7: 30-45.

ZIMMERMANN E. and PETRYKOWSKI W.v. 1983: Magische Krankheitsvorstellungen ausländischer Eltern als Problem in der Pädiatrie. *Der Kinderarzt* 14 (9): 1113-1120.

# Section 3

# Local Cultures of Biomedicine

# The Practice of Biomedicine and the Discourse on Hope

## A Preliminary Investigation into the Culture of American Oncology [1]

Mary-Jo DelVecchio Good

### Introduction

Since 1971 and the passage of the National Cancer Act, we have witnessed in the United States an extraordinary burgeoning of specialization in medical, surgical and radiation oncology and of the complementary fields of research oncology, including clinical trials and clinical epidemiology. The creation of these new specialties, the concomitant commitment of government funds to the National Cancer Institute for biomedical research in oncology, the expansion of clinical trials to international cancer 'centers' and collaboration on research protocols, and the newly developing fee-for-service cancer research and treatment programs raise numerous questions for anthropological exploration [2].

In this paper, I seek to formulate some of these questions and to address them in a preliminary fashion, drawing on research in progress, which includes interviews with fifty-one medical, surgical and radiation oncologists who maintain clinical practices and conduct research in the Harvard Medical School teaching hospitals in Boston and 55 patient interviews conducted in the radiation oncology unit of the Massachusetts General Hospital. In addition, conversations with colleagues from Italy (Deborah Gordon, Mariella Pandolfi), Japan (Tsunetsugu Munakata, Margaret Lock on Japan), Turkey, and with Marvin Zelen (Dana Farber Cancer Institute, (HMS/HSPH), a leader in cancer biostatistics and international research protocols, led me to examine the international literature on oncology and to consider variations in local meanings associated with cancer and with oncological practice and treatment. This international perspective suggests that American oncology as practiced at centers of medical knowledge and education, is a unique cultural variant of international biomedicine (Geertz 1983). An emerging culture of cosmopolitan and international oncology also appears to be challenging 'local' practices concerned with disclosure and physician-patient relationships. The nature of this challenge seems, in part, to be structured by questions raised by American oncological practice.

These observations suggest two hypotheses: first, there are local (national, provincial, regional) variants of international biomedical culture which are powerfully influenced by societal and institutional contexts and by the history of modern medicine within those contexts. (See also the collection of essays in Biomedicine Examined, eds. M. Lock and D. Gordon, 1988). The 'local' aspects of biomedical culture are particularly evident when relationships between physicians, patients and their families and patterns of clinical discourse are examined. Second, choice of treatment options and use of biomedical technology are influenced by (1) patterns of disclosure and patient-physician relationships, and (2) by the extent to which a society's biomedical culture is dominated by the specialties of international oncology [3].

In addressing the concerns of this conference, I would argue that the American culture of oncology is infused with a particular discourse on 'hope'; it is grounded in popular American conceptions of the relationship between mind and body, or psyche and soma, and it patterns modes of disclosure. Patterson (1987) identifies this discourse as "the message of hope" which provided popular support for National Cancer Institute research funds and which was carried to a large degree by the American Cancer Society activities. Indeed, the American emphasis on the 'message of hope' in the 'war on cancer' has not only shaped and funded the development of the subspecialties in oncology but it also provides central symbols in debates within the oncological specialties about the efficacy and meaning of hope in actual clinical practice. This American discourse on hope incorporates both popular and professional dimensions in our culture of biomedicine. Its emphasis on 'will' - if one has enough hope, one may *will*, through control of mind and psyche, a change in the course of disease in the *body* - articulates fundamental American notions about personhood and the relationship of body to mind. Townsend (1978) in his comparison of German and American popular and professional ideologies of mental illness and treatment (in contrast to psychiatric knowledge) noted that Americans not only attribute mental illness to environmental forces, but also that they emphasized personal effort or 'will' in both causing and curing mental illness. In contrast, German professionals and the lay public held two clearly distinguished notions of mental illness, one that was curable by therapy and attributed to a precipitating event, and the second which was untreatable (insanity, mental illness) and inherited or genetic. Jenkins, in her analysis of conceptions of schizophrenia comparing Mexican-Americans and Anglo-Americans found that Anglo-Americans emphasized individual will in overcoming disease (1988:1241). I would argue that these fundamental American notions about the 'will' and the relationship of will to somatic and psychological manifestations of disease and cure appear to influence how disclosure and hope are expressed in clinical discourse and practice in the American context.

In an effort to begin unpacking the relationship between popular American culture and American biomedicine, three dimensions of the culture of oncology are explored in this paper: (1) physicians' concerns with disclosure of diagnosis, prognosis and discussions regarding treatment; (2) physicians' concerns regarding the maintenance of hope - for themselves as well as their patients and their families; and (3) physicians' engagement with clinical challenges and the bioscience and technology of oncological practice.

## Truth-Telling and Disclosure

The past two decades have witnessed a radical shift in American physician attitudes about disclosure of the diagnosis of cancer. In the 1950's, the majority of physicians surveyed indicated that general practice doctors never or seldom told patients the diagnosis of cancer. Reluctance to break bad news, even for physicians who were inclined to share the diagnosis of cancer with their patients, arose when physicians feared patient responses or were discouraged from 'truth-telling' by family members (Novack et al. 1979, Radovsky 1985). More recent studies have indicated a major shift in physician attitudes; Novack et al. compared a university physician sample surveyed in 1961 in which 90% of physicians reported that they generally did not inform patients of diagnoses of cancer (Oken 1961) to a 1970 sample in which 91% of physicians frequently or always informed patients, to a comparable university sample, surveyed in 1977, in which 98% of physicians indicated that disclosure was their usual policy and 100% felt it was the patient's right to know the diagnosis (Novack et al. 1979). Novack and colleagues noted that the physician respondents' written comments suggested that the current policy of telling was "accompanied by increased sensitivity to patients' emotional needs", and they argued that "there is some evidence that telling is the best policy" (1979: 900). Remarkably, Novack and his colleagues felt that many questions remained about what actually constituted the content and mode of the telling and what influence modes of disclosure had on patient outcomes. The subtleties of disclosure could not be assessed in the survey. Radovsky, in an occasional note for the *New England Journal of Medicine* (1985), grappled with the more complex issue of distinguishing disclosure from informed consent, arguing for a processual model rather than a point-in-time model. Taylor, in a break with standard survey methodology exploring physician attitudes about disclosure, observed patterns of telling the diagnosis of cancer of 17 surgeons in clinical practice in Canada; Taylor identifies the telling as a routinized process (1988). This processual model of disclosure emerges in our discussions with oncologists, especially with medical and radiation oncologists who see patients through to cure, recurrence, or death.

The Italian and Japanese approaches to discussing diagnosis and disclosure contrast sharply with the current American pattern. Until recently, both Japanese and Italian physicians masked the diagnosis of cancer, in what was considered to be the most caring of approaches to protecting the patient (Munakata 1986, Long and Long 1982, GIVIO 1986) [4]. In a recent article in the British Journal of Cancer (1986), the GIVIO research group from Milan, addressed the question of what doctors tell patients with breast cancer about diagnosis and treatment (GIVIO 1986). In this study, 39% of Italian physicians gave thorough information on diagnosis to their patients, and 11% of patients received detailed information on the surgery. After controlling for patient characteristics, these researchers found that physicians who were most likely to give thorough information were associated with oncology research groups and practiced in major medical centers. A remarkable component to GIVIO's analysis was that physicians who told breast cancer patients that they had a "benign tumor" were considered to be delivering "partial information" rather than what I would perceive (from an American perspective) to be 'misinformation'. This analysis suggests that the Italian pattern of dealing with diagnosis and disclosure may be in a state of flux and that a challenge arises from the international biomedical and oncology research groups. The labeling of 'benign tumor' as partial rather than misinformation also suggests that there are indeed local variants of international oncology; the meaning of 'benign tumor' in both lay and medical culture in Italy may be deliberately vague and suggestive of malignancy, whereas in the American research context it would be perceived as misinformation and would not convey the appropriate cultural cues. However, in American clinical practice, 'partial information' might include initially telling patients that they have a 'tumor' otherwise undefined.

## American Oncologists and Cultural Documents

In this essay, I draw primarily upon material gathered in interviews with 51 oncologists, 14 from radiation oncology, 14 from surgical oncology, and 23 from medical oncology. All practiced in the teaching hospitals of Harvard Medical School. The mean age of those interviewed was 40.1 years, illustrating the newness of these subspecialties (34 years for medical oncology, 40 years for radiation oncology, and 45 years for surgical oncology). An open-ended but structured interview was followed for all physicians. An additional set of closed questions were given to each participant to complete at the conclusion of the interview. Because my analysis is based upon interview material rather than observations of daily practice and interaction with patients, the research text, the cultural document, emphasizes how oncologists conceptualize and explain what they do. The interview texts give us access to the ideologies of these subspecialties and the participants' explanations of these ideologies. How these explanations and ideologies structure clinical practice is posed as the research question for the next phase of the study.

## Issues of Disclosure and Maintenance of Hope

Issues regarding disclosure take on a different cultural form for specialists in medical and radiation oncology than for internists and general surgeons who often are the first to break 'bad news' of diagnosis to patients. Disclosure in the context of oncology is about prognosis and treatment and not primarily about diagnosis. Thus the process of disclosure colors clinical discourse between physician and patient and families of patients; it appears as an evolving process and is not bounded in one point of time. In our study of radiation therapy patients, overwhelming concerns as expressed by patients were first about prognosis, second about disease process, and third about treatment side effects, options and fundamental information. For the clinicians, these concerns of patients are managed 'in time' and over time. The clinicians set a pace for disclosure, in part a pace attuned to the phasing in of biotechnical interventions of surgery, radiation therapy, chemotherapy or some combination of the three.

Patterns of disclosure are also tied to increasing certainty of outcome. Although physicians claim that they individualize disclosure for patient and family, there is an alternative ideology and justification toward clinical norms or routinization (as in Taylor's findings, 1988). The balancing of the individualized response (and the physician's own individuality in shaping the response) with the demand of the specialty, and thus standardization and routinization, provides a tension in oncologists' narratives relating the process of disclosure.

Physicians frequently frame the first step in disclosure as an exchange *from patient to physician* rather than vice versa (as patients struggling with their specialists often wish it to be). The following quotations suggest how subspecialists conceptualize this process; there is some difference between surgeons and medical oncologists and radiation therapists.

*Surgeon* (male, age 47)

Q.: How do you determine whether or not a patient knows what their diagnosis is?
Response: Well, I ask them how I can help them. It's the only question I ask to begin an interview and they have to spin it out from there.
Q.: If a patient comes to you not knowing their diagnosis...
Response: I tell them that we are concerned it may be a tumor... I never use the word cancer, ah, it is just not in my vocabulary.
Q.: Why?
Response: Because that is a word that patients assume and don't want to hear you say until they work up the courage to ask "Is the tumor malignant?" or "By tumor do you mean cancer?". So I give them the control to decide when they want to phrase the question that way. They all do - but ... some ask immediately and others - it will be a second or third visit before they'll ultimately ask if the tumor I'm talking about is a malignant one.

*Surgeon* (male, age 65)

Q.: How do you determine whether or not that patient is aware of his disease?
Response: Just by asking them, talking to them. And we have it differently here because this is a cancer hospital. ... It's very different from a general hospital and things are very different from what they were 20 years ago in this area. The main change is that 20 or 25 years ago, there was a strong tendency, particularly among surgeons, to be reassuring and to be less up front with diagnosis and to try to be more protective of patients. Some of us have always felt - or for many years - one should be rather more up front, and that it made it easier to deal with a patient who did have cancer. And I think it is the medical oncologist who has taken the lead, just out of necessity in discussing the issue of whether the diagnosis was cancer or not ... Gradually, as the population became educated and as physicians and surgeons became educated, we have all begun to tell the patient more candidly what the diagnosis was. I think physicians have changed because medical oncologists have changed; ... they naturally had to, they couldn't discuss in most instances, advance cancer treatment without using the word cancer.

*Medical oncologist* (male, 41)

Q.: How do you determine whether or not that patient is aware of his disease?
Response: ... I might say - I have some information from Dr. X who referred you and I am not sure exactly what he has told you, - just so we are on the same wavelength, tell me in your terms what your understanding is ... getting the person to tell what their understanding is, not only what they understand objectively, in terms of the facts, but you also have to understand what the disease means to the patient. Because what it means to the patient, and what bothers the patient most

about the disease, is often very different than what bothers the doctor about the disease.

*Medical oncologist* (male, 43)

Q.: How would you determine whether or not a patient knows what their disease is?
Response: My standard first question when I talk to them after taking my own history or when it's time to talk turkey is "What did Dr. Bloggs tell you?" I would never stumble into a trap of cutting off somebody's defense.

*Medical oncologist* (female, 45)

Q.: Are there issues you consider as you are deciding what or how to tell someone about their diagnosis?
Response: ... I think that patients let you know what they want to hear ..., in the way that they talk to you. (pause) I think one has to listen to that; I think knowing something about a patient is terribly important before you go blurting out something like that.
Q.: How explicit do you find you are ... using the word cancer, and do you find it helpful to be more or less direct?
Response: I almost slip the word in somewhere ... and even when you say cancer, the patient sometimes hears it and sometimes doesn't. So I tend to be very direct. I think it is very important for the patient to think you are capable telling things because if they don't know you are capable of bad news they can never believe the good news.

*Radiation oncologist* (female, 45)

Q.: I have some questions about how you discuss diagnosis, prognosis and treatment of cancer with a patient.
Response: I usually find it wisest, having read whatever I have been handed about them, in charts or letters, to ask the patient what they understand first of all. What they have been told, because it is a mystery to me how much information they have received themselves or assimilated. And coming in at the back end of all their consultations, as one usually does, it is a fallacy that they will be well-informed about the disease ... And I often meet quite a bit of divergence in what I expect they know and what they do know.
Q.: Can you give an example?
Response: I think that I ask them, for example, do they know why the doctor asked you to come and see me? What is your understanding of the surgery you just had done? You can get into this strange box right away, where they will say I've got a tumor and then you need to find out what they mean by the word 'tumor'. Because they may blithely understood that to mean a benign situation. I think sometimes people just problably think is is benign. It would come out after the patient had come home, it acutally hasn't been a very good transaction about it and the patient has gone on their way with "I hope it is a benign tumor" is what they have retained, heard and been told ... probably told. And then there haven't been the appointments fixed up very fast, too much is done on the telephone to avoid dragging the patient in again. *And so you are presented with somebody who is actually only part of the way through the conversation.*

Staging disclosure about prognosis and treatment is part of this ongoing 'conversation' and emerges as a major clinical phenomenon in the world of medical and radiation oncologists.

*Oncologist* (male, 43)

"It's like when you climb a mountain, and it looks so enormous, the only way you can climb it is one step at a time. And I tell them that the only way we are going to deal with this is one step at a time. The first step is going to be to control the pain and then we are going to work into these various options. Because then people put their mental energies onto something and they are not so overwhelmed."

*Radiation oncologist* (female, 45)

"Yet in general, I don't believe you can make one statement. But you are taught to approach people very cautiously and develop the dialogue or the discussion over time. The first time you meet somebody it is not necessarily the smartest time to talk turkey. And we are taught to stage things and I think that has stayed with me, because I have seen how easy it is to jump in assuming a level of reception that isn't there."

The staging of disclosure and the flow of information and clues to elicit disclosure between patient and physician is contingent on the type of 'partnership' that the specialist constructs with the patient. In American oncology, the *primary unit of treatment* is the patient, rather than the family of the patient; and although patient families are considered in the staging of disclosure and movement to 'the next step', the ultimate focus is on the patient, the *contract* is with the patient. At times, patients are perceived by clinicians as 'telling you their mandate' - setting out the goals of partnership that involve the type and extent of information which they wish to elicit from the physician. However, the staging of disclosure on prognosis and treatment coincides with the evolution of the partnership, and the content of disclosure is equally influenced by the form the partnership takes as treatment proceeds apace.

A radiation oncologist, commenting on this process, remarked:

"There is a lot of housework, the boring part of getting people into radiation therapy. All those little appointments and time schedules and things. All of that occupies the trivial time when you circle around each other perhaps."

A medical oncologist, commenting on partnership, noted:

"What you do get from talking to people - the questions they ask and the way they conduct themselves - is how much the person wants to know. We want patients to have a good understanding, especially if we are going to be treating them. We need them to be kind of partners in what we are doing. So although we don't bludgeon people with the truth, I mean ultimately you would like each patient to have the fullest understanding of the disease that they can tolerate ... Initially some people may give you very clear messages that they want some information about the rest of it (beyond diagnosis), so this may develop over time. You may decide that you want to get a few key things worked up and out on the table the first time and you can build on that the second or third time you see a patient. So that by the end of the month you can really get up to speed."

A radiation oncologist, commenting on the evolving partnership, noted in response to a question asking if he gave information to patients that he did not give to family members:

"I might ... I do other things ... I'll do little things to give the person the feeling that the relationship is really between ... you and I ... and you as the patient. And although I am willing to tell your relatives everything, it is really us and I want to keep it that way."

A medical oncologist: "My commitment and bond is to the patient."

The image of the specialist constructing a 'partnership' with the patient, which is salient in the majority of these interviews, is complemented with stated ambivalence about 'paternalism' in structuring disclosure of information. In addition, these oncological specialties have a very clear mandate to instill and maintain hope in patients. Thus oncologists engage in a balancing act which requires weighing disclosure of information about the disease, about prognosis and about treatment *options*, in the context of considerable biomedical uncertainty, against the demands of 'partnership' which emphasizes disclosure and the mandate to instill and maintain hope. Disclosure becomes a contested domain especially when patients request numbers - "our national mania" remarked one physician who had moved from Europe during training years. Revealing numbers (putting odds on longevity, recurrence, cure) undercuts 'hope', emphasizes uncertainty, and may, in the view of many clinicians, undermine the 'partnership' and the inherent hierarchy of the professional-lay arrangement. Yet, withholding of information can be threatening to the partnership as well (DiGiacomo 1987).

## Instilling Hope

Although all physicians in the study responded to questions as to whether they thought that instilling hope in their patients was of critical importance with emphatic expressions 'absolutely', 'certainly', 'essential', their understanding of what was meant by hope and how it may be best conveyed in clinical practice was quite diverse. In addition, the specialists' definition of what constitutes hope and of its efficacy in patient care, is at times at odds with much of the popular lay and medical literature on the importance of hope in producing immunological and thus beneficial physiological responses.

Physicians in general and oncologists and clinical trial researchers in particular are intrigued with the question of whether psyche (hope, a fighting spirit) has any major impact on the immune system and the biological outcome of cancer; and the nature of this relationship is debated, sometimes hotly, in esteemed professional journals. The paucity of quality data adds to the ideological ferment and is reflective of popular cultural notions and in tension with them (Angell 1985, Cassileth et al. 1985, Cassileth et al. 1988, Cousins 1985, New York Times 1986). And yet, despite serious questions about the biological efficacy of hope, of the psyche's ability to prolong life and to have an impact on physiological processes in cancer, all those interviewed expressed the mandate of instilling and maintaining hope.

Curiously, hope is Janus-faced; the oncologist's vocation is to instill hope in patients - but as captured in the following comments - hope or an optimistic stance - is also at the essence of practice.

*Radiation oncologist* (male, 42)

"I don't think it (hope) makes a biological difference, but it certainly makes a difference in your outlook and the way you live the rest of your life. And ... I think if you don't have hope, that what you're going to do is going to make a difference, then you really - this is probably the wrong field ... I just think it's such a fundamental thing to have that I couldn't imagine somebody not having it and being an oncologist, to tell you the truth."

*Medical oncologist* (male, 43)

Q.: How important is it for you to instill hope in your patients?
Response: Essential. They have to have hope and they have to have structure. Otherwise they have a clock ticking in their head. Doctors who don't provide hope, facilities that don't provide hope, or at least some positive outlook or support, the patients don't stay. Now, I'm not saying that this is a business where you are trying to attract them like the Pied Piper of Hamlin, but on the other hand, you have to give people something to wake up for in the morning, otherwise they might as well take some cyanide ... you can't take away all those things ... look - the glass has to be one tenth full rather than nine tenths empty.

Q.: Do you think a patient's level of hope alters how they will progress in their disease?
Response: Not objectively. There are studies to that effect, self-assessment... But it certainly relates to how hard they will try in maintaining their weight, how compulsive they will be to take their medications on time, how they are going to act through all these things. Yes, that it does.

Q.: Would you say hope helps organize the way in which you relate to patients?
Response: I try to make patients feel better after they have seen me. I'll always focus on the positive. If they look good, I'll tell them they look good. If their weight has gone up an ounce, I'll focus that their weight has gone up an ounce. Whatever it is. Because in part I'm a cheerleader. I've got to be their cheerleader.

*Surgeon* (male, 47)

Q.: How important is it for you to instill hope in your patients?
Response: Absolutely.

Q.: Do you think it alters the course of disease?
Response: Well, of course we can never know that. But it is my impression that it does, yes. Yes. I think people eat better, work harder to live, and feel better about themselves if they are encouraged and given hope. I think I saw it more clearly when I was in training and had the opportunity to see patients of many different doctors - some of whom were pretty pessimistic. And what they thought - they were brutally honest with their patients and their patients would often give up. And others who were able to let their patients know that they'd be with them through whatever happened, and they'd support them. Whatever happened, they didn't have to worry about pain or discomfort. Those patients were able to go on living for whatever time they had, much better. And I think, often a bit longer for it, but it's hard to prove.

Q.: Do you have a sense of how instilling hope becomes an integral part of your communications - how it works or happens?
Response: I think you have to love your patients. And I think that if you love people and you truly love your patients - we all respond to love and - patients respond to it... you build a relationship with a person that is mutually enjoyable. ... It's not a sense of hope that they are going to live for a long time or forever. It's the sense first of all of reminding them of the uncertainty with which we all live everyday from the time of birth. And I often go with them through Freud's writing. (Discusses Freud and fear of death as a primal drive). Working through their own mortality is often a very beneficial and helpful thing. Realizing that someday they may die and it may be soon and it may not be, gives a greater color and richness to every day that they'll live thereafter. So I encourage them to take the benefit of this disease which is to challenge them with their own mortality. And to evaluate their life in terms of enjoying their family and their friends and their job and the beauty of a Fall New England day in a richer way than they would have otherwise ...

These comments and the themes that emerge in each of the interviews record the complexity of the meaning of hope to these oncologists. Hope is not simply defined in terms of attitudes which have a measurable impact on the biology of cancer, although many physicians consider wistfully the potential link between optimism and immunological responses. As with disclosure, hope is staged, given in calibrated, achievable and realistic bits. Hope to enjoy the richness and color of daily experience, hope to control pain, hope to be secure in the doctor-patient relationship are at minimum mandated. Hope to affect the quality of life, perhaps to influence the magnitude of side effects, to give treatment options their best chance are at the maximum of the mandate within the profession.

The diversity of specialists' opinions about 'hope' and about instilling 'positive' or 'optimistic' attitudes in patients was captured not only in our interviews but also in a series of closed questions. The following tables summarize these closed-ended responses; scores are based on a 6 point scale (1= not at all, 6= most definitely).

**Table 1: Oncologists' perceptions of the importance of positive attitudes***

I. Do you think a 'positive attitude' affects the outcome in early stages of cancer? (i.e. Stage II Breast Cancer)

| Specialty Affiliation | Responses | | | |
|---|---|---|---|---|
| | Not at All (1) | Somewhat (2-4) | Definitely (5-6) | Mean |
| All Physicians | 23.5% | 50.9% | 25.5% | 3.0 |
| Radiation | 35.7% | 50.0% | 14.3% | 2.6 |
| Surgery | 35.7% | 28.6% | 35.7% | 3.3 |
| Medicine | 8.7% | 65.2% | 26.0% | 3.1 |
| p=.10 | | | | |

II. Do you think a 'positive attitude' affects the outcome in later stages of cancer? (i.e. Stage IV Breast Cancer)

| Specialty Affiliation | Responses | | | |
|---|---|---|---|---|
| | Not at All (1) | Somewhat (2-4) | Definitely (5-6) | Mean |
| All Physicians | 23.5% | 45.2% | 31.4% | 3.2 |
| Radiation | 35.7% | 42.8% | 21.4% | 2.9 |
| Surgery | 28.6% | 14.3% | 57.1% | 3.9 |
| Medicine | 13.0% | 65.2% | 21.7% | 2.9 |
| p=.04 | | | | |

* Sample: n=51; 14 radiation oncologists, 14 surgeons, 23 medical oncologists

III. Do you think a 'positive attitude' affects the patient's experience of complications?

| Specialty Affiliation | Responses | | | |
|---|---|---|---|---|
| | Not at All (1) | Somewhat (2-4) | Definitely (5-6) | Mean |
| All Physicians | 2% | 13.8% | 80.4% | 5.1 |
| Radiation | - 0 - | 28.5% | 71.4% | 5.0 |
| Surgery | - 0 - | 21.4% | 79.6% | 5.1 |
| Medicine | 4.3% | 8.6% | 86.9% | 5.1 |
| p=.69 | | | | |

IV. Do you think a 'positive attitude' affects how the patient views the doctor?

| Specialty Affiliation | Responses | | | |
|---|---|---|---|---|
| | Not at All (1) | Somewhat (2-4) | Definitely (5-6) | Mean |
| All Physicians | - 0 - | 11.8% | 88.2% | 5.3 |
| Radiation | - 0 - | 7.1% | 92.9% | 5.4 |
| Surgery | - 0 - | 7.1% | 92.9% | 5.5 |
| Medicine | - 0 - | 17.3% | 82.7% | 5.1 |
| p=.81 | | | | |

V. Do you usually make an effort to change your patient's attitude to a more optimistic view?

| Specialty Affiliation | Responses | |
|---|---|---|
| | No | Yes |
| All Physicians | 12% | 88% |
| Radiation | 7% | 93% |
| Surgery | 14% | 86% |
| Medicine | 13% | 87% |

VI. Reasons noted for encouraging patients to have an optimistic view, to instill hope. Percent of yes responses.

| Reasons | Specialty Affiliation | | | |
|---|---|---|---|---|
| | Radiation | Surgery | Medicine | All |
| Tolerate Adversity | 71% | 50% | 65% | 63% |
| Live Longer | 14% | 21% | - 0 - | 10% |
| Live Better | 71% | 79% | 74% | 75% |
| Easier to Manage | 29% | 7% | 30% | 24% |
| Enhance MD relationship with patient | 29% | 21% | 39% | 31% |

* Sample: n=51; 14 radiation oncologists, 14 surgeons, 23 medical oncologists

Although over one quarter of all physicians stated that they thought a positive attitude influenced outcome in both early and later stages of cancer, when asked about why they attempted to instill optimism in their own patients, only ten percent remarked that it was so that patients would live longer. This apparent contradiction, in part produced by the two modes of measurement (a 6 point scale versus a yes/no response), is also reflected in the interviews. Almost all physicians rejected conveying to patients statistical information on longevity - as if conveying negative information would have a major impact on psyche - on a patients' attitudes, and thus on soma - and hasten the progression of the disease. Most were unwilling to clearly take a position on the relationship between psyche - or hopeful and optimistic attitudes - and soma; their stance was uncertainty. Although these oncologists would speculate on the possibility of a relationship between a positive attitude and the progression or physical experience of cancer, clearly, they are most comfortable in asserting that 'hope' or 'optimistic attitudes' have most effect on the non-biological or non-somatic aspects of cancer, on the social experience of cancer. Many expressed discomfort at the popular medical literature which gives legitimacy to the notion that individualized *will* can have an impact on the progression of the disease (i.e. Siegel 1987, Simonton et al. 1978). Although the biological world is perceived as dominant and resistant, in the long run, to influence from *will*, it is the specialist's personal sense of 'hope' in being able to 'make a difference' that frames commitment to research and clinical practice in oncology.

**Bioscience and Clinical Care**

Among the clinicians in our study, 92% were or had been involved in laboratory research; most frequently enrolled patients in clinical trials; all participate in the academic functions of teaching residents and medical students to varying degrees. Thirteen were fellows in oncology at the time of the interview. Thus, as specialists in oncology at major cancer research and treatment centers, these 51 individuals participate in the production of medical knowledge and standards of practice for the field and in the education of the next generation of oncologists.

In exploring why these physicians are involved in oncology, two themes emerged in the interviews which are associated with the psyche-soma dichotomy. Oncologists express their commitment to the field in terms of challenge - the challenge of engagement in biomedical research and the creation, trial and utilization of aggressive treatments. Clinicians place themselves in different positions; some see themselves as more engaged in the existential issues of the patients, in managing the patients' psychological and social experience of the disease; others see themselves as deeply committed to research or to devising new modes of treatment. Many identify themselves as being in both camps, deeply involved in patient care through the use of aggressive treatments and biotechnology and through seeing patients through the process of treatment and the progression of the disease. Caring thus becomes conveyed through the offering of treatment options, be they 'curative' or 'palliative'.

The sense of confronting the biological frontiers of disease and commitment to altering those biological frontiers emerge in the way oncologists speak about and describe their work.
A radiation therapist, in describing the dual dimensions of her work, clarified a statement she made to me on "cutting close to the bone".

"I mean anatomically. Very often I am working to the tolerances of normal tissues. How close am I coming to the spinal cord. I am going to skim it by a couple of inches. How reliably can I do that? Am I being too daring? ... Our morbidity is not the instant variety like surgeons... Radiation oncologists are incredibly fixated on critical organs - are we polishing off an eye, a spinal cord, a kidney. There's a different series of anatomic considerations that don't have the quick finesse of the surgeon's knife - but we have this long term debt you can't take back."

"Aggressive treatment is the only way you cure the majority of tumors. So if you are pussyfooting around, you might just as well sprinkle holy water on it and go home. Most patients only get cured with our mode of treatment ... Vigorous therapy centers take the worst cases on... your are going for broke. ... We have a superb program (describes experimental treatment program for skull tumors) ... it is dangerous ... We are right up to what we can get away with. It's exciting, but it's a different brinkmanship."

"Patients may take all of that quite calmly, what they actually get out of the transaction is where they are talking - that's a separate mode and I don't think they collide really. It' s a schizoid way in which you operate. One is talking to people and the other is figuring out the best way to go for it."

Medical oncologists also discuss the 'challenge' of the profession, of working with patients who are facing the frontier between life and death. In describing why he turned to oncology, one physician remarked:

"Scientifically, ... it's very demanding because you are dealing with diseases of all the body organs and all the complications of the diseases ... I have everything from a brain tumor to synovial sarcomas of the toe, all with different complications of therapies. You have to be very broad based as an internist. Also, the diseases are very threatening to patients and you are really dealing with important stuff. And I like that. You have to be a little bit of an old fashioned doctor, a little bit of a psychiatrist, little bit of high tech - like the bone marrow transplant - all rolled into one. And you are dealing with serious illness. I didn't want to deal with zits. Or people who couldn't sleep, or have low back pain and are depressed. *Everyone who walks into my office might die*. And although I don't wish them all to die, I kind of get more of a buzz out of dealing with something. Everyone of these cases is up against it ... There is this competitive side of it too - I always try to be just a little bit better. Everybody loves the case that comes that has been elsewhere and they say that nothing can be done for me - and here I am, sort of the fifth opinion, and there will be another year for that patient because of something that somebody (else) didn't think of. So there are a lot of positive things."

The interview texts suggest that for many oncologists 'hopefulness' appears to be grounded primarily in the biomedical dimensions of their work. Caring for patients is conveyed through the treatment process, on offering therapeutic options, and on holding out hope for the development of new treatments on the cutting edge of biology and medicine. Many of those interviewed noted developments in what they could offer today that could not be offered in the recent past, and in several interviews the most difficult part of oncology was perceived as 'not having anything to offer' patients with advanced or recalcitrant diseases. And although the medical oncologists in particular speak about seeing patients through to the end of their lives, offering control of pain and the 'security of the doctor-patient relationship', and view themselves as better able to care for dying patients, the cultural texts provided by these interviews strongly suggests that the 'hospice' role does not dominate any of the three subspecialties of oncology.

## Conclusion

In the introduction to this paper, I proposed that the expression of hope and modes of disclosure in American oncology are powerfully influenced by popular American notions of psyche and soma, that individualized *will* can influence bodily processes. A review of responses to questions on disclosure and hope suggests that in this arena of uncertainty surrounding the biology of cancer, oncologists, while skeptical about the relationship between psyche and soma, still seek to enhance the *will* of patients under their care through instilling hope and through 'educating patients in their disease'. Thus, disclosure of diagnosis is seldom at issue (the rare cases noted include the foreign born, non-English speakers, and children). This contrasts with dominant biomedical traditions in many European societies, in Japan and the Middle East, where relationships with patients are not only hierarchical but paternalistic and 'protective', and frankness in disclosure of diagnosis is at least contested as a clinical goal if not considered cruel and inappropriate. (Some Japanese physicians see the American pattern as protective only of the physician (Feldman 1985), see also New York Times 1989, Boston Globe 1989, Wall Street Journal 1988.)

The specialists in our study argue that the biotechnology of American oncology encourages a certain degree of disclosure and frankness about disease processes and treatment. In part, disclosure is viewed as essential to building partnerships with patients, and constructing the partnership is viewed as a necessary component of treatment, of getting patients to accept chemotherapy or radiation, of making patients 'work' or be 'responsible' for eating well, taking their medications, keeping treatment appointments, participating in a therapeutic endeavor which is often highly toxic. And yet, oncologists voice the tension they experience in how much and what to tell patients about their disease and treatment, in particular their prognosis. Almost all the clinicians in our study expressed caution at discussing statistical odds about cure and longevity, although all noted that some patients push them to be explicit about chances for cure, for information on prognosis and process. These clinicians conveyed various ways in which they manage this challenge for 'numbers' and for explicit information, and argue that controlling this information is essential to maintain or instill an optimistic attitude in their patients. Indeed, when asked to comment on how they would organize disclosure for three hypothetical cases, the extent of details, including numbers, that clinicians claim they would convey to patients was highest for Hodgkins Disease - a disease with a high cure rate, 'good' numbers and outcome, information that would encourage patients and give them hope [5]. Although these oncologists argue that aggressive treatment requires disclosure, total frankness about prognosis and treatment is not the norm; disclosure is still a contested domain. Information is tailored to the clinical task, to perceived capacities of individual patients, to maintaining the superior - subordinate structure of the 'partnership' and to the efficacy of biomedical treatment.

Is American oncology a unique cultural variant of cosmopolitan biomedicine, of international oncology? Are variations in relationships between physicians and patients and in customs of disclosure shaped by the cultural ground in which oncology is practiced? Do societal concepts of individual will and popular notions of psyche-soma breed different forms of biomedicine, of clinical practice and research in oncology? This brief foray into American oncology suggests that biomedicine is indeed shaped by local cultures. Although it is common to speak of cross-cultural variation in psychiatric practice and scientific 'knowledge', we are less inclined to consider biomedicine from this perspecitive (granting heteroglossia to the field) and more inclined to view 'it' as monolithic (institutionally) or hegemonic (as a dominant mode of knowledge and practice). And yet, biomedicine, and in particular the subspecialties of radiation and medical oncology and oncological research are organized, both culturally and institutionally, in cosmopolitan as well as local worlds [6].

We are left with questions for the next phase of research - to what extent are we witnessing the emergence of a potent cosmopolitan and international form of oncology? How influential is the American variant in shaping the clinical and research domains of this specialty? How are forms of clinical practice and research translated into local variants of biomedicine? Do clinicians and researchers in non-American cosmopolitan centers resist modes of biomedical practice grounded in American cultural notions of will, psyche and soma, and hope and in American notions of truth-telling and disclosure? Does the production of biomedical knowledge and technology, and the very utilization of this knowledge and technology, challenge local variants of clinical practice, and vice versa, do local variants of clinical practice shape the production and utilization of knowledge and technology? Oncology, as an international specialty, has a short history, remarkable for radical changes in clinical practice and investment in research, less remarkable for its effect on mortality. Our conference reflections on the lengthy history of psychosomatic medicine and its use in the cross-cultural paradigm may enlighten the analysis of this new field.

## Notes

[1] I would like to acknowledge the contributions of my colleagues in the conceptualization of this research endeavor: Byron Good, Ph.D., Harvard Medical School, Deborah Gordon, Ph.D., Florence, Italy, and Stuart Lind, M.D., Massachusetts General Hospital and Harvard Medical School, Department of Medicine, Hematology/ Oncology. Cynthia Schaffer, HMS 1990, participated as a student researcher on the project, assisted in conceptualization, and conducted the majority of the interviews with clinicians under my guidance. This project constitutes the first phase of a longitudinal study of the culture of American oncology and its relationship to international biomedicine. The study of radiation therapy patients was conceptualized and implemented by Stuart Lind (MGH), Byron Good, Tom Csordas, Steve Seidel (HMS 1989) and myself. I would also like to thank the physicians who participated in this research and gave generously of their time; the conveners, Arthur Kleinman, Thomas Maretzki and Beatrix Pfleiderer, of "Anthropologies of Medicine: A colloquium on West European and North American Perspectives" Hamburg 4-8, 1988, and the participants in this colloquium. This research was funded in part by small grants from the Department of Social Medicine, HMS, the Milton Fund, HMS, and by BRSG S07 RR 05381-27, Biomedical Research Support Grant Program, Divison of Research Resources, National Institutes of Health.
Portions of this paper were published in *Culture, Medicine and Psychiatry* 14: 59-79, 1990, In Good et al. "American Oncology and the Discourse on Hope"
[2] See Patterson (1987) for an historical analysis of the meaning of cancer in American society in the past century and an analysis of social and governmental movements which led to the creation of the American Cancer Society, the establishment of the National Cancer Institute, and the variations in the popular, scientific and governmental "war against cancer". For a discussion of the recent development of fee-for-service research and treatment programs and a questioning of the ethics of these endeavors, see Lind (1986) and Oldham (1987).

[3] A third question concerns the political economy of specialist/ research oncology: how does it articulate with the culture of "hope" and of "disclosure" in specific societies. An examination of the American historical and contemporary material on the war on cancer suggests that the move towards frankness in disclosure of diagnosis and discussion of treatment is intimately associated with social movements (the American Cancer Society activities, alternative health movements) and with government research investments in and through the National Cancer Institute. The elevation of cancer researchers to positions of high prestige, the magnitude of investment in the research and treatment programs of the National Cancer Institute, have given oncology a major boost as a specialty. Recent dissatisfaction with the results of such an "expansive" research war (as noted in the New York Times, 1987) and the success of environmental efforts which have focused on etiology (the anti-smoking endeavors in particular) may have clouded the prestige of oncologists and basic science cancer researchers. And yet, Americans continue to finance the seeking of cancer cures (to the extent personally feasible); the latest private endeavors utilizing basic NCI research produce tailored-made treatments and are financed via fee-for-service care (Lind 1986).

[4] The Japanese press was admonished not to use the term cancer in discussing Emperor Hirohito's terminal illness, suggesting that in certain public contexts the discussion of cancer is considered inappropriate (Wall Street Journal 1988, New York Times 1989).

[5] Each specialist interviewed was asked to discuss how they would convey information regarding diagnosis, prognosis and treatment to patients in three diseases: Hodgkins Lymphoma IIA (high response, high survival), Resectable Colon Cancer, stage C2 (low to moderate response, low to moderate survival). Over 85% of Hodgkins patients are curable using radiotherapy or a combination of therapies. Five year survival rate for colon cancer is approximately 50% with surgical treatment. For Gastric cancer, less than 15% of patients survive 5 years. The specialists were much more likely to say that they would use numbers and give explicit and detailed information to patients in cases when physicians believed in the efficacy of the treatment; extent of frankness and disclosure declined precipitously with survival chances as measured in these three cases,

[6] Several oncologists commented on variations in treatment and research ideologies across cancer centers within the same local medical community. Thus, the meaning of "local" and cultural variation can ever be refinded.

## References


ANGELL M. 1985: Disease as a reflection of the psyche. *New England Journal of Medicine* 312: 1570-1572.

BOSTON GLOBE 1989: Cancer: in Japan, a word rarely heard. January 12.

BUCKMAN R. 1984: Breaking bad news: why is it still so difficult? *British Medical Journal* 288:1597-1599.

CASSILETH B., LUSK E.J., MILLER D.S. et al. 1985: Psychosocial correlates of survival in advanced malignant disease? *New England Journal of Medicine* 312: 1551-1555.

CASSILETH B.R., WALSH W.M. and E.J. LUSK 1988: Psychosocial Correlates of Cancer Survival. *Journal of Clinical Oncology* 6: 1753-1759.

COUSINS N. 1985: Unpublished response to Cassileth et al. and Angell M. (1985). Courtesy of the author.

CLINICAL CANCER LETTER 1987: *Optimism Plays No Definitive Role in Survival of Patients With Advanced Prognostically Poor Cancers,* 10: 1-2.

DIGIACOMO S. 1987: Biomedicine as a cultural system: an anthropologist in the kingdom of the sick. In: BAER H. (ed.), *Encounters with Biomedicine.* New York: Gordon and Breach Science Publishers.

FELDMAN E. 1985: Medical Ethics the Japanese Way. *Hastings Center Report* 15: 21-24.

GEERTZ C. 1983: *Local Knowledge.* New York: Basic Books.

GIVIO (Interdisciplinary Group for Cancer Care Evaluation) 1986: What doctors tell patients with breast cancer about diagnosis and treatment; findings from a study in general hospitals. *British Journal of Cancer* 54: 319-326.

JENKINS J. 1988: Conceptions of Schizophrenia as a problem of nerves: a cross-cultural comparison of Mexican-Americans and Anglo-Americans. *Social Science and Medicine* 26: 1233-1243.

KIRKHAM J.S. 1985: Talking to patients about cancer. *Lancet* Nov. 30.

LIND S. 1986: Fee-for-service research. *New England Journal of Medicine* 314: 312-315.

LIND S. et al. 1989: Telling the diagnosis of cancer. *Journal of Clinical Oncology* May.

LOCK M. and G. GORDON (eds.) 1988: *Biomedicine Examined.* Boston: Kluwer Academic Publishers.

LONG S. and B. LONG 1982: Curable Cancer and Fatal Ulcers: Attitudes toward cancer in Japan. *Social Science and Medicine* 16: 2101-2108.

MUNAKATA T. 1986: Japanese Attitudes toward mental Illness and Mental Health Care. In: LEBRA T. and W. (eds.): *Japanese Culture and Behavior.* Honolulu: University of Hawaii Press.

NEW YORK TIMES 1985: *Debate Intensifies on Attitude and Health.* October 29.

-- 1989: *A fear of cancer means no telling* January 20.

NOVACK D. et al. 1979: Changes in Physicians' Attitudes Toward Telling the Cancer Patient. *JAMA* 241: 897-900.

OKEN D. 1961: What to tell cancer patients: a study of medical attitudes. *JAMA* 175:1120-1128.

OLDHAM R. 1987: Sounding Board: Patient-funded cancer research. *New England Journal of Medicine* 316: 46-47.

PATTERSON J.T. 1987: *The Dread Disease: Cancer and Modern American Culture*. Cambridge, Mass.: Harvard University Press.

RADOVSKY S. 1985: Bearing the News. Occasional Notes. *New England Journal of Medicine* 313: 586-588.

SIEGEL B. 1987: *Love, Medicine and Miracles.* New York: Harper and Row.

SIMONTON O.C., MATTHEWS-SIMONTON S. and CREIGHTON J. 1978: *Getting Well Again*. New York: St. Martins Press.

TAYLOR K. 1987: Informed Consent: The Physicians' Perspective. *Social Science and Medicine* 24: 135-143.

-- 1988: Physicians and the Disclosure of Undesirable Information. In: LOCK M. and GORDON D. (eds): *Biomedicine Examined* (pp. 441-463). Boston: Kluwer Academic Publishers: (Also published in *Sociology of Health and Illness* 10: 109-132.)

TOWNSEND J. 1978: *Cultural Conceptions and Mental Illness: A comparison of Germany and America.* Chicago: University of Chicago Press.

WALL STREET JOURNAL 1988: *Japan appears at standstill as concerns over Hirohito's health preoccupy nation.* September 27.

curare Anthropologies of Medicine Special Issue 7/91, 137–156

# Culture, Cancer, and Communication in Italy

Deborah R. Gordon

## Introduction

It is still the dominant practice in Italy, but one definitely changing, *not* to inform cancer patients of their diagnosis, particularly if the outcome is uncertain. A medical practice, such as telling or not telling a diagnosis of cancer, is infused with social and cultural understandings and implications. It is not just a neutral technique nor does it belong exclusively to the realm of medicine. This presents an interesting situation, for increasingly practices current in North America find their way to Italy. Informing the patient of his or her diagnosis is exactly one of them. Bioethics is another. While informing already takes place more frequently in major cancer research and treatment centers in Italy (GIVIO 1986), the ideology of telling is spreading [1] and one can expect that normative practice will eventually follow.

This momentum towards change is due to a number of factors. One is the internationality of the field of oncology (see Good et al. 1990), where research and treatment protocols are standardized and shared. Such protocols require informed consent by the patient, and informed consent should require knowing one's diagnosis. Other important reasons include: the increased ability to diagnose, treat, and cure cancer in general; a public increasingly educated about cancer; changing relationships between physician and patient, reflected in and stimulated by the changing structure of health care (to include more private health insurance); the diminishing role of the family and the church; an increasing concern for the individual and his or her rights; and the increased internationalization of the Western world and Italy's participation in it.

This moment of transition provides special access to understanding the traditional practice of not telling and not knowing, offering a window into aspects of Italian culture that become more visible as they are called into question, both in Italy and through cross-cultural comparison. In this paper I will first explore the social and cultural logic of not informing and not knowing and the structure of hope in the context of a cancer diagnosis with uncertain prognosis. Then I will move into the background of this practice to consider what other practices and cultural understandings, typical ways of being in the world, operate within it.

My data are based on a number of studies conducted primarily in Florence, Italy: of terminally ill hospitalized patients with gastrointestinal cancer, their families, and health professionals involved in their care regarding the topic of information and communication [2]; of the experiences of women affected by breast cancer [3]; of the attitudes and practices of 'healthy' women regarding cancer and its prevention [4]; of family members, patients, and health professionals regarding needs and problems during the terminal phase of illness [5]; and more specifically of lay persons, professionals and patients regarding the meanings and practices related to disclosure of a cancer diagnosis [6]. While all the physicians (N=20) interviewed work with a significant number of cancer patients, only 7 were oncology specialists and/or work in a major treatment center. This is of relevance when considering cross-cultural comparison.

Finally, behind these studies are years of living in Italy and personal experience with cancer patients and some of their deaths.

Throughout the discussion, American culture will be an explicit and implicit partner in the dialogue, based on comparative analysis with other studies, particularly that of Good et al. (1990) and the abundant popular and professional literature, and on my being an American and my personal and professional knowledge of that culture. This comparison will be, however, with only one strand of such a complex culture, albeit one increasingly dominant.

### The Cultural Logic of not Telling and not Knowing and the Structure of Hope

In order to understand not telling and not knowing in Italy, we must first consider some of the cultural and social logic involved in this practice, what may be considered the structure of hope in this cultural context. While the perspective of patients and lay persons are incorporated in the following account, the views were often explored more openly with health professionals, primarily physicians, and family members who make the decisions. How much people who have cancer actually share these perceptions when they become severely ill is less certain.

Essentially, the logic regarding not informing patients of a cancer diagnosis with an uncertain prognosis is as follows:

(a) a diagnosis of cancer = certain death and hopelessness
awareness of one's diagnosis = hopelessness and death in the present.

(b) according to physicians and family members, patients don't want to know.

(c) a good death is living as normally and serenely as possible until one dies without being aware of it, ideally suddenly.

(d) telling the diagnosis is regularly interpreted by the physicians interviewed as giving a prognosis in terms of a specific time frame.

(e) the physician/patient, patient/family relationships are based on authority and protection. Telling the diagnosis is often equated with 'giving a sentence'.
Information and decisions regarding therapy are more the prerogative of the authority - physicians - and the family, less the patient.

(f) not telling allwos for hope and for life to continue; telling 'removes hope' and leads to death.

Let us consider these points more in detail. The few quotations presented were chosen only to exemplify statements repeated by others.

*(a) Cancer = death and hopelessness. Hope cannot coexist with an awareness of one's eventual death. Awareness of one's eventual death = hopelessness and death in the present:*

Cancer still is equated with a certain death by many patients, family members, and physicians alike. In fact, patients indicated greater willingness to talk about death than about cancer. For most, a diagnosis of cancer (with uncertain prognosis) fuses with a terminal prognosis, such that hearing one has cancer is lived or assumed to be lived as meaning a certain death (Gordon et al. 1989, Gordon 1990). A remarkable number of physicians interviewed also indicated that a diagnosis of cancer was essentially a death sentence leaving little hope, suggesting that the cancer = death association is still strong among them as well as some nurses.

"The word *tumor* is associated with the word death; that is, it doesn't give hope, it gives only certainty of death, of dying now, within a year, within a month, or to go ahead among a thousand difficulties, all things that are poorly accepted." (Chief, General Surgery, male, in 50's)

"So at the point when the professor or whoever tells me to my face, "you have cancer", I have no more hope, I can no longer think about life." (Head nurse, Pulmonary Department, male, in 50's)

In fact, slightly over half of 13 physicians questioned said they would like to be informed of the diagnosis if they had a tumor, the same if they had a serious prognosis. Nurses were much more interested in being informed on both accounts, suggesting gender differences worthy of further exploration.

Hope, on the other hand, equals life:

Q.: Hope, how important is it for you in your practice?

A.: "Hope for cure represents everything. If you take that away from a person, the hope for cure, then it's only a revolver that he's left with." (Chief, General Surgery, male, in 50's)

Living with knowledge of one's eventual death kills the present and essentially spells the end. It would effectively be all over:

"The confirmation (of a suspected diagnosis) is difficult to give because it closes everything, it closes possibilities, hopes, the hope to participate in life, everything ... *if the patient knows, there is no more hope.* But this happens in a situation in which there is no more therapy to offer." (Assistent, Pulmonary, male, in 40's; emphasis added)

More specifically, the expected responses most often cited were committing suicide, going crazy, throwing oneself out the window, becoming very depressed and refusing treatment, losing the will to live (Gordon et al. 1989). One head nurse predicted:

"If patients were to be indiscriminantly told the truth ... how many of these patients perhaps would throw themselves from Giotto's Tower (in Florence) or commit suicide with a revolver, or whatever other means they would find to kill themselves?"

Similarly, physicians, such as the following, frequently cited such possible consequences:

"Patients are not culturally prepared ... the patient can go and meet with suicide ... it happened a few times ... To say to one of our sick people that they have an illness of this type means to destroy them completely."

Q.: "What do you imagine would happen if you told a patient that he had cancer?"

A.: "Reactions of anger, of great depression ... in other words, he comes home and says, "I am a dead man, it's already finished for me..." and sometimes, from the point of view of therapy, some still have chances that they can use, and instead ... many times these people lose these chances..."

Q.: "Would you personally want to be informed if you had a serious prognosis?"

A.: "Well, maybe in this case with more difficulty (than knowing the diagnosis), because I believe that to have a sentence is not a thing about which you can rejoice ... to have a precise sentence." (Associate, Pulmonary, male, in 50's)

Living with consciousness of one's impending death, accepting death, is viewed as very difficult, if not impossible:

Q.: "Do you believe one can arrive at an acceptance of death, after the initial shock?"

A.: "No, I don't agree, not even for the 75 year old person. I understand that one must die, of this there is no doubt, and we need to help a person as much as possible. But it's not telling them that they must die in a month, or in a day or in three hours that we give them an advantage. It's rather being close to them, giving them the cures ... " (Head Nurse, Pulmonary)

Q.: "Do you think that one can accept that one will die, and within this fact have hope for example of living one more year or of not being abandoned?"

A.: "I believe that the thing that influences most positively is the hope of cure, to the very end; then, perhaps in the last days certainly, a different attitude enters." (Chief, General Surgery)

And there seems to be little difference, at least from a physician's perspective, between a terminal prognosis of one month or one year:

Q.: "But if one knows that one may die within a year, not tomorrow, may one 'live' (meaningfully) with that knowledge?"

A.: "And with one year? One goes to the seashore this year with one's children, next year I will no longer be able to go to this beach with my children, I will no longer have them, how can I live this year, in a tremendous spasmodic waiting for death? ... Death is not an acceptable situation, unless, I repeat, one has a particularly solid structure by moral preparation, by culture, by ideology, by religion ..." (Chief, Pulmonary, male, in his 60's)

As we know from people who have faced this experience, one theoretically can live the last year with much greater intensity and joy, knowing it may be the last.

*(b)Patients don't want to know, argue many physicians and family members:*

The belief that patients don't want to know is sustained by many health professionals and family members, even when the patient bombards them with questions regarding the diagnosis, treatment, and prognosis. These questions are rarely interpreted as evidence that the patient wants to know, but instead as requests for reassurance and denial (Gordon et al. 1989). A version of this can be heard by the following physician:

Q.: "When a patient asks for information regarding the possibility of having cancer, how do you usually respond?"

A.: "I enter in dialogue with the person in order to understand what the patient *really* wants." (Associate, Gastroenterology, male, in 40's)

These assessments by physicians and family members, made less frequently by nurses, do not generally concord with the expressed and implicit statements of patients. For example, approximately three quarters of the 23 hospitalized patients interviewed with terminal cancer indicated they wanted more information on their illness, treatment and/or prognosis. Eighty-five % of the 23 patients interviewed indicated that they would like to be informed if their situation took a turn for the worse, many mentioning the possibility of death and their desire to know. On the other hand, only one of 12 family members questioned felt that their relative would want to know his or her diagnosis, and only one fifth that the patient should to be informed about a change for the worse (Gordon et al. 1989). Similarly, in a national study of 31 hospitals, the research group GIVIO found a significant number of women wanting more information on their diagnosis and treatment (prognosis was not discussed) than that offered (Moscone et al. 1990).

*(c)A good death is to live as normally as possible until the end, not knowing ahead of time when that end will be:*

This aspect came out very clearly in speaking to family members who managed the death of someone at home. It was considered a real achievement if the person died without being aware that he or she was dying, and was able to continue living as normally as possible till the end. It was also reflected in the number of people, including those who knew their diagnosis and prognosis, who 'chose' not to be aware that they were actually dying at the end. This view was also heard repeatedly from health professionals. For example:

Q.: "If I understand you correctly, the ideal is for the person who is dying to live his or her life as normally as possible up to the moment of death?"

A.: "Absolutely ... the idea that one can live one's life knowing when one will die is, according to me, unacceptable ... but fortunately, in each of us there is the hope that we will still go with our children for a walk ... Life is made in a way that one doesn't know when one is born and when one will die. If the subject knows exactly when he will die, I don't know if his life, so ordered, is worthy of being lived." (Chief, Pulmonary)

Two people (one physician, one nurse) recounted positive stories of the death of their fathers from cancer. The fathers were never informed, and in fact great care was taken to act normally. But very shortly before dying, they both broke from the situation, one to announce his will in front of all present, the other, the night before going in coma, saying, "now I'm beginning to go away. I salute you all".

Q.: "No other mention was ever made that he was dying."

A.: "There's no need to say that .... Otherwise it becomes almost a desire to put a finger in the wound. But why? It's absurd."

Q.: "In the United States, there's a sense that people involved with someone dying have things to say to one another, to express their love, to say goodbye. If I understand you correctly, it's not necessary to do so with words?"

A.: "Absolutely not. I believe that words are not necessary ... to the contrary, a look maybe is better ..." (Head Nurse, Pulmonary, male, in 50's)

Q.: "What percentage of your patients would you say know they have cancer?"

A.: "I would say almost all of them. They don't say so, because they don't have the courage, but they are aware."

Q.: "But you say that almost all know, then is it like a theatre?"

A.: "Yes, it is a theatre, a fiction that both parties play. We pull ourselves ahead in this way, with this fiction. It's in their interest that it is thus ... For example, those who go for chemotherapy, naturally they know that there is a tumor, because they lose their hair, they know, but no one tells them they have a tumor. However, I know that they know, they know very well that I behave in this way because they have a tumor, and we go ahead all together, but we joke, we laugh, and we face the things with a certain fantasy, together." (Associate, Pulmonary, male, in 50's)

This recalls the ritual of mutual pretense described by Glaser and Strauss (1965).

No quality was mentioned more than serenity as the ideal state for a patient, particularly for the person with a terminal prognosis of cancer. Obviously a serene patient allows for a smoother hospital and family life. But it may also have a religious meaning associated with a life free from sin, the victory of the good over the bad. A tormented death, as one can have with a tumor, may suggest that a person is morally in conflict or being punished. Furthermore, how much serenity before death is considered a prevision of life after death is not clear, but it may symbolize the final judgement of the patient.

*(d) Telling the diagnosis is regularly equated with giving a prognosis in terms of a particular time period:*

As we have already heard in the above examples, telling is usually equated with giving an exact time frame. Whether this is a tactic used by physicians in particular in discussion to rationalize the practice of not telling by presenting telling in its worst form, or whether it is what many imagine telling to be, in part because they have had little experience with communication of the diagnosis and may discuss the prospect in abstract terms, or because they may in fact believe in the numbers, is not clear.

*(e) The physician/patient, patient/family relationships are based on authority and protection. Telling the diagnosis is often equated with 'giving a sentence'. Information and decisions regarding therapy are more the prerogative of the authority - physicians - and the family, less the patient:*

Patients supposedly participate in treatment because the physician orders it. They, as a rule, are not involved in the decisions. Motivation is not needed for compliance necessarily, better trust and respect. We hear this traditional view stated in the extreme in the following:

"On the choice of therapy, I believe the patient never must say anything, of this I am convinced. The choice is to be made by the physician... if he knows how to do his profession, whatever therapy he proposes to the patient, he will accept. If you say to a person, "You must cut off your arm, then your leg must be cut off because it is important", this patient first has his arm cut off, then his leg." (Assistent, Pulmonary, male, in 40's)

"One wonders how the patient believes the incorrect information or misinformation he is frequently offered by physicians. Not only does he want to believe this information, the physicians maintain, but, "he follows our reasoning, has trust in us. The relationship is one of trust." (Chief, Pulmonary).

*(f) Not telling allows for hope to live and thus for life to continue. Telling essentially destroys hope, leads to a collapse and a shutdow:.*

Not telling allows one to continue with one's life as normally as possible. The non-confirmation, even if one suspects, allows room to hope (this is one of the differences between people 'knowing inside' without being told). Because with the pronouncement, one can hope no longer.

"The difference between knowing oneself (that one has cancer) and being told is that one can not accept it (the diagnosis), one can know but refuse to know, and thus one can remain attached to life ..." (Head nurse, Pulmonary)

## Not Telling and Not Knowing as Ways of Being in the World

Within this cultural logic of not telling and not knowing are more pervasive ways of being in the world that embody particular interpretations about life, death, suffering, consciousness, being a person and morality. It is to these we turn now, considering in particular how the world - the existential world - is given in this Italian context.

Let me begin by way of a personal anecdote that I observed and experienced among Italian friends and that brought to light some of the following analysis.

True to the traditional practice of not disclosing a diagnosis of cancer to patients in Italy, when a 41 year old friend was operated on for possible ovarian cancer, her surgeon, never having mentioned the likelihood of cancer to her before the operation, explained her diagnosis afterwards as "tuberculosis of the ovaries". Despite being a psychologist and having worked in health care settings, despite having been informed by her close medical friend, Dr. A., before the operation that ovarian cancer was a possibility, the woman initially accepted this diagnosis, and when she learned that she had lost her ovaries and her uterus she was not only distraught but furious. Dr. A. thus decided that it was necessary to tell her the real diagnosis, saying that rather than it being tuberculosis as her surgeon had said, there was a very strong likelihood that it was a tumor, they would have to wait for the histological report for confirmation (even though it had already been studied and clearly documented as carcinoma, third stage, diffuse, during the operation). The anguish and confusion and distress that followed was expectedly tremendous. One of her first reactions was the desire to throw herself out of bed. In subsequent days the woman waivered between the two diagnoses, challenged her surgeon about it being a tumor, and was temporarily calmed when he gave her a detailed explanation of how in fact it really was tuberculosis, all the while telling her she would have to begin chemotherapy treatment. As time went by the woman torturously grappled with the possibility that it could be cancer, referred to as a "tumor". After some time she was informed clearly by Dr. A. that it was in fact a tumor. This information she kept a secret from most people, including her mother.

Dr. A., working in a traditional medical setting in which patients are not informed of a cancer diagnosis, and thus not accustomed to working with informed patients, later commented to me about how much more distressful it was for the woman, and of course for everyone involved, that she knew the diagnosis. How much more suffering there was.

This statement bewildered me, as I thought to myself, of course there is more suffering, but the suffering is already there in the fact that there is cancer. For me, and given my similar age the situation had particular relevance, I felt there simply would be no choice: if I had cancer, that would be part of my life - suffering would not enter the picture regarding knowing or not. As a fact of my life, as a part of who I am, it would be something I would have to know and to deal with. Consciousness would be essential. Not knowing would be not living my life to some important extent.

I was also stunned first by the possibility of speaking about tuberculosis of the ovaries in 1989 in the same breath as chemotherapy and second, by the willingness, albeit waivering, of this educated and already semi-informed woman to believe this diagnosis. I was struck not only by her active desire but also her willingness not to know.

Having studied the issue of telling and not telling a cancer diagnosis in this Northern Italian context and knowing the commonness of this situation, I imagined myself to be more North American in my thoughts and feelings - by no means exclusively so - not because I would want to 'control' and direct my care, as is usually emphasized, but in feeling that the fact about whether I have cancer, or whatever illness, is a fact about myself that constitutes part of who I am, is part of me. Nobody especially wants suffering, but something much too fundamental intrudes first. It's not a matter of choice to avoid, because it already exists. And the idea of someone 'sparing' me this knowledge makes little sense.

The image of Jesus Christ came to my mind, the image of a person absorbing the suffering and sins of others. I wondered how much this image and more generally the Roman Catholic religion played a role here.

How North American these feelings are, how Italian are the others, is no simple dichotomous matter. Many people in both countries share similar orientations and experiences. Denial of the diagnosis even after being informed is well documented in North America; and many patients in Italy feel the need to know, not just for pragmatic reasons, but because 'it is their life'. In fact, a number of studies document patients' desire for more information (Gordon et al. 1989, Moscone et al. 1990, Falchini 1989). I think, however, that currently the dominant approach in each country still differs and is culturally grounded in different ways, even though the ideology and gradually the practice in Italy is moving closer to the North American pattern.

We may ask, how could the communication of a diagnosis of tuberculosis of the ovaries be culturally possible, socially and psychologically credible and even temporarily accepted? What other understandings and practices does this practice of not telling and not knowing presuppose?

In beginning to answer this question we will note that in Italy there appears a world in the face of illness and suffering, but also more general than that, in which one's being and one's reality is very much defined in the present, in the actual being, rather than through abstract, cognitive parameters, a world defined more 'socially' than 'objectively', a world that embodies the values of connection, continuity, and protection, a world amenable to human wishes. And the social here means, primarily, the family.

We may, for highlight, contrast this world to a common situation in America, where a value is that one must live in the 'objective', 'real' world, must 'face the facts', a world indifferent to not only the individual but to humans in general, a 'disenchanted' world. This world is supposedly autonomous from humans and individuals, it is not socially given [7]. It is even oppositional to the self. Autonomy and equality, and, in turn, self reliance and control are dominant ideals in this context.

It appears that while this 'objective' world of course exists in the Italian context, for many people it is secondary to the more socially given, mediated one, that is more immediately accomodating to human wishes. Much as the 'objective' world celebrates and proposes the autonomy and control of the individual from nature, culture, and society (see note 7; also Gordon 1988), this more socially given world stresses hierarchy and the dependence of the individual on others (Dumont 1977), the social porousness of people, the dedication of the family collective to the individual, faith in authority, and the limits of controlling the objective world. Continuity and connection are stressed on the one hand (Gilligan 1982), confrontation and rupture avoided.

Here we will begin to triangulate a few dimensions of this world through examples, at times comparing them to America, and reflecting on some possible explanations involved in differences noted. This analysis must be considered preliminary, more a stimulus for further research than sound findings.

### *How the world is presented and defined*

The world is presented and defined more through actual situations lived in the present than abstract, mental descriptions that anticipate a future situation. Preparation, plans, schemes play relatively little role in structuring this world. Rather experience unfolds in the present, experience after experience, with few cognitive handles to hold onto. The individual takes his or her definition more situationally than from abstract definitions. For example, one finds the following cultural practices:

(1) Patients are rarely warned of what is to come: women who pass through a series of diagnostic tests in order to ascertain whether they have breast cancer, women who can tell they have parted from the routine, will be told nothing, and denied information if they ask about the *possibility* of their being a problem, even by doctors who tend to inform patients of a *certain* diagnosis. Denial will follow denial, words will diverge from behavior as long as possible. In the end, women are often presented the certainty that they must have surgery, for example, or that they in fact have a tumor.

(2) Many women who found a lump in their breast followed the same pattern, often not telling their spouse of the discovery until an operation was scheduled.

(3) Women are rarely warned ahead of time as to the kind of surgery and the kinds of treatment and physical and emotional experiences they may expect after surgery (there is no institutionalized 'informed consent' *in practice*). Many are told instead, "we'll operate and we'll remove the problem". Afterwards: "you're like new". Only step by step do many discover that they have begun a new beginning rather than reached an end. (This contrasts, for example, with the tremendous time and consideration given to pre-operative and post-operative patient teaching by registered nurses in some parts of the United States [Gordon 1984]).

In fact, to plan ahead for an unwanted eventuality, according to one common line of thought, threatens to bring about that very eventuality, as if thinking about it can bring it on. As we will discuss later, this belief in the power of thinking usually works more for the negative rather than the positive.

Situations, then, are continually defined and revealed in the present, in activity, images, and experience, less often through cognitive schemes, plans, or abstract parameters.

We may interpret that underlying this extreme situational approach is a background understanding that life is a stream of unexpected events that can at best be creatively and spontaneously mastered *ad hoc*. For no matter how much you prepare, the situation will always have its unpredictability and defy the best plans and efforts to control it. This contrasts with a common orientation in the United States that seeks maximum preparation possible in order to better control the situation, the expectation being in this case that events are relatively predictable and understandable, and thus controllable.

One is brought to consider the relative histories of the two countries as an operative context for these different approaches: Italy, living through a sweep of invasions since the time of the Romans, changing dominations, economic and political precariousness in contrast to America, deliberately pioneered by immigrants who sought to take their destinies in their own hands, and who have a history of social domination rather than subordination, with relative economic security.

Another way to describe this world is as 'socially mediated realities' instead of supposedly individually mediated ones. It is the family and the physician who traditionally mediate between the person and the illness; the illness is given less through objective parameters, through objective information, but more through how they present it to the person. It is explicitly socially constructed. 'Reality' or the situation is mirrored in the other. Thus if the physician or the family are tranquil, the message is that the situation is tranquil and the patient can thus be tranquil. On the other hand, if family members cry or show upset, this cues the patient to the severity of the situation. Thus people 'read' each other to understand what the situation is. And people in turn contrive themselves in order to present a particular reality:

"If she (the patient) sees you tranquil ... "Well," she thinks, "if he is so calm why should I have to worry myself!"" (breast surgeon)

"It was extremely difficult to present myself as calm with her (his dying wife). But I learned it helped her when she told her aunt, "I know that I'm feeling bad and the situation is not good, but I see my husband, and I see him tranquil, and if he is tranquil, that means that the situation is not so bad for me." (husband recounting the death of his wife from breast cancer)

If reality is socially defined, one important way of denying an unpleasant reality is by socially defining a different one.

Another way in which we may say this world is socially defined is in the power attributed to the sayer, obviously, not any sayer, but one with authority, be it the physician, a family member, a priest. Truth is then socially defined, not necessarily 'out there'. The emblematic example relevant here is the sense described by all that the person who tells the diagnosis, the physician, is the one who 'gives the sentence'. He who names, sentences. Note that the 'sentence' is not attributed to the existence of the illness in the body of the person.

Some parallels in the culture at large come to mind, such as the traditional Roman Catholic church which traditionally mediates between a person and God (in contrast to a more direct relationship in Protestantism), as the priest interprets the scriptures. These mediators don't just mediate, they can take on and solve the problems of the others. They embody the others in themselves. In fact, Holland and colleagues found that countries dominated by Catholicism were less inclined to inform patients than Anglo Saxon countries, where presumably Protestantism plays a greater role (Holland et al. 1987). The role that religion plays in the not informing, not knowing story, however, clearly merits further investigation that is beyond the scope of this paper.

While truth may reside in the powerful sayer, truth can also be understood without being put into words. Many things can and best be left unsaid. This is particularly true with a dying person, as described earlier, as it is often admitted that the dying person does in fact know, but there is no need to confirm it explicitly with others. Nor does there seem to be much expressed need or desire to 'say goodbye', as one hears frequently in the death and dying literature in America (Kübler-Ross 1969) and where feelings or understandings, to be considered real and true, best be explicitly expressed (Arney and Bergen 1984) (the romantic "I love you"s of American films strike many Italians as humorous or embarrassing). Words are not taken on face value and much symbolic communication takes place together with much effort to grasp the real meaning of what the other has said (Munakata 1989).

This difference in explicit and implicit communication may reflect what Hall calls *high context* and *low context* cultures (Hall 1976), the former being those in which many things can be taken for granted and left implicit, the latter being those in which the context provides little information. America, a country of immigrants, is generally a low context culture, and this may be another example of this.

Finally, we see the dominance of the social over the 'objective' in that a truth really takes on its reality when it becomes social. This is best exemplified by the apparent contradiction that many people maintain that most cancer patients know their diagnosis and possibly a bad prognosis, even if they are not told, while adamently insisting that a person would effectively die if he knew. Given that the most important reality appears to be the social reality, not admitting the diagnosis into this important reality means it can be lived as less real. By not telling, the hope that the reality was not really real could live on.

Similarly, sharing the information of a 'terminal' diagnosis in the social context endangers bringing about a social death right then and there, as it begins the separation between those who will be leaving from those who will be staying (Gordon 1990).

### *Ways of living painful emotions and suffering*

Communicating and learning a diagnosis of cancer inevitably touch very strong emotions and cause suffering for others, give what women called *dispiacere*, an upset, to another. Practices and understandings about these emotions and how best to handle them enter the communication picture.

It is considered a strong obligation in social relationships to avoid causing displeasure to others. To protect someone from suffering, to avoid hurting someone, by not giving them bad news, for example, often takes moral precedence over other more long-range goals or values, such as growth, control, freedom, independence, or honesty. In fact, many practices work to avoid and prevent suffering and more generally, to minimize opposition, while promoting harmony and tranquillity. For example:

(1) As we heard in the above example - the diagnosis of tuberculosis of the ovaries - physicians say they end up telling the most incredible things to patients, but that "they (patients) will believe anything!". Several recount stories of chiefs of oncology departments who denied to their death that they had cancer.

(2) A large number of women described the reactions of their family members and friends to the news of having to be operated on for a possible breast tumor with the following *leitmotiv*: "Be tranquil!", "it's nothing!", "but no, what are you talking about, it's nothing!", "you're exaggerating!" - what the women called 'minimization'.

(3) The tendency to deny bad news is so strong that the denial can continue even after the patient has concrete proof of a diagnosis of cancer. One patient described for example how her family physician, who had been following her for years, refused to admit that a tumor was involved, even after she said that she had read the letter addressed to him in which the diagnosis "carcinoma" was clearly written. He even bet her a dinner that it would not be true.

This avoidance of giving displeasure in part explains the non communication of the diagnosis by physicians, by no means exclusively for Italian physicians:

"Without a doubt, there is a difficulty on my part, and I think it is true of everyone, in giving the sentence. I have difficulty, perhaps, in the upset that I could cause in this person in telling him. In other words, there is always on the part of the physician the need to make the person feel better ..." (Assistant, Pulmonary, male, in 30's).

It also partly explains the tendency of informed women with breast cancer not to inform their family members. They, as women, are supposed to be the *protectors* from pain, particularly for the young or the old, not the cause of it.

Suffering can be taken on by another. It is not the property nor does it necessarily 'belong' to an individual. It can be collective (familial). Of course with this goes the social authority to decide for the other. The body, and more generally the life of the person, is not so much exclusively his or her body or life but also belongs to a collective.

In fact, protecting others from suffering is a major cultural practice, particularly paradigmatic of the role of mothers. We hear this clearly in women who developed breast cancer themselves:

"When I see a person in need, I don't count, I don't exist."

"I was taught from childhood ... that women must be brave, capable, must be strong, must undertake all the problems of the family."

"I don't know how to ask, I'm not used to the fact that others help me, I've always had to arrange alone ... I'm the cuddling mother, the protective wife, I have to protect the others. I would put them all under a bell-jar in order that they are not treated badly, that they don't get tired, that they don't have any problems."

Yet this image of the strong, protective, selfless mother presents a major problem for women having a potentially fatal illness and needing help. What to do with their suffering when it is they who are supposed to absorb the suffering of others? They not only feel guilt over this, but desperation over who will help them and where to put their own feelings:

"Thus, as I see my husband who is crying, my children who are crying ... so you have to play the heroine, you have to say "it's nothing, I'm alive, I'm here."

"I always had fear of bothering someone, fear of making them uncomfortable, I always tried to understand that others have problems ... first I want to try alone... I always try to spare the person."

By not telling their loved ones, as a good number of these women did not, they could maintain there ideal as protectors while avoiding a dreaded role of the bearer of pain and suffering. This contributed to a sense of a world divided between *outside* and *inside*, with painful feelings lodged on the *inside*, as we discuss below. Some of the women's expressed reasons for not telling their loved ones about their illness were:

"Naturally, in my house I didn't say anything to anyone because I thought; "If I do that I'll put them all in alarm, and we still don't know what it is, so let's just leave the things as they are" ... and I carried this weight upon me."

"I didn't want to make them sad, it's my thing."

"I didn't want to give them more worries; it's not right."

"I didn't tell anyone in my house, I kept it all tranquil. I didn't want to give problems to others, so I said nothing."

"I try not to have my family feel all this of mine that I have inside.... I keep all inside. It doesn't seem right to share it with my son (adolescent) or my husband."

This taking on the suffering of others again recalls the impact of the Roman Catholic religion and the major cultural paradigm, not only of Jesus Christ, who absorbs the sins and suffering of others, but of the Madonna as the protector of others, and the role of the Great Mother in Italy and Mediterranean society in general (Bernhard 1969). However, again, the pervasiveness, meaning and complexity of these figures and practices require an in-depth study beyond the scope of my present understanding.

The other side of the value of not causing suffering to others is a cultural acceptance of either avoiding suffering oneself or of the ability to stand the suffering without putting it on others. For example:

Q.: "If I have understood your reasoning for not telling, there is no real purpose in suffering, and thus it is better to avoid it?"

A.: "Absolutely no purpose. And this is the point - it should be avoided. The desire, the will to try and believe that it is something else (other than cancer) is something that takes away the suffering, because if I think that all is finished, I feel bad, I suffer. If, on the other hand, I live in uncertainty, I can extend myself, stretch myself out."

Q.: "Suffering is not something that can make you stronger?"

A.: "I feel strong when I am serene, at peace. I don't feel strong when I am worried or when I am suffering." (Head nurse, Pulmonary)

Patients say they are often 'tranquillized' by the words and the calm of the physician (and others) and place their trust in him. Some acknowledge allowing themselves to be seduced by this tranquillity even though 'inside of them' they 'knew' the situation to be different. This tranquillity, however, is not always long lasting and requires regular boosting. Furthermore, many women said they appreciated their physicians' telling them the diagnosis but in a way that 'sugared the truth', made it softer and more palatable.

One doesn't hear particular value placed on an ability to 'face the truth', for example, the 'cold' truth, but more a belief that there isn't necessarily one truth, there can be many, and they certainly can be manipulated and made 'warmer' and 'softer'. Contrast this with an American moral strand that stresses that one should live without illusions, one 'should' face the truth, face the facts, and not 'regress' to denial or self delusion. This moral strand is likely indebted not only to the Enlightenment, but also to the influence of Protestantism on American culture (Fox 1989).

The different approaches to suffering (and to death) may also be explained by the fact that suffering is already in the background of the Italian context as something already lived through historically, as inevitable and unavoidable, thus the attempt may be to avoid it when it comes and search for serenity. Perhaps pessimism about suffering lies close below the surface here. Does this explain, in fact, the tendency to 'collapse' into despair, to succumb to hopelessness, to crumble under the situation that is a common and expected initial response to learning bad news, such as a diagnosis of cancer (Gordon et al. 1989)? This seems not only to be a common way of living bad news, but also that the 'news', the definition of reality, seems to be taken as an absolute truth, more than a challenge to beat, to defy, to fight against, as is apparently more common in the United States (Simonton et al.1978, Siegal 1987, Cousins 1979, 1989, Good et al. 1990). Further, in America one is struck by the dominance of the pursuit of happiness, the attempt to *eliminate* suffering and the belief that this is possible through maximum control of one's environment (Payne 1988) and self.

*Consciousness and information*

Saying and thinking are lived as very powerful, both saying the good (such as 'I'm very healthy') but worse, saying or thinking the bad or the unwanted. Four examples:

(1) The most often cited anticipated consequence of informing a patient that he or she has cancer offered by health professionals and family members was that death would begin right then and there: "It would be as if it is all over." Similarly, the reason many 'healthy women' most frequently gave for not wanting to be informed if they had breast cancer was that they would die earlier, that *knowing* they were going to die would bring on the death sooner.

(2) One of the dreaded dimensions of cancer by 'healthy women' is the "consciousness of one's own death".

(3) The phrases, "I don't want to think about it", "I try not to think about it", "I can't think about it" return and return in interviews with healthy women regarding cancer.

(4) In the face of cancer or a difficult or potentially dangerous situation, over one third of healthy women say they live by the motto, "the less I know the better I am".

Knowledge and consciousness, then, are lived as very powerful. But interestingly this power is considered primarily negative and thus potentially dangerous. Consciousness is manipulated through denial, more than manipulated for instrumental action. The 'power of positive thinking', the belief in mind over matter, an approach common and growing in North America (Siegal 1987, Cousins 1979, 1989, Good et al. 1990), is used in this Italian context more as a technique to *avoid* thinking about or *denying* the bad than as a serious means to *bring upon* the good through mental power. Will power is less considered a powerful resource to cultivate in order to use for purposes of cure (Good et al. 1990), rather it is more that the power is in the hands of the physician and the medicine, and where this no longer exists, in the caring family.

The belief in fate or destiny and/ or in God's power to influence events, specifically health events, was found to exist among over 50% of the Florentine sample of healthy women (Gordon et al. 1990). Very likely these beliefs are touched when the presence of cancer is discovered, and may also help explain why people referred to the communication of a diagnosis of cancer as "giving a sentence" to a patient or "condemning one to death".

This may play an important role in the differential approaches to disclosure in North America, in contrast to Italy and Japan. Not only are the latter two cultures more traditionally family rather than individual dominated - and thus it is the family who manages the information and the illness experience of the person during at least the period of illness (Munakata 1989) - but further, sensing the ability to *challenge* a diagnosis of a fatal disease presented by an authority figure such as a physician is also a sense fostered by individualism. In this moral context, the ideal is that the individual understands himself as separate and autonomous from others and as sovereign as possible in determining his or her fate, and in fact, in overturning the fate written by others. Defiance not only in the face of the illness, but in the face of the physician, requires belief that the individual can determine his or her own fate.

This defiance is less apparent in Italy. In many respects this reflects the authoritarian structure of Italian relationships, on the one hand, and a history of foreign domination on the other. The individual appears to experience himself as less powerful to define his or her own fate. Furthermore, it is holism, with others, with one's environment, perhaps even with one's illness that is culturally stressed more than opposition. Perhaps this contributes to bad news being more frequently denied than defied (Cousins 1989).

In sum, many Italians expressed concern over the negative power of consciousness of a cancer diagnosis in the Italian context and that it could accelerate the process of dying more than as something positive and potentially therapeutic. It is not seen to provide the means to *change* the objective situation, but is manipulated in fact in order to avoid change or rupture.

Consciousness or knowledge in the American context on the other hand is considered helpful and even essential for treatment and even cure (Good et al. 1990). Knowledge is an important armament necessary to fight the enemy (the illness), to defy, to attempt to control the illness by maximizing one's health care and more recently, by exerting mind over body.

We should emphasize here that neither country appears to have a monopoly on 'magical thinking' or rationality, but that the different countries emphasize different ways of coping with what we may call hopeless situations. While the Italian way tends to be through denial, normalization, and dependence on others, a common way of coping in the United States seems to be to take one's case in one's own hands and try to control the disease. Whether this improves the fate of the disease and changes the outcome is not yet clear. It is clear, however, that the belief that it can is strong (Good et al. 1990) and that enacting this belief may improve the experience of the illness.

Furthermore, consciousness for its own sake does not appear to hold any particular symbolic or moral meaning in and of itself in the traditional Italian context. In contrast, in the United States, one could argue that being aware of one's 'objective' condition offers the symbolic and moral gain of being an autonomous, rational person, and master of one's fate.

Finally, information does not appear to necessarily 'belong' to the person about whom it refers in Italy, even though there is an increasing ideological recognition of the 'right' to information. Information has a long history of being the prerogative of the authority, and in the physician-patient relationship, in fact, the physician still communicates very little information to patients, whatever the diagnosis (Gordon et al. 1989); he is expected to use it at his or her discretion for the benefit of the sick person.

Information as a 'right' of the individual is also quite implemented in practice in America. Information regarding one's person, body, and life is understood to belong to the person about whom it refers, not to the physician nor to the family. This undoubtedly reflects the different extent to which one's body, one's suffering, and one's life is considered to be the possession of a self that is primordial and separate from the collective (Sandel 1982).

### *Personhood, illness and emotions*

Illness and feelings appear to be on the external edges of a person and, for the most part, come from outside the person or the self. Feelings can be caught as well as manipulated by others. For example:

(1) Women, both healthy and affected by breast cancer, use the following expressions repeatedly: "I had the illness *upon* me"; "I don't want to *weigh down* others"; "they threw the fear *onto* me"; "one feels the *weight* of age"; "he (the doctor) removed the illness *from upon* me".

(2) Cancer is nearly always described as something external to the self entering an internal, coming from without and ruining the good within. The most common cause given for breast cancer (in a sample of 144 women) is to receive a blow, a trauma, on the breast.

(3) A large number of healthy women described themselves as being "very impressionable" to what others say and do and experience. This works both as contagion on less pleasant emotions, such as anxiety, depression, as well as more "tranquil" states of being. Just hearing the word "cancer" or the situation of another put some in a terrible crisis.

(4) Women with breast cancer often repeated this internal/ external directionality, such as in the following explanation of the disease in themselves: "The illness I see as a thing in the air, that is there ready for everyone, not that it's for me ... in a certain moment in my life I was really disposed ... that day must have been a particular one, it entered in me and began to plant roots in this part of my body, and I see it now as a thing that absorbs everything that is good and healthy in my physique ... it takes it and I lose it ... slowly slowly it becomes owner of the body and then gives me the final blow ..."

(5) Healthy women frequently referred to cancer as if it was in the air, ready to strike, using such terms as, "this one here", "these here".

The expression of strong painful emotions is generally avoided. This too may be explained in part as an avoidance of disrupting a harmonic social fabric. After all, with people so porous, so open to each other, difficult, painful emotions "weigh down" others, cause them pain and discomfort. This leads one to consider how much the silence of many terminally ill patients is their way of protecting their family members and physicians from suffering by controlling their emotions, as the Longs (1982) suggest is true for Japan.

When people cry, and this ranges from women newly diagnosed with cancer to newborn babies, a very frequent expression is "*su, su*", "up, up!" (in morale, in mood). It is not generally considered therapeutic to express one's sadder, more depressed emotions and to regard this expression as a passage to more hopeful and happier emotions, or as a natural reaction to a situation. Again, with the world existing effectively in the present, these emotions become everything.

Not surprisingly, in counterpoint to the social field in which the unwanted, painful, non-harmonic emotions are denied, is another that is called "inside of me", "*dentro di me*". The certainties that are acted out in the public arena often reverberate as uncertainties "inside", as, for example, the doubt many women with breast cancer expressed about their actual health status, what they can believe or not. Likely the non-confronting, accomodating and harmonic public world is parralleled by a suspicious, aggressive, and perhaps paranoid world that is placed "inside" the person. Hear three women affected by breast cancer:

"Even though the physician said it was not a tumor, I knew inside of me it was."

"Inside of me a bomb exploded, but the splinters scattered all about."

"I didn' t make a tragedy of it, but inside of me, *dentro di me*, I was upset."

What we may say that all these phenomena have in common is that a large part of reality is defined in the present, in the lived context, and more specifically, by social others. It is as if a large part of people is lived very much open and susceptible to, and interacting within, a given social space. Being what is usually called "a self" here is very much a process, given more in social terms than abstract objective ones.

Responding, complementing, and reflecting this ideal public world is the sense of a private, inside world, a repository that allows women to meet their ideals.

### Care and Hope

Care consists of anchoring a person in a world. The anchors and the worlds may differ culturally, and while one apparently universally recognized need is for hope, offering hope is a form of caring that will be grounded in a cultural context. In Italy, where the social dimension is paramount, not only is the situation socially defined, but caring is expressed through protection, and through maintaining the social bonds, by not letting the person part company, by giving the message, "nothing has changed, we are here with you as before, you are all right, there is nothing wrong, you have not gone away from us".

In America, where the "reality principle" is more the so-called "objective" reality, we are not surprised to hear that many patients prefer that their physicians "level" with them, "give them the facts", speak "matter-of factly" (Cassileth et al. 1982). Information is an important part of the person as it *defines* that very person. And for many, it is the "objective" diagnosis of cancer that defines that they have cancer, less, perhaps, the person who tells the diagnosis.

Providing the "facts" in a concrete way may also be a way to give people hope (Trillin 1981, Good et al. 1990). For where the values of autonomy and self reliance are so strong, one type of caring is through fostering a sense of control. This, as Cassell notes, is the means to offering hope, "by keeping patients as much in command of themselves, their symptoms, and their situation as possible" (Cassell 1984: 148).

What else does the physician have to offer the patient with a terminal prognosis? As mentioned, knowing one has cancer with an uncertain prognosis and hope are considered by many Italians to be mutually exclusive. Many do not share a great faith in technology, neither physicians nor patients. And while religious faith certainly can be called upon, it is significantly weak. Rather, an Italian physician (again, not necessarily the oncology specialist in a major treatment center) offers hope by offering the security and tranquillity that *he* will take care of the situation, by protecting the patient, and by offering reassurance, often denying the facts and the gravity of the situation through "tranquillizing words" and actions. By keeping the patient uncertain he leaves the door open to hope. Whether this actually is experienced as tranquillizing and hopeful by patients is less clear.

One tendency in America on the other hand is to offer hope in the battle against mortality and perhaps for immortality through technological intervention, and we hear many physicians aggressively working to stave off death by pursuing uncertainty and thereby constructing a sense of "chance" (Muller and Koenig 1988, Payne 1988, Powles 1973, Good et al. 1990). The bias, the magic here, if you will, is placed in the instrumental, in technology (Powles 1973, Gordon 1987a, b). It is in this way that many express their caring to patients (Muller and Koenig 1988, Good et al. 1990).

While manifestly there is less public discourse about death in Italy than in America, and with it more apparent denial than acceptance, a more likely hypothesis is that Italians accept the inevitability of death to a greater extent and thus do not attempt to stop it frantically with technological bombardment [8]. If death cannot be stopped, and if knowing it is coming causes suffering, then there is no real use in knowing. Furthermore, possibly death is lived more "on the skin", in the sensation, in the imagination, and in the present than cognitively.

On the other hand, perhaps death is considered less inevitable in America, as the battle is to continue to defy it (Cousins 1989). Underlying this approach is likely a deeper optimism - verging on magical? (Parsons 1951) - about the inevitability of death, and about the ability to control it. Knowledge and information are essential for control. Objectification and depersonalization are likely used more in America, such that things are managed more with the mind, as much of the popular literature encourages, than experimentially lived.

## Conclusion

In this paper I have emphasized the cultural grounding of the practice of not telling and not knowing a diagnosis of cancer with an uncertain or poor prognosis. In doing so I have clearly stacked the deck by emphasizing cultural differences in comparision to the United States. It must be repeated that while there still remains a great deal of cultural consensus in Italy around not telling and not knowing, particularly regarding a terminal diagnosis, there is also significant dissatisfaction and call for change, particularly by patients and professionals.

The fact that these practices are currently undergoing change, then, raises some interesting questions. If in fact the practice of disclosure is so imbued with other social relationships and cultural understandings that partake of the larger social and cultural context as well, should we expect that other changes must precede, accompany and/ or follow an increase in the practice of informing?

Certainly the tremendous public and professional discussion of the argument, "to tell or not to tell the truth to cancer patients", the abundant conferences, seminars, articles, and television programs on the topic suggest that people are grappling with a significant change and what it implies. For the current internationalization of the Western world in which Italy participates, the notable influence of American culture, together with the international nature of oncology, together with changes internal to Italy and its own history suggest that the change to informing will eventually take place. How much it will done and be done in an Italian way, and how long it will take remain to be seen. But we may expect that within the general framework of informing there will be different cultural styles and that the same practice will not necessarily be practiced or lived the same way in different cultural contexts, such that non informing in the American past may not necessarily equal the current non informing in Italy or Japan, or that informing in Italy may not necessarily have the same meanings it has in America.

## Acknowledgements

The author gratefully thanks the many people who participated in the various research projects, the organizers of the stimulating conference, "Anthropologies of Medicine" and the editors. Allaman Allamani, Mariella Pandolfi provided fruitful exchanges for understanding Italian ethnography, Mary Jo DelVecchio Good for the conceptualization of cross-cultural comparison. The studies cited were supported by the Regione Toscana, the Italian League for the Fight Against Cancer, Florence branch, and the Italian Association for Research in Cancer. Part of this paper was presented at the American Anthropology Association meeting in Washington, D.C., November 1989.

## Notes

[1] The right to information is already specified in the 'Rights of Sick People'. The Association of Physicians recently voted to require that all patients be informed of their diagnosis and treatment (Negro 1989).

[2] A study of hospitalized patients with gastrointestinal cancer (N=23), their family members (N=20), physicians (N=11) and nurses (N=19) (compared to a contrast group of patients with "terminal" cirrhosis of the liver, N=17) regarding information and communication concerning the diagnosis, treatment, and prognosis of cancer. Not one of the

21 patients communicated to the researchers that they knew his/ her diagnosis (two patients with gynecologic cancer however did). None of the five physicians interviewed working in the division in 1987 in which the patients were hospitalized inform the patient directly of the diagnosis. None of the 11 physicians interviewed inform of a "bad" prognosis. It is rather the family who is informed, and invariably they together who choose to withhold the information from the patient. These findings are presented in Gordon et al. 1989.

[3] An interdisciplinary study of approximately 20 women with breast cancer (members of an association called "*Donna Come Prima*" "Woman as Before", jointly sponsored by the Center for the Study and Prevention of Cancer and the Italian League Against Cancer, Florence branch. Data were collected through indepth group interviews (10 weekly meetings, 2 groups, 20 women) that explored the womens' experiences from before the diagnosis to the present. The women ranged in age from 37 to 63 years old, and were mostly of middle class backgrounds. They constitute a special sample in their willingness to communicate and relive their experiences. This research is presented in Paci and Venturini (1989), including an article by the author: Gordon (1989). Several physicians involved in their care were also interviewed, see Mucci, Muraca and Tanzini 1989.

[4] A study of the attitudes and practices regarding cancer and its prevention among a sample of 200 "healthy" Florentine women between the ages 50-70 (N=144), using semi-structured interviews. Progetto Firenze/ Donna, Center for the Study and Prevention of Cancer, Florence, Italy. Findings presented in Gordon et al. (1990).

[5] A small pilot study of the expectations, problems, and needs of terminally ill cancer patients, based on interviews with patients, family members and health professionals (N=12): "Terapia del Dolore" Project, sponsored by the Italian League for the Fight against Cancer, Florence branch (cf. Cardoso et al. 1989).

[6] An anthropological cross-cultural study conducted by the author on the cultural meanings of not telling and not knowing the diagnosis of cancer in Italy (Gordon 1990). This study also forms part of an international study including the United States (Mary-Jo DelVecchio Good) and Japan (Tsunetsugu Munakata).

[7] This of course is the ideology, not the reality. For a compelling argument, see Taylor (1985).

[8] This suggestion was made by Mariella Pandolfi.

## References


ARNEY W. and BERGEN B. 1984: *Medicine and the Management of Living*. Chicago: University of Chicago Press.

BERNHARD E. 1969: *Mitobiografia*. Milano: Adelphi.

CARDOSO P. et al. 1989: *Per un progetto sperimentale di assistenza continuativa ed integrata al malato oncologico*. Firenze: Lega Italiana per la Lotta Contro i Tumori.

CASSELL E. 1984: Autonomy and Ethics in Action. *New England Journal of Medicine* 297: 333-334. Cited in ARNEY and BERGEN (1984: 148).

CASSILETH B.R., ZUPKIS R.V., SUTTON-SMITH K. and MARCH V. 1982: Information and Participation Preferences among Cancer Patients. *Annals of Internal Medicine* 92: 832-836.

COUSINS N. 1979: *The Anatomy of an Illness*. New York: Bantam.

-- 1989: *Head First: The Biology of Hope*. New York: E.P. Dutton.

DUMONT L. 1977: *From Mandeville to Marx*. Chicago: University of Chicago Press.

FALCHINI S. 1989: "Perche' non collaborano?" In: PACI E. and VENTURINI A. (eds.) 1989: *Dall esperienza della malattia una nuova cultura. (pp. 109-114). Firenze:* Comune di Firenze, La Lega per la Lotta Contro i Tumori, Centro per lo Studio e la Prevenzione Oncologica, Firenze.

FOX R. 1989: Comments on D.R. Gordon's paper "Not to know, not to tell: Cancer and communication in Italy". American Anthropological Association Meeting, Washington, D.C., November, 1989.

GILLIGAN C. 1982: *In a Different Voice.* Cambridge, Mass.: Harvard University Press.

GIVIO (Interdisciplinary Group for Cancer Care Evaluation, Italy) 1986: What Doctors Tell their Patients with Breast Cancer about Diagnosis and Treatment. *British Journal of Cancer* 54: 319-326.

GLASER B. and STRAUSS A. 1965: *Awareness of Dying.* Chicago: Aldine.

GOOD M.-J., GOOD B.J., SHAFFER C. and LIND S.E. 1990: American Oncology and the Discourse on Hope. *Culture, Medicine, and Psychiatry* 14 (1): 59-79.

GORDON D.R. 1984: *Expertise, Formalism, and Change in American Nursing Practice.* PhD Dissertation. San Francisco/ Berkeley: University of California (Medical Anthropology Program).

-- 1987a: Aspetti magici in biomedicina. *Psichiatria e Psicoterapia Analitica* VI (I): 93-98.

-- 1987b: Magico-religious Dimensions of Western Medicine: The Case of the Artificial Heart. *Acts of the First National Conference of the Cultural Anthropology of Complex Societies.* Rome (in press).

-- 1988: Tenacious Assumptions in Western Medicine In: LOCK M.D. and GORDON D.R. (eds) *Biomedicine Examined* (pp.19-56). Boston: Kluwer Academic Publishers.

— 1989: "Vivendo questa nostra storia": La voce delle donne. In:PACI E. and VENTURINI A. (eds.) 1989:*Dall esperienza della malattia una nuova cultura.*(pp. 17-57) *Firenze:* Comune di Firenze, La Lega per la Lotta Contro i Tumori, Centro per lo Studio e la Prevenzione Oncologica, Firenze.

— 1990: Embodying Illness, Embodying Cancer. *Culture, Medicine, and Psychiatry* 14: 273-295.

GORDON D.R., BELLONI L. and ALLAMANI A. 1989: *Informazione e comunicazione nell' assistenza al paziente con neoplasia e alla sua famiglia. Report finale, Progetto Oncologico.* Regione Toscana, Firenze.

GORDON D.R., VENTURINI A., ROSSELLI DEL TURCO M., PALLI D. and PACI E. 1990: What "healthy" women think, feel, and do about cancer, prevention, and breast cancer screening: A study of women in Italy. Currently under review for publication in the *European Journal of Cancer.*

HALL E. 1976: *Beyond Culture.* New York: Anchor.

HOLLAND J. et al. 1987: An international survey of physicians' attitudes and practice in regard to revealing the diagnosis of cancer. *Cancer Investigation* 5: 151-154.

KÜBLER-ROSS E. 1969: *On Death and Dying.* New York: Macmillan.

LONG S.O. and LONG B.D. 1982: Curable cancer and fatal ulcers.*Social Science and Medicine* 16: 2101-2108.

MOSCONE P. et al. 1990: Informazione sulla diagnosi di tumore: medici e pazienti a confronto. *Federazione Medica* (in press).

MUCCI M., MURACA M.G. and TANZINI O. 1989: A colloquio con i medici. In: PACI E. and VENTURINI A. (eds.) 1989: *Dall esperienza della malattia una nuova cultura.* (pp. 115-136) *Firenze:* Comune di Firenze, La Lega per la Lotta Contro i Tumori, Centro per lo Studio e la Prevenzione Oncologica, Firenze.

MULLER J.H. and KOENIG B.A. 1988: On the boundary of life and death: the definition of dying by medical residents. In: LOCK M. and GORDON D.R. (eds.) *Biomedicine Examined* (pp. 351-374). Dordrecht: Kluwer Academic Publishers.

MUNAKATA T. 1989: The socio-cultural significance of the diagnostic label "neurasthenia" in Japan's mental health care system. *Culture, Medicine, and Psychiatry* 13: 203-213.

NEGRO F. 1989: Nasce per il medico un nuovo codice di comportamento. *La Nazione*, July 18, 1989.

PACI E. and VENTURINI A. (eds.) 1989: *Dall esperienza della malattia una nuova cultura. Firenze:* Comune di Firenze, La Lega per la Lotta Contro i Tumori, Centro per lo Studio e la Prevenzione Oncologica, Firenze.

PARSONS T. 1951: *The Social System.* New York: Free Press.

PAYNE L. 1988: *Medicine and Culture.* New York: Henry Holt and Co.

POWLES J. 1973: On the Limitations of Modern Medicine. *Science, Medicine and Man* 1: 1-30.

SANDEL M.J. 1982: *Liberalism and the Limits of Justice.* Cambridge: Cambridge University Press.

SIEGEL B. 1987: *Love, Medicine, and Miracles.* New York: Harper and Row.

SIMONTON O.C., MATTHEWS-SIMONTON S. and CREIGHTON J.L. 1978: *Getting Well Again.* New York: Bantam.

SWINBANKS. D. 1989: Japanese Doctors Keep Quiet. *Nature* 339: 8: 409.

TAYLOR C. 1985: Introduction. In: *Philosophical Papers I and II.* Cambridge: Cambridge University Press.

TRILLIN A.S. 1981: Of Dragons and Garden Peas: A Cancer Patient talks to Doctors. *New England Journal of Medicine* 30: 699-700.

# Images and Interpretations of Severe Illness:
## Ethnological Aspects of Dealing with Cancer

Jutta Dornheim

About the time I was conducting the present study, the Cancer Research Center in Heidelberg began receiving an interesting series of letters. In these letters (received from 1977 to 1979) people from every social class aired their views about the cause and cure of cancer. No one had solicited the correspondents' opinions; they had written on their own initiative. Among the causes of cancer mentioned were infection from the clothes of cancer sufferers as well as from "sexual intercourse", "sleeping around" and "licentiousness". One writer thought that the pathogen orbits within the atom of a cancer cell; another that it lies in the nucleus of a negatively-charged entity consisting of gas or cancer salmonellae. Therapeutic proposals extended from dietary measures, such as fasting or the intake of fresh cucumber juice, to religious counsel and magic rituals.

These letters document quite graphically the starting point of my study of notions, attitudes and forms of behavior in everyday dealings with cancer. However, I first learned of them when they were reported in a medical journal in 1981 (cf. Schmähl 1981). My interest in this subject was awakened in another way.

### Starting Point and an Empirical Approach

In 1976 an interdisciplinary work group was founded at the University of Tübingen called "Dealing with Cancer Patients". The initiative stemmed from several physicians at the University Gynecological Clinic. These doctors saw themselves confronted in their daily work with problems which, in their opinion, went beyond the sphere of medical competence. Through an exchange with professionals in other disciplines they hoped above all to understand better the fears and inhibitions that interfered with their own dealings with patients.

The work group soon became interdisciplinary in the broadest sense: members included nurses, doctors, social workers, and hospital chaplains, as well as teachers and students of psychology, sociology, and *Empirische Kulturwissenschaft* [1]. I belonged to the last. Each participant attempted to find approaches to his or her own fears and the fears of the patients on the basis of his or her specialization (see e.g. Braun and Hardin 1979).

At first, I considered looking into traditional research in folk medicine, an area which has for about half a century belonged to my discipline (earlier known as 'folklore' [2]). It soon became clear, however, that the conceptual basis and research methods used in previous studies in folk medicine were inadequate. I cannot go into these problems in detail here, but they are illustrated by the form that the discussion began to assume within the work group. The hospital chaplain and social worker reported, for example, that husbands of several cancer patients had moved out of the marital bedroom. Other women maintained that in their village church people had drawn away from them in the pews - presumably from fear of infection. What caught my attention was the remarkably secure context of understanding in which all these reports were received. We asked ourselves how far these accounts may have been influenced by the patients' subjective perceptions. But not a word was said, for instance, about the 'fear of infection' which several patients had themselves suggested as the explanatory clue.

Only very gradually did I realize that our unquestioning 'understanding' was itself part of the problem. I then became interested in how the reported experience of the patients was connected with the common stock of information about cancer and further, how this collectively available information about cancer cohered with other everyday experiences. I finally arrived at the working hypothesis that behavior towards cancer sufferers is co-determined by notions and attitudes whose ingredients belong to the common property of our time and culture; and that these 'images', as I would like to call them, have a history independent of single individuals - even when, as it is likely, each person individualizes them in the course of a life-long appropriation. Earlier collective experiences with sickness probably play a part in the development of these images, experiences which repetition and transmission have gradually consolidated. In this way, apparently spontaneous behavior could be determined by patterns of which the individual is quite unaware.

Since many patients in the Tübingen Gynecological Clinic came from villages in the Swabian Alb, it seemed reasonable to begin the study there, with interviews of healthy people. Thus in winter of 1976/77 I conducted individual interviews of several hours, each time with ten men and ten women, in a small Alb village. The interviews (which were non-standardized) were concentrated on the subject of cancer, and were recorded with the permission of the speakers. A physician in our work group arranged the first contact with a woman in the village. After that other participants were found through her and through a sort of snowballing-system.

The results of these exploratory interviews suggested further studies, especially in urban environments. There followed a project entitled *"Medical Culture and Cancer"* [3], subsidized by the German Research Society and conducted by myself and my colleague Wolfgang Alber. Within the scope of this project we interviewed not only healthy people but also cancer patients and relatives of de-ceased, acutely ill and cured cancer patients, as well as professionals, that is, physicians, nursing teams, hospital chaplains and social workers. We further conducted group discussions with members of self-help groups. Nearly all the patients and relatives with whom we spoke lived in urban or suburban settings.

In the following, I present selections mainly from the Alb study [4]. The presentation follows the actual course of the research and will thus provide the necessary background as it unfolds.

## Images and Interpretative Patterns in Linguistic, Life-World, and Historical Contexts

I will present an extremely narrow selection from the interviews conducted in the Swabian Alb, chiefly those passages related to very specific structures of interaction with illness. These passages often deal not only with verbalizations of notions and attitudes, but also with whole chains of association or interpretation as well as their precedent modes of perception and experience. I will first discuss the 'images' of illness (notions, attitudes), then the interpretative patterns. Since the actual borders between images and interpretations are fluid and a strict analytical distinction would require an elaborate discussion, I will distinguish the two only when it is necessary for understanding. Most of the time I will speak simply of "images and interpretative patterns".

### *Images and interpretative patterns of contagiousness and heredity as concretions of diffuse concepts of infection*

Taking my lead from the difficulties with others that cancer patients had mentioned, I directed my attention first to contacts between the sick and the healthy. When I asked a 48-year-old farm woman how often other people behaved towards women who had undergone operations for cancer, whether and how others visited them, I received the following answer:

"Well - it's not that you're exactly afraid that - that you could catch it or something. I don't believe that, that it's contagious." [5]

This interview was the first one I conducted in the village and I had not yet mentioned the word contagious. But my interlocutor had immediately perceived the question about visits within the context of infection and reacted to what, for her, was implied in the catchword visit.

The fear of infection surfaced again later very strikingly in the personal experience of a 56 year-old woman whose brother had died a few years earlier from stomach cancer. This woman described first the close relationship to her brother which she had had until his death, and then continued:

"But - ... when he wanted to let you drink from his glass -, well, you didn't want to do that. No - I couldn't do it."

When I asked her whether she otherwise felt an aversion to drinking from the glass of another person, she said that she did not and also that she had had no inhibition to do so with her brother before he fell ill with cancer. Asked to describe the change in her feelings, she said that she was probably afraid of "infection". At the same time, she stressed that she did not regard cancer as contagious. At this point in the interview, she noticed herself that her opinion contradicted her former behavior.

Thanks to this description of the concrete situation, a notion about contagiousness was conveyed to me in its everyday, practical context, a context which I would probably not have been able to anticipate at my desk. I therefore redescribed this situation in all subsequent interviews and asked my interlocutors to tell me what they would have done. Naturally, statements about such hypothetical situations need not correspond to actual behavior in real situations; but I was chiefly interested in the associations which the description would evoke. In this respect, the reasons given for refusing to drink from the glass of a cancer patient were of particular interest. They went from the statement that one would do so only out of love for a person very close to one, to the feeling

"I'd have some inhibitions",

to the admission that

"I'd have to think about it. If it was contagious - and I'm not saying that I believe it is - well, then the danger for me would be greater."

It is also worth noting that many of the interviewed felt obliged to describe similar situations which they had themselves experienced. A 38-year-old man, for instance, said that he was scared of talking with a neighbor who had cancer of the mouth because "germs could flow out". Another thought that medical instruments used in early recognition examinations "spread diseased cells". These are only some of the situations bound up with images and interpretations of contagiousness.

It is not these images and interpretations as such that are significant, but the connections in which they must be seen. When those interviewed were asked directly whether they regarded cancer as contagious, not one answered in the affirmative. Nonetheless, 'dangerous' situations such as those above were described again and again. Moreover, many of the interviewed said:

"*Others* say that cancer is contagious".

This seems to mean the following. As well as other people in our society, the people in this village are provided with the information made available by modern medicine. One piece of this information is that cancer is not contagious. But, at the same time, contradictions exist between the informed mind and deep-seated fears which sometimes determine behavior more strongly than does the mind.

The discrepancy between the fact that "cancer is not contagious" and the readiness to behave as though it were becomes more understandable when one considers that 'contagiousness' is not always identical with its strict medical sense. The notion of contagiousness entered the general picture very diffusely. Thus various people made a distinction between "direct" and "not so direct" contagiousness, according to which the latter was more closely identified as contagiousness "from generation to generation". This notion corresponded to the synonymous use of the terms "contagious" and "hereditary" by several of the interviewees. But whether contagiousness "from generation to generation" referred to a disposition in the strict medical sense or something quite different was unclear to those who themselves used the expression (sometimes it could also mean prenatal infection through the mother or even the father - a notion that turned up several times). In any case, it must be assumed that the interviewees have a more comprehensive and more abstract concept of contagiousness than that used in medicine. Notions about "contagiousness" and "hereditary" may be termed concretions of such abstract concepts. Yet even when "contagiousness" is conceived of in the medical sense, the contradiction between scientific information and everyday behavior is not completely incomprehensible. It is a well-known fact in the social sciences that our behavior and motives often conflict with our conscious knowledge [6].

### *TB as the most frequent association with cancer*

There was hardly any interview in which the interviewees did not themselves raise the subject of tuberculosis. Thus, for example, a 53-year-old seamstress requested - as did many other interlocutors - that mobile stations equipped with teams to conduct early recognition check-ups should come to the village itself. She then expressed the hope that one day medical science would get cancer "under control" as it had tuberculosis, and continued:

"And that was the worst thing about TB, it's been contagious ... That was the worst thing. And that's why I think they ought to smoke cancer out with some sort of Early Recognition Bus."

Although here the transfer of a characteristic of TB to cancer appears relatively inadvertent and was certainly not fully conscious to the interviewee herself, in other interviews comparisons between TB and cancer, and even the equation of the two diseases, were quite explicit. Thus a 45-year-old man arrived at the conjecture, through comparing alleged symptoms of respiratory TB and lung cancer, that it is a matter of the same disease in both cases. First he said:

"Run-away TB ... maybe back then that was really cancer. Lung cancer."

Then he explained that the symptoms of respiratory TB and lung cancer are the same:

"It's very quick", "The lungs start bleeding", "And when the lungs start bleeding, then it's all over."

This interviewee was by no means the only one who had the idea that TB and cancer are similar or identical. Another, of about the same age, said:

"I just also want to say that before, people -, were also - whole families -, you could say died out -, but back then they said 'They just wasted away'. I can't exactly picture what that means, but - it's got to be something like that also."

He then immediately proceeded to speak of a married couple who had moved to a city and whose son had just that year died of "consumption", after two daughters had already died of the same disease, and concluded:

"Now that's just the sort of family I mean - you knew something there wasn't right."

By "consumption" the people in the village meant and mean tuberculosis. The interviewee expressed his doubt that consumption is different from cancer with the hint "something like that also", by which he was referring to cancer, the subject of the foregoing conversation.

*Earlier behavior towards TB sufferers as the background for present behavior towards cancer sufferers*

Since cancer and TB were so closely connected in the minds of the inverviewees, it became important for my study how people in the village used to deal with TB sufferers - that is, I became interested in the structures of interaction and the social relations at the time when TB was still very common. The collected material contained numerous references to TB; I have selected two. A 59-year-old farm woman said she was glad that cancer isn't hereditary (by which she meant contagious), because:

"Before, it used to be bad. If someone had TB - ha! - why then a lot of people wouldn't even go into the house or anything - ... they used to say, 'You can't go in there', or, 'It's hereditary'... or 'He's got germs'."

In this statement is is not simply a matter of fear of contagiousness, but also of avoiding contact with a family of whom one knows that "something there isn't right".

That TB sufferers were shunned not only because of the risk of physical infection but also because of a sort of social danger, emerges directly from statements about marriage. Thus a 72-year-old shoemaker and farmer said that TB sufferers above all had to fear "not being able to get married".

At this point, an excursus on the earlier way of life among villagers living in rural conditions of production would be necessary in order to show the enormous importance of a healthy spouse and healthy children as laborers [7]. I will confine myself to observing that to be healthy was of particular importance for the Alb farmers with their small farms and the meagre, work-intensive soil. In these conditions, "damaged ones" ("*Angeschlagene*"), as the chronically ill were called, meant a heavy burden for their families and sometimes for the entire village.

This rough historical background makes more understandable two very specific forms of behavior which were as characteristic of the earlier relation to tuberculosis as they are of relations today to cancer. In order to describe them, I will again begin with a quotation.

A 38 year-old man, whose father was suffering from prostate cancer at the time of the interview, described how such a new case is handled in the village:

"With us, you - let's say - you mainly hush it up. Maybe you say it' s an inflammation, or this or that - ... because you don't come out with it so openly -, that it's cancer."

Accordingly, the mother spread the story that her husband had an inflammation of the bladder and the interviewee, too, held to the story. In the interviews it was often mentioned that, in the past, one learned somebody had TB only once he suffered a fatal hemorrhage, and that today cases of cancer were often "hushed up" too (more frequently, in my experience, with men than with women; see

Dornheim 1983: 144-148). Another man, in whose family there were no cases of cancer, observed about "hushing up":

"Yeah, that's normal. Yeah ... you never find out about it from the relatives." [8]

Not only actual cases of cancer are hushed up: preventive check-ups are also often kept secret. Although there is a doctor in the village who conducts early recognition examinations for men, most of the men interviewed said they go "out of town" to be examined. The reason for this emerged in the interview with the man who had spoken of "consumption". In connection with the request for a mobile early recognition station, we had the following dialogue:

"It's my opinion that if it was like with getting x-rayed - if a bus came here - well, then I'd say, it ought to be a law - that everybody has to go, like with getting x-rayed -- everybody has to go -.

"You think that would be the right way -:

"The right way, yes. I think that would be the absolutely right way. Absolutely.

"Yes.:

"Absolutely.

"But -, if it weren't a law, do you then think that when such a bus came more people would go than now?:

"Oh, no. I don't think that at all.

"You don't.:

"Because then - whoever went into the bus - afterwards -, like it is in a small town -, everybody would say, 'Look at him'.

"I see.:

"Now when somebody goes out of town --, nobody here knows that he's going to a doctor out of town because of it - because of cancer."

Spying on one another and concealment belong together; concealment is the social reaction to the expectation that one will be spied upon. Both are relics of the time when knowledge about sicknesses in other families was of great practical importance. This emerged when I asked the man who had hushed up his father's prostate cancer about the reason for the concealment:

"Well -, I'm not really sure why, there's just this dumb idea -, maybe here in the country -, maybe everywhere. Maybe it comes from back when they had TB -. People just wanted to -, because of the danger of catching it, and that's why I think that generally people are a little careful.

"You mean that the experience from before with TB is still so deep-seated.:

"Yeah. Especially with contagious diseases.

"So that now people transfer it to cancer.:

"Yeah. And you don't want to let everybody know -, because - here in the country, - here everybody knows everything about everyone -, you know the whole family history -, the kids, parents, grandparents, great-grandparents."

This man (whose own father was suffering from cancer) explains the concealment of illnesses in view of the mutual spying extends over generations in small rural communities; and he in turn explained the need for this spying because of the risk of infection which in the past often accompanied tubercular illnesses. Concealing preventive check-ups, which indicates that cancer and the behavior undertaken to detect cancer have the same undesirable social consequences, must be seen in the same historical light. To go for an examination means to go to a doctor, and in the past the villagers did that only when they were severely ill [9]. In less severe cases they simply didn't consult a doctor; and none of them would ever have sought medical help merely to find out whether he or she were still healthy. Thus when today one of the villagers who seems healthy goes for a check-up, his neighbors assume that he must have his reasons; and this means that something will "come out".

So far we have seen that

a) characteristics which could be applicable to tuberculosis are more or less consciously transferred to cancer; and

b) ways of behavior which used to be reasonable in dealing with tuberculosis (i.e. functioned to maintain the social system) are still to be met with even though they are no longer reasonable.

The further question then is: Are characteristics transferred only to the illness or also to the sick person himself?

### TB and Cancer Sufferers: Analogous Ascriptions

When it came to identifying those who held prejudices against TB sufferers, the interviewees invariably referred to "other people". These prejudices concerned women exclusively [10]. Thus "others" said that women who fell ill with tuberculosis had indulged in sexual excesses - by which a causal connection was intended. Similar statements of various degrees of directness surfaced about women with cancer. Again, when asked to identify those who had made these attributions, the same "others" were named, as in the following statement of a 39-year-old woman whose sister-in-law had, as a result of cancer, had a breast amputated.

"When, let's say, someone was a little livelier than the rest when she was young, then they say, 'There, you see. You knew something like that would happen to her. She deserved that sort of punishment.' - You heard that often enough about my sister-in-law. - Or they say, 'You see, she had a good time when she was young,' or ' I knew she would get something like that.' Now I don't think that's right either: someone can get it who -- who'd lead a perfectly respectable life - or more than respectable. That's got nothing to do with the disease."

The prejudices which this women attributed to other people appeared in some interviewees as their own unreflected opinions. Thus the same 72-year-old man whom I already quoted responded to the question how one could best protect oneself from getting cancer with the following train of thought: You really couldn't recommend anything; he wouldn't even advise against drinking and smoking, because in plenty of drinkers and smokers you couldn't find "a trace of cancer". And

"Others who led such respectable lives - so respectable -, like I said before - about women -. Older women who didn't drink - or smoke, or, let's say, who were still unmarried. -- No sex, nothing - and still they got cancer."

This means experience, contrary to expectation, teaches that sexual and other forms of abstinence do not protect against cancer. The surprising lesson is then offered as proof that sexual intercourse and other forms of behavior have nothing to do with the disease.

Both the last quotations bring to light conditions underlying notions about the causes of cancer which go beyond the experience of TB and would have to be interpreted in overarching ideological and social-psychological contexts. So much, however, is clear: these last two statements refer, as do all the interviews that were conducted in the village, to older views about TB sufferers.

### On the Function of the 'Historical View' of Severe Illness

The historical perspective available to the interviewees spans not only the most recent, personally experienced past and present, but also reaches much further back to times in which the experience of specific diseases could only have been preserved by the 'collective memory' [11], and opens out into the future. Thus interviewees often saw cancer as part of an historical series of diseases which began with bubonic plague and smallpox, and extended from cholera and tuberculosis to the present blight of cancer, although it by no means ended with cancer.

"If they found a cure for cancer ... like with TB -, then there'd be something else ... There'll always be something else -, another disease that they can't cure."

This remark of a 63 year-old farmer recurred in more or less the same form again and again. The historical perspective served to orient the interviewees in the present. With the help of familiar and comparable diseases, they tried to appropriate the unfamiliar and still incomprehensible phenomenon 'cancer'. Tuberculosis is exemplary for this purpose because many of those interviewed still remembered personal experiences with it and because its present controllability inspires a certain hope.

### The Scope of Some Specific Interpretative Patterns

Now that the historical context has made the behavior of the villagers more understandable, I will prepare the basis for a proposed typology by describing several far-reaching interpretative patterns underlying that behavior.

The example of 'mouth cancer' and 'germs flowing out' clearly shows that in assessing the risk of contagiousness, the notion of 'inner-outer' plays an important role. Thus a 54-year-old postal worker drew the following distinction between "cigarettes" and "money": he would have considerable misgivings to accept a cigarette from a colleague who had cancer, but he had none whatever in daily handling money, since in doing so he "took nothing into" his body. If one could get cancer from external contacts, remarked the interviewee, "then half the world would soon have cancer".

Against the background of this interpretative pattern, it is not surprising that several men thought one could be infected - in the medical sense - with cancer through sexual intercourse with a woman who had been operated on for cancer of the uterus. A 64-year-old tailor, for example, first said that one could catch cancer only through sexual intercourse, but then corrected himself by saying that after the woman had had radiotherapy, "the worst" of the cancer would be gone [12].

I will conclude the section on images and interpretations of contagiousness with a reference to the results of the second stage of my research, the interviews with people living in urban and suburban settings. These interviews revealed a group who had no fear of infection with cancer - in whatever sense, and even when notions about contagiousness were present. Representative of this group were several women from a village in the close vicinity of a city who described themselves as Pietists, or as having been reared as Pietists [13]. In their view, it is "clearly the teaching of Pietism" that "it actually exalts one, that it is, so to say, a sign of special favor - from God, that only those are allowed to suffer who are chosen by the Lord."

For these women, cancer is in so absolute sense a manifestation of God's will that it is wholly irrelevant whether or not cancer is contagious; all such possibilities are strictly subordinated to their concept of divine providence.

Along with the 'inner-outer' distinction (which is no doubt promising material for psycho-analytical analysis), another interpretative pattern is also of great importance. This is the pattern of ascribing 'responsibility' versus. 'non-responsibility' [14]. In the interviews, this took the form of statements to the effect that someone "didn't take care of himself", by which was meant bad eating habits, smoking, drinking alcohol, or too much work. A frequent example of bad eating habits was that of eating a cold instead of a warm dinner and of drinking cold drinks. It was striking that the responsibility for this habit was sometimes placed not on the sick person but rather - so far as it concerned men - on his wife [15].

The ascription of non-responsibility manifested itself mainly in statements such as that to get cancer is "simply bad luck", that "it's in the air", or that cancer is "sent from our Lord God". These views were voiced chiefly by people in whose families there was or had been a cancer sufferer, and probably express their refusal of responsibility which they feared or expected their neighbors would place on them. A possible variant of the 'responsibility/ non-responsibility' pattern may be seen in remarks to the effect that not the sick person is responsible for his sickness, but rather the society which produces unhealthy living conditions. Among the unhealthy conditions which could cause cancer were mentioned "foul-smelling factory air" (cf. TB), "fall-out and that stuff", "the Pill", and generally "everything artificial". By contrast, "water veins" and "earth currents" were mentioned as things for which nobody is responsible and which belong to some degree to the natural conditions of human life.

Finally, one image or pattern appeared to be applied only by sick persons themselves, or at least by some of them. One cancer sufferer was quoted as having said:

"I'm a cancer woman (*"Krebsmensch"* [16]), and no one can help me now."

When I asked my interlocutor what she thought the woman could have meant, she replied that she meant "she couldn't go see people any more", because many people think that cancer is contagious. The interviewee herself had "got the chills" from the expression "cancer woman". Whether the cancer sufferer actually herself associated contagiousness with cancer is a moot point; but the most significant connotations of her remark seem to me to be those of loneliness, isolation, and perhaps the feeling of being 'marked'. This interpretation is supported by the reaction of my interlocutor, who had reassured the woman with the words:

"You shouldn't think that -, ... I could get it today just like you, just like somebody else tomorrow. It's not like you alone are saddled with it."

These words express a sympathy and solidarity with the sick person which account for a good part of the therapeutic potential of non-professional help [17].

## Attempt at a Typology

Each of the following four classificatory types are illustrated by examples in the words of the persons interviewed. The various perspectives, though their formulation is mine, were implicit in the interviews. The list of examples and their related perspectives is far from complete and is intended only to furnish concrete instances of the four types.

**A. Origin-related or 'etiological' images and patterns**

| **Examples:** | **Dominant perspectives:** |
|---|---|
| not taking care of oneself (smoking, drinking alcohol, cold drinks, bad nutrition) | way of life; individual responsibility |
| fall-out, foul-smelling factory air, poisons in foods | environment; working conditions; social/collective responsibility |
| contagiousness and infection | proximity and distance of bodily and psycho-social contact |
| individual weak points: the germ is already present | disposition, predestination |
| heredity; whole families get it | diachronic and synchronic observation of families |
| bad blood, too little resistance | remnants of old medical theories transmitted by naturopathy |
| simply bad luck that it got you | denial of all responsibility, blaming higher powers |
| sent from God, Divine providence | religous explanation; dichotomy of 'here' and 'beyond' |

**B. Process-related or 'prognostic' images and patterns**

| **Examples:** | **Dominant perspectives:** |
|---|---|
| when you have cancer, it's all over; when you have cancer no one can help you any longer | dichotomy 'live and be healthy' vs. 'get cancer and die' |
| it often takes a long time | temporal process |
| it keeps on eating away | physical process |
| cancer is painful, a martyrdom | individual bodily experience; suffering, isolation |
| your own stamina can make a difference | it is possible to cope with cancer through individual psychological effort |
| there can be remissions | analogy with TB - there is still hope |

**C. 'Topographic' images and patterns**

| Examples: | Dominant perspectives: |
|---|---|
| in the case of mouth cancer germs can flow out | outer vs. inner illness; criterion of risk (for others): link with 'outer'; assumption of a medium |
| stomach and lung cancer are dangerous because of germs in spit | link or non-connection with 'outer' as criterion of risk; analogy with TB |

**D. Essence-related or 'ontological' images and patterns**

| Examples: | Dominant perspectives: |
|---|---|
| cancer is like the Plague, leprosy, like an epidemic | comparisons to historical diseases; analogical reasoning; tendency to metaphorization |
| cancer is a disease of the people | frequency, risk (epidemiological view) |
| cancer is treacherous | anthropomorphizing, personifying view |
| cancer can go wild, can lay in wait somewhere, can eat away | comparisons to animals; tendency to metaphorization |
| in the past no one knew what it was/that it was cancer | medical process has brought cancer to light |
| something else will always turn up, another incurable disease | cancer as link in a historical series of epidemics |
| cancer is a penance, a trial, a special favor of God | religious interpretation, often with consolatory potential |
| cancer is punishment for something done or not done, for a kind of behavior | moral-religious interpretation |
| cancer is fate, bad luck | resignative interpretation |

The images and patterns with which an individual actively operates are probably determined by a variety of factors, including experiences and modes of orientation acquired through socialization, class, gender, as well as age. But, above all, it is the biographical factor of the personal or familial impact of cancer that influences the active application of individual images and patterns.

Along with all these factors affecting individual behavior towards cancer, the active use of images and patterns, as well as the functions which they are accorded, depends on the concrete situation. Thus, for example, the same person can use the etiological pattern 'getting cancer is simply bad luck' as well as its contrary 'you get cancer if you don't take care of yourself'. Such choices are always influenced by the threat felt by the person to himself and those close to him: deciding for the explanation 'not taking care of yourself' can function as a release if he thinks that he leads a healthy life and regards the way of life as an etiologically relevant factor. Further, the active use of images and patterns is bound up with the role adopted in the interview situation itself as well as the roles adopted beyond that situation. This complex interconnection between interview situation and role within and beyond it on the one hand, and the selection of images and patterns on the other, tends to substantiate that empirical examples are dependent on the concrete situation and linguistic context, and cannot therefore be strictly correlated to any single classificatory type. This accounts for why the correlations can sometimes appear ambiguous. Several (similar) statements had to be summarized under each dominant perspective, and this required generalization and abstraction. The general conclusion that may be drawn for interpretative efforts like mine is the following: the specific function of each image can be determined only in view of the concrete situation [18].

Nonetheless, considering the whole of the collected material, three notions appear to have more extensive importance than the others. In the interviews, they came to expression in the following words:

1. "Cancer is incurable", "No one can help you any longer once you have cancer", "When you get cancer, it's all over."

2. "Nobody knows where cancer comes from", "They don't know it", "They can't smoke it out", "They haven't gotten to the bottom of it", "Nobody knows what causes cancer", "Nobody knows the why and the wherefore."

3. "Anyone can get it", "Nobody is safe", "The worst thing is nobody can be sure that at any hour, any minute, he won't get it."

Each of these groups of statements distinguishes itself according to certain common semantical features. An attempt to formulate these common denominators might be:

1. The outcome of cancer is certain; its outcome is *inescapably* pregiven.
2. Cancer is the great *unknown*, the contemporary form of the *unknowable* pure and simple.
3. Cancer threatens *everybody*, and it threatens *me* personally as well.

I call these notions - unavoidably inadequately formulated - 'basis-notions', because they were not only expressed explicitly but were also present implicitly in many specific images. This becomes apparent if we recall, for example, the statemant "I'm a cancer woman, and no one can help me any more", as well as the ensuing response.

How are these basis-notions related to my four groups of specific images and interpretative patterns? One hypothesis is that basis-notions, which represent constant and universal threats, are restricted by more specific notions, that is, are to some degree 'defused' by them. Etiological notions probably illustrate best this unburdening function; thus if all that - bad nutrition, smoking, heredity, contagiousness - doesn't apply to *me*, then I don't need to feel so threatened. Yet many of the prognostic, topographic and ontological images and patterns also have their focus in the threat perceived to oneself and one's family. That cancer sufferers and healthy persons differed most in their prognostic notions tends to support this view [19]; many cancer sufferers had even freed themselves from the basis-notion that 'cancer is the end' and often regarded themselves as healthy, or at least as using what health they had to live. Correspondingly, the images and interpretative patterns of sick and healthy persons agreed most with respect to the basis-notion that 'cancer is the great unknown of our time'. In general, basis-notions and specific images correspond to prevailing social structures [20] and to the currently achieved mastery of nature.

## Dealing with Cancer - Some Results

I will now summarize in thesis-form some relevant results of my study.

a) As a specific segment of notions about disease and behavior towards sick persons, notions about cancer and behavior towards cancer sufferers rest on collective and personal experiences of disease in general. The traditional and dominant example of disease is tuberculosis, which exerts a determining influence on dealings with cancer.

b) Fostered by our inadequate knowledge about cancer (from which medical experts as well as laymen are not exempt), interviewees ascribed characteristics of other diseases, above all tuberculosis, to cancer. This was especially evident of the characteristic 'contagiousness', which appeared in two forms: 'infectious' and 'hereditary'. Although interviewees in their statements denied (in agreement with professional medicine) that cancer is contagious, their real acceptance of its contagiousness appeared in the discrepancies between this opinion and their actual or imagined behavior. This behavior was sometimes supported by reasons, chiefly by the distinction between inner and outer contact, direct contact with the inside of one's body meaning danger. The notions of 'infectious' and 'hereditary' are bound up with historical conditions, possibly with changes in the understanding of notions of contagiousness; as 'rational' grounds, they co-operate with the motivation to distance oneself from cancer sufferers (the conflict manifests itself in the ambivalent application of the notion of 'infectious').

c) Fear of contagiousness was the dominant way in which disease was experienced. Bound up with this fear is the mutual observation of families over generations as well as the counter-strategy of concealment and, as defense against this counter-strategy, the 'smoking-out' of diseases. This pattern also occurs with cancer; sometimes it operates to protect even the cancer sufferer himself from what is thought to be untimely information (it appearing more often in the case of men with cancer than of women). It also operates with respect to 'preventive' behavior. This manifested itself in the demand for compulsory participation in early recognition check-ups; these check-ups should, it was proposed, be conducted in a mobile examination station which comes directly in the village and to which 'everybody' must go.

d) Closeness of relation to the cancer sufferer is the independent variable that determines the feeling of personal menace. One effect of this factor appears in the numerous coping-strategies and counter-strategies of the people in question. Sometimes contradictory notions and attitudes exist next to each other without the contradiction becoming conscious. This obliviousness is connected with a factually real but over-sized anxiety-potential (an assumption which does not so much exclude other approaches as makes them superfluous). The personal impact of cancer also influences the attitude towards early recognition check-ups.

e) Characteristic of cancer is an extraordinary complexity, displacement and condensation of meaning, as well as polar evaluations of etiological factors. In the complex notion of 'cancer' converge

- collective notions, sometimes empirically supported and sometimes not, anchored in everyday life through experiences of disease;
- notions belonging to a new sort of irrationality arising from the use of undigested scientific and

technical information;
- notions based on culturally specific systems of thought or belief, chiefly Christianity;
- notions based on personal experiences of early and later socialization.

f) Cancer is the object of diversely motivated attempts at interpretation. It often appears to be a foothold in the outer world by which an over-sized anxiety-potential, fed by both inner and outer sources, can be realized.

### Theoretical and Methodological Considerations

For a detailed discussion of the theoretical and methodological approach of my study, I must refer the reader to my full-length presentation and to my publications on individual aspects. Here, in conclusion, I confine myself to a few summary remarks.

Starting with the observation that interaction with cancer is determined by extreme anxiety [21], I sought a better understanding of collective behavior towards the disease by elucidating its socio-cultural conditions. My starting-point was the assumption that our dealings with cancer are anchored in unquestioned everyday experiences which constitute the life-world (cf. Schütz and Luckmann 1979) in its historical forms. Insights gained in an interdisciplinary work-group led to adopting a model of consciousness that explicitly comprised unconscious and pre-conscious attitudinal and behaviorial motivations. It thus seemed reasonable to suppose that psychoanalytical theories could contribute to understanding behavior towards cancer. Applying these theories to structures of concrete social systems highlighted particularly the inter-subjective implications of psychoanalytical concepts. This means that although my data and analytical results are based on the statements of empirical individuals (including the interviewer herself), they are not the result of an analysis of individual structures [22].

Quite apart from the supposition of such one-dimensional causation, the feed-back from images of illness to everyday social structures is unsatisfactory unless the share these images have in overarching and culturally specific symbolic systems is also taken into account. I made allowance for this in only a very preliminary way in my study. Today I would develop the hints that the interplay between discourse and the semantic system which underlies it is to be understood as a reference to immanent socio-historical premises [23]. Even if talk of the 'post-modern' has no other object than a new attitude towards things and theories about things, it still issues a challenge in the form of the thesis "that in our society signs no longer refer to a signified, but only to other signs; that our discourse never encounters something like a signification, but only circulates in an endless chain of signifiers" (Bürger 1977: 7, 7-12). This challenge makes central for studies like mine the question of how symbols mediate everyday experiences and the forms of individual socialization. Moreover, growing sensitivity to the link between knowledge and power (which has been developed mainly by Michel Foucault [24]) has placed many social scientific methodologies under a new demand for legitimation. My own confrontation with the consequences of this link for as ticklish a subject of empirical study as our dealing with sickness (and health) continues.

- translated by Jonathan Uhlaner

## Notes

[1] Empirische Kulturwissenschaft (literally: 'empirical cultural science'), the name of the subject taught at the Ludwig-Uhland-Institut of the University of Tübingen, corresponds at other universities to Volkskunde, Europäische Ethnologie, and Kulturforschung. In Tübingen until 1971 the discipline was called "Volkskunde".

[2] For the history of research in folk medicine, see Barthel (1986), Grabner (1967); Schenda (1973).

[3] The German title was "Heilkultur und Krebs - Untersuchung kultureller Determinanten und sozialer Bedeutungen von Krankheitsbildern". The project was conducted from January 1981 to March 1984 under the auspices of Hermann Bausinger of the Ludwig-Uhland-Institut in Tübingen.

[4] For a full presentation of the results of this study, see Dornheim (1983).

[5] Key to the transcriptions: - (= short pause); -- (= longer pause); ... (= omission in quoted material).

[6] For the conception of a consciousness to which its own rules remain inaccessible, see Leithäuser (1976).

[7] There is a series of socio-historical, regionally specific studies on this subject. See Jeggle (1977); Lipp/Kaschuba (1982).

[8] The village priest told me, referring to an actual case, that sometimes relatives conceal the nature of the illness even from him.

[9] I have since learned that this was bound up with the lack (until 1972) of mandatory insurance for full-time farmers.

[10] The background of this socially and subculturally modified image of women cannot be discussed here.

[11] This concept is from Halbwachs (1967).

[12] In this assessment, the personal experience of the interviewee with gonorrhoea during the war played an important part.

[13] Pietism is a Protestant religious movement which originated in the 16th century and whose goal was a renewal of religious life through strict rules of conduct. Württemberg, and especially the Swabian Alb, was one of the centers of the movement.

[14] T. Parsons has observed that the sick person has the right not to be held responsible for the state of being sick. See Parsons (1969: 57-87).

[15] This shows the villagers' traditional conception of gender.

[16] This expression is also a neologism in German. Similar neologisms such as 'cancer personality' ('Krebspersönlichkeit') and 'cancer person' ('Krebsperson') were already familiar to me from popular scientific articles.

[17] On non-professional help and self-help, see, e.g., Ferber/Badura (1983).

[18] In my research, accordingly, analysis of individual cases always preceded the identification of images and patterns as well as of classificatory types. The whole context, including the statements of the interviewer and her relation to the interviewee, were taken into account, although I have not always emphasized this. A case study illustrating the approach is presented in Dornheim (1984).

[19] The comparison between sick and healthy persons is also based on data from the project "Heilkultur und Krebs" (see note 3).

[20] See Mary Douglas' thesis that bodily perceptions within a culture stand in a reciprocal categorial relation with perceptions of social structures within that culture. Douglas (1981: 99-123).

[21] For a comprehensive medical-psychological study of anxiety and cancer, see Verres (1986).

[22] I distinguish between individual and subjective structures, by the latter of which I mean the objectively conditioned

structures of the socialized subject. Towards an analysis of subjective structures, see Lorenzer (1973); Leithäuser/ Volmerg u.a. (1977).

[23] J. Kristeva, e.g., has called for the "self-reflection" of discourse: it should regard itself (i.e., its "objects" and its signs) as ideology. Zima (1977: 7-31). (Citation translated from the German).

[24] "... that power always plugs into knowledge and knowledge into power" is one of the principal theses of M. Foucault, repeatedly developed in his work. Foucault (1976: 38). (Citation translated from the German).

## References


BARTHEL G. 1986: Volksmedizin und Volksmedizinforschung im deutschsprachigen Raum. In: BARTHEL G.(ed.),*Heilen und Pflegen. Internationale Forschungsansätze zur Volksmedizin* (pp. 14-24). Marburg: Jonas Verlag.

BAUSINGER H. 1980:*Heilkultur und Krebs - Untersuchungen kultureller Determinanten und sozialer Bedeutung von Krankheitsbildern.* Manuskript, Tübingen.

BRAUN H. and HARDIN B. 1979: Krebserkrankung und psychosoziale Belastung. Eine Analyse von Gesprächen mit Teilnehmerinnen an einem Programm der Nachsorgeuntersuchung.*Medizin Mensch Gesellschaft* 1: 40-45.

BÜRGER P. 1977: Vorbemerkung. In: BÜRGER C. and P. (eds.). *Postmoderne: Alltag, Allegorie und Avantgarde* (pp. 7-12). Frankfurt/M.: Suhrkamp.

DORNHEIM J. 1983: *Kranksein im dörflichen Alltag. Soziokulturelle Aspekte des Umgangs mit Krebs.* (Untersuchungen des Ludwig-Uhland-Instituts der Universität Tübingen, Vol. 57) Tübingen: Tübinger Vereinigung für Volkskunde e.V.

-- 1984: 'Ich kann nicht sagen: Das kann ich nicht.' Inkongruente Erfahrungen in heiklen Feldsituationen. In: JEGGLE U.(ed.),. *Feldforschung. Qualitative Methoden in der Kulturanalyse* (pp. 129-157). (Untersuchungen des Ludwig-Uhland-Instituts der Universität Tübingen, Vol. 62) Tübingen: Tübinger Vereinigung für Volkskunde e.V.

-- 1986: Zum Zusammenhang zwischen gegenwarts- und vergangenheitsbezogener Medikalkulturforschung. Argumente für einen erweiterten Volksmedizinbegriff. In: BARTHEL G. (ed.),*Heilen und Pflegen. Internationale Forschungsansätze zur Volksmedizin* (pp. 25-41). (Hessische Blätter für Volks- und Kulturforschung, Vol. 19) Marburg: Jonas Verlag.

DOUGLAS M. 1981:*Ritual, Tabu und Körpersymbolik. Sozialanthropologische Studien in Industriegesellschaft und Stammeskultur* (Orig.: Natural Symbols. Explorations in Cosmology. London 1970). Frankfurt/M.: Suhrkamp.

FALLER H. 1983: Subjektive Krankheitstheorien als Forschungsgegenstand von Volkskunde und medizinischer Psychologie.*Curare* 6: 163-180.

FERBER C.v. and BADURA B. (eds.) 1983:*Laienpotential, Patientenaktivierung und Gesundheitsselbsthilfe.* München/Wien: R. Oldenbourg.

FOUCAULT M. 1976: *Mikrophysik der Macht. Über Strafjustiz, Psychiatrie und Medizin.* Berlin: Merve.

GRABNER E. (ed.) 1967: *Volksmedizin. Probleme und Forschungsgeschichte.* Darmstadt: Wissenschaftliche Buchgesellschaft.

HALBWACHS M.1967:*Das kollektive Gedächtnis.* Stuttgart: Enke.

HERZLICH C. 1973: *Health and Illness. A Social Psychological Analysis.* London/New York: Academic Press.

HORN K., BEIER C. and WOLF M. 1983: *Krankheit, Konflikt und soziale Kontrolle. Eine empirische Untersuchung subjektiver Sinnstrukturen.* Opladen: Westdeutscher Verlag.

JEGGLE U. 1977: *Kiebingen - eine Heimatgeschichte. Zum Prozeß der Zivilisation in einem schwäbischen Dorf.* (Untersuchungen des Ludwig-Uhland-Instituts der Universität Tübingen, Vol. 44) Tübingen: Tübinger Vereinigung für Volkskunde e.V.

LEITHÄUSER T. 1977: *Formen des Alltagsbewußtseins.* Frankfurt/New York: Campus.

LEITHÄUSER T., VOLMERG B. et al. 1977: *Entwurf zu einer Empirie des Alltagsbewußtseins.* Frankfurt/M: Suhrkamp.

LIPP C. and KASCHUBA W. 1982: *Dörfliches Überleben. Zur Geschichte sozialer und materieller Reproduktion ländlicher Gesellschaften im 19. und 20. Jahrhundert.* (Untersuchungen des Ludwig-Uhland-Instituts der Universität Tübingen, Vol. 56) Tübingen: Tübinger Vereinigung für Volkskunde e.V.

LORENZER A. 1973: *Zur Begründung einer materialistischen Sozialisationstheorie.* Frankfurt/M.: Suhrkamp.

MÖCKLI von SEGGERN M. 1965: *Arbeiter und Medizin. Die Einstellung des Zürcher Industriearbeiters zur wissenschaftlichen und volkstümlichen Heilkunde.* (Schriften der Schweizerischen Gesellschaft für Volkskunde, Vol. 46) Basel: Habelt.

PARSONS T. 1969: Definition von Gesundheit und Krankheit im Lichte der Wertbegriffe und sozialen Struktur Amerikas. In: MITSCHERLICH A. et al. (eds.) *Der Kranke in der modernen Gesellschaft* (pp. 57-87). Köln/Berlin (2nd edition): Kiepenheuer u. Witsch.

PFLANZ M. and KEUPP H. 1977: A Sociological Perspective on Concepts of Disease. *International Social Science Journal* 19: 386-396.

PFLEIDERER B. and W. BICHMANN 1985: *Krankheit und Kultur. Eine Einführung in die Ethnomedizin.* Berlin: Dietrich Reimer.

SCHENDA R. 1973: Volksmedizin - was ist das heute? *Zeitschrift für Volkskunde* 69: 189-210.

SCHMÄHL D. 1981: Vorstellungen über die Krebskrankheit. Laienzuschriften aus den Jahren 1977-79. *Deutsche Medizinische Wochenschrift* 106: 246-248.

SONTAG S. 1977: *Krankheit als Metapher* (Orig.: Illness as Metaphor. 1977). München/Wien: Carl Hanser.

-- 1989: *Aids und seine Metaphern* (Orig.: AIDS and Its Metaphors. 1988). München/Wien: Carl Hanser.

VERRES R. 1986: *Krebs und Angst. Subjektive Theorien von Laien über Entstehung, Vorsorge, Früherkennung, Behandlung und die psychosozialen Folgen von Krebserkrankungen.* Berlin/Heidelberg/New York: Springer.

ZIMA P.V. 1977: Diskurs als Ideologie. In: ZIMA P.V. (ed.) *Textsemiotik als Ideologiekritik* (pp. 7-31). Frankfurt/M.: Suhrkamp.

# Emil Kraepelin and the Origins of American Psychiatric Diagnosis

Allan Young

Like every clinical science, psychiatry has a linguistic jumping off place, a bedrock of basic words and meanings that no longer attract scrutiny. It is on this foundation that writers, researchers, and clinicians build their routines and reputations.

The culture of science says that no meanings are immune from scrutiny. Within its purview, all interpretations are necessarily tentative and subject to revision. In everyday life, things are a bit different, however, and researchers and clinicians are not ordinarily inclined to scrutinize their basic words and meanings. Most of these people are pragmatists and puzzle-solvers, and their eyes are fixed on getting results - producing knowledge, making correct diagnoses, getting desirable outcomes - and not splitting epistemological hairs. In their mouths and minds, everyday words are tools for building knowledge, not weapons for subverting it. And, so, in the ordinary course of events, outside of the rare scientific revolution or coup d' état, the bedrock language changes only slowly and incrementally, without calling attention to itself.

The bedrock language of contemporary American psychiatry is encoded in *The Diagnostic and Statistical Manual of Mental Disorders, Third Edition*, an official publication of the American Psychiatric Association, and commonly referred to as "DSM-III" (published in 1980 and revised in 1987, as DSM-III-R). DSM-III is essentially a nosology, a classification of over 200 named mental disorders, intended to be used by clinicians and researchers. DSM-III is not a product of evolutionary change. Its language is fundamentally different from the words and meanings found in previous psychiatric classifications, including earlier editions of the manual (DSM I and DSM II). Its publication was a revolutionary event, vigorously but unsuccessfully resisted by the opponents of radical change in psychiatry. Critics of DSM-III continue to write about its supposed deficiencies, but the battle is over. A decade after publication, the manual provides the standard language of American psychiatry and the authorized jumping off place for problem-solvers (Williams, Spitzer, and Skodol 1985).

DSM-III is the product of many people, representing the spectrum of mainstream clinical and theoretical orientations. Fourteen advisory commitees, composed of experts in each category of mental disorder represented in the text (organic mental disorders, substance use disorders, etc.) prepared a succession of drafts. These were read and commented on by a liaison committee established by the American Psychiatric Association (APA) Assembly (which is composed of representatives from all of the APA's district branches); APA members discussed and debated the drafts at various meetings and conferences; drafts were presented to and commented on by other professional groups, including the American Association of Chairmen of Departments of Psychiatry, organizations representing psychodynamically-oriented clinicians and researchers, the American Academy of Child Psychiatry, the Academy of Psychiatry and the Law, and the American Psychological Association (DSM-III: 2-5).

It is hard to believe that so many people representing so many different points of view would simultaneously want to replace the established psychiatric language with a new one. Nor is that what happened. Although diverse views were solicited and incorporated into the drafts, the spirit of pluralism was, without exception, limited to shaping DSM-III's *content*, i.e., mainly the texts making up individual entries. On the other hand, the *structure* of DSM-III - which, as we shall see, is the feature that most distinguishes this text from previous editions - was the product of a relatively small and homogeneous circle of people, known among themselves and to their critics as "neo-Kraepelinians".

Kraepelin's approach was premised on the following string of ideas. Mental disorders are analogous to physical diseases. Medicine has made great advances in explaining, preventing, and treating diseases, but psychiatry continues to lag far behind. In order for psychiatry to move forward, psychiatrists must adopt the methods and perspectives of medicine. Medicine's first step was to distinguish among and classify the different kinds of diseases, therefore this is where psychiatry must begin, in order for it to move ahead also.

Emil Kraepelin (1856-1921), a German psychiatrist, was an influential figure in the study of schizophrenia and a pioneer in the systematic classification of mental disorders. According to Kraepelin, progress in psychiatric classification demanded fidelity to three principles. One, each mental disorder, like each disease, has a distinctive etiology. Just as it is wrong to think that one pathophysiological process can account for all diseases, so it is also incorrect to believe that there is a universal process which underlies and unifies all or most mental disorders. Two, the classification of mental disorders can be achieved only through careful empirical observation. Inferences drawn from theories which lack solid evidence (here he includes psychoanalysis) must be rejected. More concretely, Kraepelin emphasized the importance of systematically observing, recording, and comparing case histories in order to identify clusters of symptoms (syndromes) that go together from case to case and that follow a discernible course over time and lead to a predictable outcome - that is, the patterning of syndromes. Three, research will eventually show that serious mental disorders have organic or biochemical origins. While little is known about the actual etiologies of mental disorders at the present time, this lack of knowledge is not an obstacle to correctly identifying and classifying disorders. In fact, the reverse is true: classification is usually a necessary first step on the way to identifying causes (Kraepelin 1974 [1920], Spitzer and Williams 1980: 1043).

After collecting and comparing large numbers of case histories, Kraepelin proposed a classificatory scheme that, in broad outline, resembles DSM-III's system. His so-called "descriptive approach", emphasizing the clinical importance of classifying and diagnosing mental disorders, was initially popular in American psychiatry, but by the 1930's Kraepelin's ideas had been displaced by psychoanalytic views. By the 1960s, most American psychiatrists remembered Kraepelin either for his research on schizophrenia or, as Franz Alexander recalled him, as a man "looked upon by the younger generation of psychiatrists as a rigid and sterile codifier of disease categories ...", someone belonging to an "antipsychological era" that had passed into history (Alexander and Selesnick 1966: 211-214).

Alexander, like Freud before him, could find no common ground between Kraepelin's descriptive approach and his own psychodynamic views, since the two positions are not merely different, but also antithetical. Nowhere are the differences more obvious than in their contrasting conceptions of symptoms. In Kraepelin's system, symptoms are signifiers which get their meanings through being juxtaposed with other symptoms/signifiers in stable formations (syndromes), like letters within words, and patients are equivalent to the surfaces upon which symptoms inscribe their meanings. In the psychoanalytical system, symptomatic behavior gets its meaning from processes which are secretly played out beneath these surfaces. (To be more specific, symptoms are expressions of defenses against unconscious anxiety originating in unresolved intrapsychic conflict - that is, the kind of underlying unitary process that Kraepelin categorically rejects.)

Put into other words, Kraepelin sees symptoms as properties or attributes of disease-like mental disorders, while psychoanalytic writers see symptoms as polymorphic expressions of individual lives and psyches. And while Kraepelin sees the idea of diagnosis based on symptoms as being psychiatry's future, psychoanalytic writers see the Kraepelian interest in classification as essentially medical and non-psychological and, for this reason, reject it (Rycroft 1968: 35).

During this period of diminished American interest, from the 1930s to the 1970s, Kraepelin's ideas continued to have an important, even dominant, influence on academic psychiatry in Europe and, from this source, helped shape WHO's *International Classification of Diseases* (Klerman 1986). But even in the United States, Kraepelin's views did not disappear altogether. Among other places, they continued to be represented in a popular psychiatric textbook, by Meyer-Gross and his colleagues (1954), and among the faculty of the Department of Psychiatry at Washington University, St. Louis. It is to this department that DSM-III's "neo-Kraepelians" trace their intellectual roots (Robins and Helzer 1986: 410-414).

Like most turning points in the history of ideas, American psychiatry's rediscovery of Kraepelin was preceeded by important technological and institutional changes, which inclined people to see his once rejected ideas in a new light. Up to the 1950's, there was a tendency for American psychiatry to relinquish responsibility for a mental disorder once an organic or biochemical etiology was demonstrated or an effetive biological or pharmacological treatment was found. "Vitamins were discovered, whereupon vitamin-deficiency psychiatric disorders no longer were treated by psychiatrists. The spirochette was found, then penicillin, and neurosyphilis, once a major psychiatric disorder, became one more infection treated by non-psychiatrists" (Goodwin and Guze 1984: xi). Once the efficacy of barbiturates and Dilantin for treating epilepsy was confirmed, psychiatry abandoned the disorder to the exclusive attention of neurology (Klerman 1984: 539).

By purging itself of these disorders, psychiatry remained essentially psychological, and researchers and clinicians continued to have little incentive for raising questions about diagnosis and classification. In the early 1950's an important change occured, when chlorpromazine (Thorazine) and other psychopharmacological agents of demonstrated efficacy and specificity became available. Unlike earlier occasions, when responsibility for treatments and disorders were transferred to other medical specialties, the new drugs were incorporated into psychiatric practice, despite some resistance. By the following decade, psychiatrists were employing four classes of drugs, each specific for a different category of disorder: psychoses, depression, anxiety, and manic-depressive disorder.

The adoption of these drugs had important implications for advocates of the Kraepelinian point of view, since clinical experience now showed a strong correlation between patterns of response to the drugs and the descriptive classification proposed by Kraepelin. Just as Kraepelin had originally argued, disorders of affect (depression) were shown to be distinct from disorders of thought (schizophrenia), and the so-called neurotic conditions were shown to consist of discrete and discontinuous types of disorders: phobias, obsessive-compulsive conditions, anxiety, and panic states (Klerman 1984: 539-540).

These findings underlined the importance of being able to correctly diagnose a patient's symptoms as a precondition for making the right treatment decision. Here there was a problem, however, since American psychiatry lacked a standard nosology of mental disorders that was compatible with the recent psychopharmacological findings. Although DSM-I had appeared in 1952 with the imprimatur of the American Psychiatry Association, it clearly did not fill the bill. Its system of classifying disorders was strongly shaped by the views of Adolph Meyer, an influential psychiatrist at Johns Hopkins University. Meyer's position was that diagnostic groups represent quantitatively different manifestations of a unitary process: the personality's reaction to changes and challenges in its environment, i.e. combinations of interacting psychological, social, and biological factors. In contradistinction to Kraepelin's ideas, this "reactive" or "environmental" view sees mental health and mental illness as different points along a single gradient of mental functioning (Weissman and Klerman 1978: 707, DSM III: 1). In addition to being at odds with the recent psychopharmacological findings, DSM-I failed in a second respect, the mission of providing psychiatry with a standard language. In the absence of any compelling reason for abandoning their previous notions and practices, many psychiatrists and psychiatric institutions simply decided not to adopt the manual's nomenclature (Spitzer and Williams 1980: 1045).

DSM-II (1968) departed from the first edition in some important respects, partly because it was intended to be compatible with the system used in the *International Classification of Diseases* (ICD-8). In the eyes of the neo-Kraepelians, however, two serious problems remained. While Meyer's notion of "reaction" had been effectively dropped from the new nomenclature, the term "neurosis" - an underlying process of intrapsychic conflict resulting in symptom formation that serves unconsciously to control anxiety (DSM-II: 39) - continued to be used throughout the manual. Second, according to Robert Spitzer, DSM-II's "classifications were severely flawed by their failure to provide formal criteria for determining the boundaries of their diagnoses". Without these criteria, clinicans were still forced to rely on "global descriptions of disorders that frequently entailed etiological assumptions" (Bayer and Spitzer 1985: 188).

In 1974, the board of trustees of the American Psychiatric Association passed a resolution calling for a revision of DSM-II and appointed Spitzer as head of the task force responsible for drafting the new manual. Spitzer asked for a clean slate, arguing that it would be difficult to make substantive progress if his task force included the same people who drafted DSM-II (Millon 1986: 29). The board agreed, and Spitzer chose as his collaborators a "group of psychiatrists and consultant psychologists committed to diagnostic research and not to clinical practice", with "its intellectual roots in St. Louis instead of Vienna, and with its intellectual inspiration derived from Kraepelin, not Freud" (Bayer and Spitzer 1985: 188).

Spitzer's intention was to produce a diagnostic system based on the following principles: (1) Each disorder needs to be identified by means of criteria that are amenable to systematic empirical observation and measurement (known in the psychiatric literature as "operational criteria" [American Psychiatric Association 1980b: 79]). (2) Psychiatry does not satisfactorily know the etiologies and underlying processes of most mental disorders, and clinical impressions and conventional interpretations about pathogenesis are not enough. Theories must be judged according to "the principles of testability and scientific verification". Because few theories have so far met this condition, most classifications must be based entirely on "shared phenomenological characteristics", i.e. the patterning of syndromes to which Kraepelin had called attention. It is for this reason that the term "neurosis", referring to processes at work beneath the surface of appearances, can no longer be included among the diagnostic criteria of any disorder. (3) Because the new classification is based on atheoretical and operational criteria, it will be possible for clinicians and researchers with different theoretical perspectives on etiology and process - psychodynamic, biological, behavioral, and so on - to adopt uniform diagnostic practices. At the most concrete level, they will be talking and writing about the same set of disorders (Bayer and Spitzer 1985: 188). More generally, they will occupy their respective places - whether competing or complementary - within a single language community

(Talbot and Spitzer 1980: 27-28).

Spitzer's third principle notwithstanding, he is reiterating the Kraepelinian project, and it should come as no surprise to learn that psychoanalytic circles vigorously objected to his conception of DSM-III. Spitzer's critics made four main points. First, they rejected Spitzer's claim that the manual's system of classification is atheoretical. They argued that there is no value-free observation language - despite what neo-Kraepelians appeared to assume, not even in science - for the sorts of things psychiatrists are interested in. Further, many DSM-III entries are based on culturally- or socially-specific presuppositions (Faust and Miner 1986: 965). Second, DSM-III's "simple summation of itemized data" tends to give each symptom or diagnostic feature equal weight "even though there is no evidence that all have equal importance". Critics referred to the maual's "cook-book approach", saying that it threatened to "deprive the diagnostic process of ... an integrative gestalt based on clinical experience", i.e. the psychiatrist's privileged access to clinical meanings (Frances and Cooper 1981: 1200). Third, DSM-III risks confusing symptoms with disease, by failing to distinguish between the sources of a person's distress and his responses to the distress, i.e. sources and responses are given equal places on lists of diagnostic features. For instance, the manual does not distinguish between anxiety arising from things a patient perceives as threatening his intrapsychic equilibrium (= his source of distress) and his efforts to defend himself, express feelings of dysphoria and despair, cry for help, etc. (= his response to distress) (Vaillant 1984: 544-545). Fourth, some critics saw the publication of DSM-III as being equivalent to an atempted coup d' état, hatched by a small circle of conspirators. "The task force that forged DSM-III was not representative of the interests, the values, or the theoretical diversity of the profession. It was composed of an invisible college that is only one college in the university of American psychiatry" (Michels 1984: 549).

In articles, conferences, and conversations, the neo-Kraepelians rejectetd each of these claims. They argued that DSM-III's entries are objective, atheoretical, and neutral, because they are both a scientific product, i.e. based on techniques for scientifically assessing the validity and reliability of the manual's diagnostic criteria, and a collective product, i.e. the work of people and associations representing all of mainstream psychiatry and clinical psychology. Second, the psychoanalysts' point about DSM-III being ethnocentric bias is largely beside the point, since the manual is intended to be used primarily by American clinicians. Even so, the manual has been translated into various European and non-European languages and psychiatrists in foreign lands say they have used it to good effect. Calling DSM-III a "cook-book" distorts the manuals's mission, which is to establish a scientific nosology. To this end, it is essential to have diagnostic criteria that are sufficiently explicit and amenable to observation so as to eliminate any future need for diagnosis based on tacit knowledge, privileged interpretations, global appraisals, and undemonstrated intrapsychic processes.

Finally, Spitzer argued that if psychoanalytic psychiatrists are dissatisfied with the work of the "invisible college", they had only themselves to blame. According to his personal account of the events leading up to the publication of DSM-III, Spitzer had made special efforts to solicit their comments, but organizations such as the American Psychoanalytic Association had shown themselves unwilling to "confront forcefully the challenge of DSM-III". Moreover, he suggested, their attitudes towards DSM-III and their attempts to obstruct its adoption were only partly doctrinal. Opposition was also strongly rooted in economic self-interest. "Psychoanalytic practitioners ... feared that a change in psychiatric nomenclature might result in a challenge by third-party reimbursement sources seeking to limit payment to patients receiving long-term therapy". It was "no coincidence", Spitzer added, that the Baltimore-District of Columbia Society for Psychoanalysis was conspicuous in its opposition to the final approval of DSM-III, since large numbers of federal employees live in the capital area and are entitled to "generous coverage for psychotherapeutic treatment". Adoption of a standard nosology would put psychoanalysts head-to-head with their more cost-efficient rivals (Bayer and Spitzer 1985: 191-192).

To an extent, the two sides were arguing about two different issues. The psychoanalysts' most serious objection was to the structure of DSM-III, i.e. its distinctively Kraepelinian foundation. But Spitzer has made it clear that, from the time he took command, he and the task force never considered this issue to be negotiable (Bayer and Spitzer 1985: 188-189). What Spitzer was willing to compromise on was the content of the manual, e.g. identifying specific diagnostic criteria. To accept Spitzer's condition was, for some critics, simply to give up the game, since their key concepts, especially "neurosis", cut across any division between structure and content.

By 1990, this dispute is ancient history. The epistemological space in which meaningful questions can be asked and in which psychiatric disorders are perceived and materialized - through diagnostic tests, therapeutic interventions, outcome measures, and so on - has shifted. The neo-Kraepelinians are correct when they write that the "historic shift from an anti- to a pro-diagnostic stance among members of the psychiatric community" is now completed, despite "pockets of resistance to the idea of categorical diagnoses that still remain among both social scientists and psychodynamic psychiatrists" (Robin and Helzer 1986: 409). Outside of psychoanalytic circles, nearly everyone routinely speaks and writes DSM-III's language. And although it is not the mother tongue of analytically oriented therapists, many of these clinicians have also adopted it as a means of communicating with other psychiatric communities, albeit often through only superficially shared meanings.

## References

ALEXANDER F. and SELESNICK S. 1966: *The History of Psychiatry.* New York: New American Library.

AMERICAN PSYCHIATRIC ASSOCIATION 1952: *Diagnostic and Statistical Manual of Mental Disorders.* Washington, D.C.: American Psychiatric Association.

-- 1968: *Diagnostic and Statistical Manual of Mental Disorders* (second edition). Washington, D.C.: American Psychiatric Association.

-- 1980: *Diagnostic and Statistical Manual of Mental Disorders* (third edition). Washington, D.C.: American Psychiatric Association.

BAYER R., and SPITZER R. 1985: Neurosis, Psychodynamics, and *DSM-III. Archives of GeneralPsychiatry* 42: 187-196.

FAUST D. and MINER R. 1986: The Empiricist and his New Clothes: *DSM-III* in perspective. *American Journal of Psychiatry* 143: 962-967.

GOODWIN D. and GUZE S. 1984: *Psychiatric Diagnosis* (third edition). New York: Oxford University Press.

KLERMAN G. 1984: The Advantages of *DSM-III. American Journal of Psychiatry* 141: 539-542.

-- 1986: Historical Perspectives on Contemporary Schools of Psychopathology. In: MILLON T. and KLERMAN G.L. (eds.) *Contemporary Directions in Psychopathology: Towards the DSM-IV.* New York: Guilford Press.

KRAEPELIN E. 1974: Comparative Psychiatry. In: HIRSCH S.R. and SHEPHERD M. (eds.) *Themes and Variations in European Psychiatry.* Charlottesville: University of Virginia Press.

MICHELS R. 1984: A Debate on *DSM-III. American Journal of Psychiatry* 141: 548-551 and 553.

MILLON T. 1986: On the Past and Future of the DSM-III: Personal Recollections and Projections. In: MILLON T. and KLERMAN G.R. (eds.) *Contemporary Directions in Psychopathology: Toward the DSM-IV*. New York: Guilford Press.

ROBINS L. and HELZER J. 1986: Diagnosis and Clinical Assessment: The Current State of Psychiatric Diagnosis. *Annual Review of Psychology* 37: 409-432.

RYCROFT C. 1968: *A Critical Dictionary of Psychoanalysis*. Harmondsworth: Penguin Books.

SPITZER R. and WILLIAMS J. 1980: Classification in Psychiatry. In: KAPLAN H.I., FREEDMAN A. and SADOCK B.J. (eds.) *Comprehensive Textbook of Psychiatry* (third edition). Baltimore: Williams and Wilkins.

TALBOT J. and SPITZER R. 1980: An In-Depth Look at DSM-III: An Interview with Robert Spitzer. *Hospital and Community Psychiatry* 31: 25-32.

VAILLANT G. 1984: The Disadvantages of DSM-III Outweigh its Advantages. *American Journal of Psychiatry* 141: 542-545.

WEISSMAN M. and KLERMAN G. 1978: Epidemiology of Mental Disorders. *Archives of General Psychiatry* 35: 705-712.

# Section 4

# From Patients' Complaints to Cultural Narrative

# The Love-Lorn Consumptive:
## South Asian Ethnography and the Psychosomatic Paradigm

Francis Zimmermann

This paper comes within the perspective of symbolical anthropology and semiotics. It heavily relies on religious ethnography in South Asia and Sanskrit textual studies. Its surmise is that some of the most essential concepts of South Asian psychiatry are not to be gained from clinical studies in mental hospitals, but from a careful reading of Sanskrit love stories. The first part is an attempt to characterize this perspective, which challenges the dominant paradigm of North American medical anthropology. The second part of this paper gathers a few scraps of religious ethnography that are relevant to the study of consumption, the wasting away of vital fluids. The third part sets out the Hindu medical concept of consumption as a *mal d'amour* and places this disease in the context of South Asian culture and society.

### The Mourning Paradigm

Medical anthropology partly is an applied subdiscipline of anthropology bound to meet psycho-social needs in our affluent societies. In so far as medical anthropologists are employed as consultants or civil servants, one can easily understand that they work in hospitals, studying patterns of help seeking, the medical bureaucracy, and the impact of ethnicity upon the social construction of illness. In the context of applied research, however, the received methods of anthropology have undergone drastic changes. The most conspicuous drift into new methods is the disappearance of fieldwork, for which clinical interviews of help-seeking patients have been substituted. Traditionally, the ethnographer visited a foreign culture, learned the language, and then described the various aspects of society from kinship to politics through religion and ethics. In contrast with this global approach, we observe today both a withdrawal of anthropology into the industrialized West and a specialization of ethnographic accounts to only one dimension; medical ethnography, for example, is a one-dimensional ethnography which ignores kinship, religion, and so forth. Cultural patterns which hitherto fell within the province of religion or ethics, for example, such as the Hindu social sterotype described below – a love-story coming to a bad end – of which the pathological syndrome (fever, delirium, exhaustion, menorrhagia) is only one aspect, have been completely medicalized.

Ethnographers and clinicians, nevertheless, tend to share a common belief in the primacy of experience and involvement (Kleinman 1988: 231-232). Dispassionate as he may pretend to be, the ethnographer actually entertains ethical responsibilities to his informants, just as the clinician has a therapeutic mandate demanding that he act in behalf of his patients. Medical anthropology is built upon this dual involvement with informants as patients. Our elders, the ethnographers of far-away cultures, had made it a rule to share the native's point of view. Similarly, being perhaps too prone to confusions between the native, the patient and the marginal, we hold today the gratifying view that medical anthropology is always on the fringe's side. We offer critical views on the evils of medicine, and we address ourselves to the formidable task of doctoring illnesses and managing sorrow. We assume that illnesses are universals. We argue that, whether or not a particular society treats neurasthenia as a disease, for example, the syndrome of chronic exhaustion is a ubicuitous illness behavior that can be described and interpreted in all sorts of situations and contexts (Kleinman 1988:

104). Therefore, the task of anthropology in a clinical context is to interpret illness meanings. The patient's body idiom, beyond the physical pain, may be expressing the pain of failure, the pain of loneliness, eliciting love and support and warding off distressing thoughts. A new mood has become widespread among medical anthropologists, pessimistic and typically *fin de siècle.* Our books, journals and conferences are all resounding with cries of stress and distress, sadness and depression, the antiquated words of acedia and neurasthenia, the doctoring of nerves and patients' demoralization. This is the new dominant paradigm of medical anthropology; let us call it the *mourning paradigm*.

This mood certainly fits the pursuits of a clinically applied medical anthropology. My only claim to relevance in such matters is that there are other forms of medical anthropology that escape the fascination of mourning. We cannot take great pride in that escape, though. It simply means that we are not doctors.

Taking pain and distress seriously is indeed the call of a physician. In the compelling style which makes his books so gripping and inspiring, Arthur Kleinman has convincingly shown that the management of suffering is a central task. However inadequate it may be, "the very act of inquiring" into pain and distress, he says, "breaks the tunnel vision of biomedicine and its veterinary tendency". I am citing a shorter text of his, dating from 1981, which stresses the practical significance of explanatory models for the clinician.

"Interpretation of illness meanings is essential for effective clinical work and it is the very essence of liaison psychiatry and clinically applied medical anthropology. The patient with chronic pain may be expressing a distressing relationship, the pain of failure, the pain of loneliness, or eliciting love and support, seeking compensation, warding off depressing thoughts and feelings that threaten to overwhelm him .... The Explanatory Models approach ... situates [the clinician] in the patient's personal and social world, and makes unavoidable the human antecedents and consequences of illness, thereby humanizing medicine by anthropologizing it .... Explanatory Models are no ways of thinking or systems of thought but practical statements about particular illness experiences. They are expressed guides to help seeking and clinical decision making " (Kleinman 1981: 375)

There is no ambiguity in this approach to pain and sorrow. Anthropology, thus 'clinically applied' in a lucid, generous and thoughtful manner, does not involve any cross cultural comparison nor generalization; in other words, it does not entail any philosophical decision on the question whether depression is a universal process or not.

As soon as the clinician-ethnographer, however, embarks on cross-cultural studies to assess the universal significance of illnesses like depression, the strand of clinically applied anthropology goes astray. New explanatory models are introduced, based on a syncretism of clinical and ethnographic data, that are no longer practical statements but tend to explain the formation and expression of syndromes. American journals are replete with 'interactionist models', built by piling up various levels of reality: mind, self, language, and *emotions*, the magic concept that aims to coagulate the whole. The very business of concocting *models* and advertising *paradigms* is indeed a new version of the grand American tradition of pragmatism. Then, clinically applied anthropology is so much in demand on the market of health care that it has determined a new casting of roles on the anthropological scene itself. For example, one of the assumptions that originally dictated the programme of the Hamburg colloquium, Anthropologies of Medicine, was that in comparing different ethnographies, exotic to one another, we could *transform the psychosomatic paradigm into ethnographic research.* In other words, ethnographers were invited to interpret their data in the light of a concept invented by clinicians. As if the clinicians of square *A1*, on the chess-board of Table 1 below, were imposing their own explanatory models upon the ethnographers and historians of squares *B2* and *C3*.

**Figure 1. Matrix of possible approaches in medical anthropology**

| | 1. Subjective world, expressive attitudes | 2. Social world, attitudes complying to the norms | 3. Objective world, objectivating attitudes |
|---|---|---|---|
| A. Clinically applied medical anthropology | interpretation of *illness* meanings, explanatory models, illness narratives, etc. | | study of *disease* |
| B. Ethnographic research | | symbolic anthropology, ethnosociology, 'anthropologie de la maladie', etc. | |
| C. Anthropology of medical traditions, including textual studies | love stories studied as deviances or diseases, etc. | | study of Galenism, of Ayurveda, 'anthropologie historique'. etc. |

It is only fair to say that the Explanatory Models school has truly revolutionalized liaison psychiatry and clinically applied anthropology. But one approach cannot tackle all the three worlds of life, I mean, the Weberian set of three spheres of action – objective, social, and subjective [1]. In constructing the concept of illness and focussing on what they call an innately human experience of suffering, which is actually a Western cultural construct, researchers of square *A1* have explored the sphere of expressive action: the body idiom, emotions, illness narrative. However, the interpretive approach to illness meanings, which originally was meant to fight the bluntness of biomedicine, has now been raised to the status of a dominant ideology. It espouses and inflates the general drift towards emotionalism and dramatization in anthropology, which Richard Shweder once labeled "anthropology's romantic rebellion" against rationalism (Shweder 1984: 27). It tends to blur the limits between the three spheres of action. The very business of constructing models and paradigms is fundamentally at odds with the intellectual habits of type *B2* and *C3* researchers, on the matrix designed above, whose approach is to let the culture they study speak for itself and unfold its own categories.

How was I to play the part of the indologist for which I was cast in the concert of lamentations? I was going to tell all you ever wanted to know about depression in India, with a few Sanskrit words to authenticate my account and various models of body-building by combining Mind, Self and Emotion, not forgetting the body politic and somatization to put myself straight with the current academicism, when suddenly I was seized by the writer's block. Then I said to myself, why not defend alternative views on medical anthropology, and try to subvert *the mourning paradigm* from inside? Let us be supremely romantic, in raising a love-story to the status of a cultural paradigm.

Hindu love tales and songs are chanted in the streets at night. The emotions and passions that are thus enacted are teaching a sophisticated knowledge of pain, the physical pain of loneliness which involves ecology, language, and religion. Ecology, since deserts and thorny shrubs, dust winds and the monsoon have some bearing on physical consumption. Language, since illness meanings are embedded in words and verse. And religion, in the way I shall tell you now.

### The Bad Death of a Virtuous Wife

Madame Marie-Louise Reiniche, who teaches South Asian anthropology at the Ecole Pratique des Hautes Etudes, once inquired into ideas underlying the tradition of erecting memorial stones to the memory of people who had died a violent death. This custom did not concern the ordinary cases of violent death, such as the death of a child, death from a wound, a mishap in childbirth, an epidemic, an assassination, a suicide. The deceased was normally excluded from the ritual of second burials and condemned to remain a wandering soul, haunting places, bothering the living. However, there were also cases of 'prestigious' bad death, so to speak, which were commemorated by erecting a stele, or sometimes, a shrine. Two categories of deceased appear to have been prestigious in traditional India, if we go by the innumerable steles erected in their honor (Reiniche 1971: 162): the male category of (Sanskrit) *vīra*, 'the hero', who died a noble death on the battlefield, and the female category of (Sanskrit) *satī*, 'the virtuous wife', who followed her husband in death. The *suttee*, in the strict sense of the word, was the widow who immolated herself on her husband's funeral pyre. The forthcoming book by Catherine Weinberger-Thomas on this institution of suttee and the politics hidden behind it will certainly bring new light on the much narrower cultural stereotype evoked in the present paper [2]. In the most general sense of the word, a *satī* is a paragon of faithfulness in love. A different but comparable category of prestigious death is recorded in the ethnographic literature. It is the category of unmarried young women who had been deified as 'Virgin' after committing suicide, or after their absconding for ever in order to shirk the parents' authority and avoid, for example, an arranged marriage. They might have thrown themselves into fire, thus imitating the suttee sacrifice, they might have fasted to death or they might have "disappeared miraculously" (Reiniche 1971: 186). Stones or shrines were erected in Tamilnad, for example, in memory of these girls who were deified by their suicide, so to speak. They were addressed in worship as (Tamil) *kaṉṉi* or (Sanskrit) *kanyā*, 'the [pure and chaste] Maid, the Virgin Maid'. The *kanyā* is to be compared both to the female character of *satī*, 'the virtuous wife', and to the male character of *vīra*, 'the hero' [3]. Their voluntary death is prestigious, because it represents a free realization of human duties and goals, and contains the seed of free will.

Georges Dumézil has shown that sprouts of "loneliness and free will" (*solitude et liberté*) were rooted in the heroic nature of the warrior (Dumézil [1956] 1985: 72-29). In that respect, I would like to suggest that *kanyā* and *satī* both represent the warrior's feminine counterpart. Their faithfulness to a certain idea of love, which they think worth dying for, is a token of loneliness and free will, within a cultural environment which tends to smother them. For lack of time, I cannot but allude to the physical and moral dimensions of *kāma*, 'love', in the Hindu ideological framework, and to love's connection with what was described by Dumézil as "the second function", that of violence. The noble heroes and the virtuous heroines stand at the same level in the scale of human goals. Their very essence is Passion, that is both a strong feeling and a value, both a physical emotion and a moral call, which nurtures or else burns life out. If the word *paradigm* were understood as designating a set of mutually substituable categories or characters, which is its definition in semiotics, one could say that the Virgin Maid's story and the Faithful Wife's story belong to the same paradigm. Their suicide sets a traditional pattern of behavior which is still to be observed. Furthermore, the psychiatrist and the medical anthropologist do have to take this cultural pattern into account. The story of the virgin maid's suicide, the story of the virtuous wife's sacrifice represent *heroic* versions of a fundamental pattern of narratives in South Asia, for which there are also *medical* versions. We shall come back later to the semiotics of text, and elaborate this idea of a narrative pattern. For the moment, let us keep to the ethnographic contents. The illness of the love-lorn consumptive represents one of the various medicalized versions of the same story. Be it through the torments of illness or through the malediction of death, the maid, the wife and the love-lorn are witnessing to the human goal of *kāma*, 'love'.

The sexist distribution of roles in this kind of stories should not be overlooked. In the Ayurvedic texts as well as in folklore and classical poetry, 'consumption' (Sanskrit *yakṣman*) results mainly from 'separation' *(viraha)* and the pain of loneliness in the case of women, while men are the victims of irrepressible 'desire' *(rāga)* and overindulge in sex, which exhausts their vital fluids. This is the conventional etiology of consumption; it reads like a parable, and it goes far beyond the boundaries of medicine. Most of the data collected in the field are about men and revolve around the so-called culture-bound male syndrome of semen loss (Carstairs 1957, Bottéro, forthcoming). However, I shall concern myself here with the women's side of things. The question whether women also have semen has traditionally been debated by the masters of Hindu medicine. Although the more widely received theory is that the embryo results from the combination of male semen with menstrual blood, reasons are adduced to think that the fecund part of menstrual products is not blood but some sort of female semen [4]. Therefore, both sexes may experience the semen loss syndrome – drastic weight loss, sexual fantasies, night emissions and urine discoloration. However, there are several levels of speech to be taken into account. 'Semen' is a metonymy for 'vital fluid', and 'semen' is also synonymous with 'virility'. Let us say that the wasting away of vital fluids is experienced by both sexes, even if its explanation as semen loss is sexist.

The village gossip all over India abounds in cases of anorexia, cachexia, depression or Stripagol (Sanskrit *strīpāgal[atva])*, 'women's fits', or 'hysterics', which correspond to folk narratives recorded by travelers when, noticing a memorial stone erected along a country road and inquiring about it, they are told the story of an unfortunate woman who let herself die from love. The connection will elude us, however, if we restrict fieldwork to interviews of help-seeking patients.

The question is whether the diagnostic approach can capture and encapsulate the wealth of affects, events and values revolving around the indigenous conception of love. The preceding discussion suggests that certain affects, events and values, and the social conditions that produce them, are crystallized into a fundamental narrative pattern that allows for different readings. Once told in ontic or existential terms (the maid's suicide, etc.), it ceases to be an illness. Once told in medical terms (fits, consumption, etc.), it ceases to be a cultural pattern. To cite Gananath Obeyesekere's excellent phrase, "the same constellation of social conditions and affects can produce at least two (and perhaps more) cultural definitions, as an ontological problem [here, love and fate] and as an illness [consumption]" (Obeyesekere 1985: 147). *Kāma*, 'love', is both defined as a human goal (in Hinduism) and as a disease. In that respect, we should compare medical treatises with the *Kāmasūtra*.

In the *Kāmasūtra*, there is a description of *kāma* as a disease developing gradually to death: love of the eye, attachment of the mind, constant reflection, destruction of sleep, emaciation of the body, turning away from objects of enjoyment, removal of shame, madness, fainting, and death – these are the ten phases of Love defined as a fatal disease [5]. It is an emergency. The poor victim is granted one of those indulgences called *āpaddharma* in Sanskrit,"[an emergency mitigation of] Brahminic rules in case of distress". Thus adultery is permitted to the virtuous man, "for the purpose of saving a man deadly wounded in his soul and body" *(ātmaśarīropaghātatrāṇārtham)*. The concept of *kāma* as disease occupies a similar position in the Ayurvedic texts. Love is the first of mental diseases. "The series [of mental diseases] beginning with Love *(kāmādi)* and the series [of somatic diseases] beginning with Fever *(jvarādi)*," Caraka says [6], "sometimes when lingering become deuteropathies of one another". The fatal wound of love brings about fever and consumption, and reciprocally, chronic fever is accompanied by fits and fantasies.

## Disease, Passion and Narrative

The question I am addressing myself to is the following. If passions are diseases, if Love is the mental disease *par excellence* – as it is explicitly said to be in Ayurvedic medicine – from which angle is the clinical gaze of the physician able to define passions in order to treat them? To this question my answer will be two-pronged. First, I would like to show, on sample clusters of diagnostic traits in Ayurvedic medicine, that passions are reckoned in, among the various facets of humoral diseases. Then, I would like to trace the sources of this medical theory of passions as loss, the wasting away of vital fluids, back to the Indo-European ideology. My purpose is to delimit the territory of *what is medical* by surrounding it and taking aim at it from neighboring territories, namely, the two territories of ethnoscience and philology.

However, this maneuver cannot be properly carried out within the framework adopted for this colloquium, because we were invited to focus on pragmatic themes. I am pleading for the recognition of the *ecological* and *philological* dimensions of medical anthropology. One may well ask here what is meant by ecology. I said I would select from my ethnographic material sample clusters of diagnostic traits that include passions. But these clusters are written down in pharmaceutical recipes, which themselves are embedded in a complex tableau of local flora, landscapes and the seasons. Just as heat and dust are essential to the torments of Love, ethnobotany and meteorology are essential dimensions of our domain. Then, why to advocate also philology and textual studies?Sterotyped lists of passions are laid down in the texts. They represent a subclass of adventitious diseases:

> Diseases caused by demons, poison, storm, fire,
> wounds in fighting, etc.,
> among humans, are adventitious diseases;
> wisdom (*prajñā*) in them is at fault (*aparādhyati*).
> Desire, grief, fear, anger,
> conceit, hatred, etc., which are
> mental diseases, belong also with them.
> All adventitious diseases
> are born from a breach of wisdom (*prajñāparādha*) [7].

Not only passions as diseases result from a misdeed, but even curses and casualties are the results of one's own deeds. The precise meaning of *prajñāparādha*, corroborated by innumerable glosses, is 'misdeed, breach of wisdom'. It is related to fate as the consequence of deeds in a previous life, and to free will and action. An accurate and exhaustive reading of the Sanskrit sources shows that what is at fault in *prajñāparādha* is not 'thinking' nor 'insight' nor 'intellect', contrary to what is found in the secondary literature on the subject, but 'action'. The nuance is essential, because it determines the interpretation of passion as disease. The theory of passions operates at two levels, that of Passion (singular) as a fundamental principle of delusion and misdeed in human existence, and that of the (various) passions or psychic diseases resulting from misdeeds. The classic texts and the learned practitioners are agreed to teach that *prajñāparādha* is of three kinds, *violence, untruthfulness, and passion,* because there are three levels of action, *body, speech, and mind.* We shall come back to this later. Just a remark, for the moment, on behalf of the philologist. Modern philology is able to decipher this so-called medical doctrine, and to show that it closely resembles the Indo-European scheme of the warrior's three sins (Dumézil 1985: 92). Philology helps to erase the boundaries of medicine.

Let me raise the question again. From which angle does the Ayurvedic physician assess a psychosomatic disease like that of the love-lorn consumptive? First, to an Ayurvedic physician, an illness narrative is always *told in the context of a landscape*; consumption is not simply emaciation and blood-spitting, but also heat and dust, in the deadly end of summer, etc. Then, the contextual framework of pathology comes to the physicians's memory in the form of *clusters, garlands of disease-names.* To study Ayurveda is to embark on a journey through deserts and forests, and gardens of spices. Landscapes, the flora, the waters and the winds are mirroring the humoral constitution of the inhabitants. Onto the landscapes and the hidden transactions of saps and humors a taxonomy is grafted, which involves thousands of stereotyped sets of names. Several hundred Sanskrit recipes are still in use nowadays for compound medicines available on the market in India. The underlying principle is that of polythetic classification; in other words, the physician never addresses himself to one disease but to a complex syndrome, the name of which is a compound. I said, a cluster, a garland of disease-names. I meant, in the semiotic sense of the word , a paradigm, and a paradigm couched in the form of a syntagma. For example, a very famous ghee (or clarified butter) medicated with dozens of plants and called 'The Ghee with the Bitter Plants and Bdellium' *(Gulgulutiktakaghṛta)* is said to cure *yakṣmārucíśvāsa* - [...], "consumption, anorexia, dyspnea . . . [the long nominal compound continues, including desiccation, anemia, madness, etc.]" [8]. We can read it both ways. Either it is a cluster, a paradigm: it cures consumption, *or* anorexia, *or* dyspnea ... which compose a family of similar diseases. Or we can read it as a syntagma: it cures a polythetic syndrome made up of consumption *plus* anorexia *plus* dyspnea . . . The second reading is the more Hindu of the two, it conveys the well-known *inclusivism* of Hindu thought. Furthermore, the rhetoric rules for composing these syntagmas are laid down in the medical texts themselves, along with an elaborate theory of the dialectic of 'idiopathy' *(svīya)* and 'deuteropathies' *(anubandha)*, that represent complications and new facets of a given syndrome. This mode of thought is wonderfully fitted to humoralism. Its plasticity also allows for passions to be featured in the system. For example, there is an oil-bath commonly prescribed to the love-lorn consumptive and in similar cases of severe psychosomatic disorders. This 'Oil with Lac' *(Lākṣāditaila)* is medicated with a number of bitter plants (against bile) and a few aromatic spices (against wind). "It cures all fevers, emaciation, madness, dyspnea, hysterics, and sin" [9]. *And sin*, that is misdeeds and passions. A more elaborate analysis would expose the reciprocal implication of fever, consumption, madness and passion, and their subsumption to idiopathies of bile with deuteropathies of the humor wind.

Fever, desiccation, emaciation, wasting away of strength and tissues – the conception which informs and conditions the clinical gaze is that of a fundamental reserve of strength that is gradually burnt out. Precisely, *ojas,* 'the quintessence of vital fluids', represents this reserve of strength. A famous article by Georges Dumézil ([1957] 1969: 92) revealed the Indo-European connections of the Vedic concept of *ojas,* which was later taken over by the Ayurvedic tradition to express the idea of a quintessence of the body tissues. Often defined as a specific fluid, *ojas* may also be assimilated either to the humor phlegm, responsible for the body's unctuousness and suppleness, or to the most precious of the seven tissues, namely, the semen. Caraka's description of consumption through semen loss emphasizes the central image of a reserve of 'unctuousness, [the quintessence of] saps' *(rasa),* located in the heart, which gets vitiated and dries up [10]. Overindulgence in sex and the very violence of sexual desire bring about semen decay and retention; retention entails rotting; then the wind comes to vitiate blood and semen; hemorrhage and semen loss follow; all the other fluids in turn get rotten and wasted away until the whole body disappears. This is the physiological version of an ontological illness narrative.

The idea which guides me in this approach is that all illness narratives about consumption are medical versions of a more general pattern, which includes the virgin maid's suicide, the virtuous wife's sacrifice and the various literary accounts of Love in abandonment, which depict the heroine

**Figure 2. An Ayurvedic classification of narrative patterns**

| spheres of action | Sanskrit categories | the three causes of diseases | subject matter of the narrative | things and beings cast in the narrative |
|---|---|---|---|---|
| the Physician's art | *yukti,* rational combination | (1) inappropriate conjunctions of the sense organs with their objects | therapeutical interventions of the physician (*) | food,utensiles, manipulations, drugs, etc. |
| the Patient's responsibilities | *karman,* deeds | (2) breaches of wisdom | illnesses | love, passions friends, family, deeds (**), pains, symptoms, etc. |
| Environment, natural settings | *kāla*, time | (3) the maturation of time | the cycle of seasons (***) | flora, fauna, landscape, weather, etc. |

Asterisks refer to examples of fixed sets of three terms that subdivide a given topic or sphere of action, e.g., (*) the three parts of medicine: (1) dietetics, (2) surgery, and (3) pharmacy, (**) the three levels of action: (1) body, (2) speech, and (3) mind, (***) the three main seasons: (1) winter, (2) summer, and (3) the rains.

awaiting her beloved at the deepest of summer heats, burnt from outside as well as from inside. The levels of speech and semantic domains are different, of course, but the basic pattern is the same.

Actually, the classic medical texts have provided the learned physician with clues to the determination of different styles or genres of case histories. These clues are laid down in the form of fixed triads – the three causes of disease, the three parts of medicine, the three levels of action, and so on. A comparison of these fixed sets of three terms generates a sort of matrix of narrative patterns, which might prove useful in our attempt to delineate the medical domain.

For clarity's sake, I have retained only one basic division in constructing the grid (Figure 2).

"There are three causes of disease: (1) Inappropriate conjunctions of the sense organs with their objects, (2) breaches of wisdom, and (3) the maturation of time, which are again subdivided into three [11]. I skip the subsidiary divisions, the many fixed sets of three terms which are crisscrossed with the fundamental one, and I cannot fully justify my resorting to the Sanskrit categories of *yukti*, 'rational combination', *karman*, 'deeds', and *kāla*, 'time'. The point I wish to make is that all the three worlds are implicated in a full-fledged narrative, I mean, all the three spheres of action – the cognitive *(yukti)*, the ethical *(karman)*, and the aesthetic. Three viewpoints must be compared, to tackle a given medical case, namely, the physician's art, the patient's responsibilities, and, so to speak, the weight of external things. Let us note the relatively limited domain of *illness narratives* proper; they are not likely to cover the third sphere, that of ecology and the maturation of time, whereas love tales on the contrary abound in descriptions of natural settings. Taking into account the weight of external things, when writing the history of a disease, requires that we escape the artificial settings of hospitals and consultation rooms, and come to grips with ethnography.

## Conclusion

Let me conclude with one of those love stories and evoke the exemplary figure of love-lorn Damayanti. The edition and translation of the Nala and Damayanti tale by Franz Bopp (1819, 1832, 1838) coincided with the Oriental Renaissance in Europe. The elegy as a genre fitted the new Romantic age, and it placed Love at the root of burning grief, both as a value and as a malady. Nala loses his kingdom in playing dice, and abandons his faithful wife Damayanti. She wanders in the forest, consumed by burning grief. "I am alone, miserable, thin from grief, searching for my husband .... I have been looking days and nights, consumed by grief .... Burning with grief, the fool woman, covered with half a skirt, constantly weeps she with that same sorrow .... " (Van Buitenen 1975: 337, 343, 349). Later poets elaborated upon the same motif. With his usual insight and elegance, Charles Malamoud recently cited a passage from the *Naiṣadhacarita* of Harsa (Malamoud [1987] 1989: 300), painting Damayanti tormented by the fiery nature of Love. She laments, invoking *Smara*, another name for Love which means also Memory: "Alas! In my heart you take refuge, you who have no body. Then why do you set it afire? When you will have burnt this fuel out in a moment, this fuel that is you, this fuel that is me, where will you be? O poor *hatāśa*, my hope deadly wounded, who acts like *hutāśa*, the fire devouring me!"[12].

To which extent should anthropology be 'medical'? In focussing on illness, we might well miss that part of the ethnographic material which eludes the clinician. Should not we come back to a generalist approach in anthropology, and consider the 'medical' reading of a given cultural pattern as one reading among others, while ethnographic research would remain the first priority? This was the methodological question underlying this presentation.

## Notes

[1] These are the three spheres of action respectively oriented towards cognitive, ethical, and aesthetic values: See Jürgen Habermas, *Theorie des kommunikativen Handelns* [The Theory of Communicative Action], I, 2 para. 3-(2), a systematic analysis of Max Weber's celebrated *Zwischenbetrachtung* (Theoretical Parenthesis). Reading Habermas is the best antidote to the Mourning Paradigm, which I take to be a medicalized version of postmodernism. The tags on Figure 1 are meant to suggest a variety of approaches in medical anthropology at large. I admit, the selection is subjective. An introduction to *ethnosociology* is found in Marriott (1990). *Anthropologie de la maladie* alludes to the approach developed by Marc Augé and colleagues in Paris. For an illustration of *anthropologie historique*, one may refer to Emmanuel Terray's brilliant essay (1990) on medicine and politics in the time of Hippocrates.

[2] Catherine Weinberger-Thomas's recent paper, which announces her forthcoming book, offers further reference about the relationship between *satī* and *vīra* (1989: 12).

[3] The comparisons made until now have emphasized the concept of (Sanskrit) *śakti*, '[feminine] energy', and the transfer of *śakti* to the deceased hero through the wife's sacrifice. My perspective is different here, when I cite Dumézil on the concept of free will.

[4] I am grateful to Dr. Alain Bottéro, MD for communicating to me his forthcoming paper, which recounts the current views of Ayurvedic doctors on this particular question in Ṣeraikella, a village on the borders of Bihar and Orissa. One of these views. which is also found in the *Suśrutasaṃhitā* (*Śārīrasthāna*, II, 36), is that the female semen is discharged at the time of orgasm.

[5] *Kāmasūtra*, V, 1, 3-5. Professor Stanley Insler (Yale) has an important book forthcoming on the ten phases of Love, a theme that recurs constantly throughout Sanskrit literature.

[6] *Carakasaṃhitā, Vimānasthāna*, VI, 8.

[7] *Carakasaṃhitā, Sūtrasthāna*, VII, 51-52 (verse).

[8] *Aṣṭāṅgahṛdayasaṃhitā, Cikitsāsthāna*, XXI, 61c.

[9] *Sarvajvarakṣayonmādaśvāsāpasmārapāpmajit, Aṣṭāṅgahṛdayasaṃhitā, Uttarasthāna*, II, 56cd. *Pāpmajit*, 'cures sin', is the authoritative reading, according to the Astavaidya tradition of Kerala, although the Bombay edition ignores this reading and prints *vātajit*, 'cures wind'. This is a typical case of dulling and ironing out of Sanskrit manuscripts.

[10] *Carakasaṃhitā, Nidānasthāna*, VI, 8.

[11] *Ity asātmyendriyārthasaṃyogaḥ, prajñāparādha, pariṇāmaś ceti trayas trividhavikalpā hetavo vikārāṇām, Carakasaṃhitā, Sūtrasthāna*, XI, 43.

[12] *Naiṣadhacarita*, IV, 75. We cannot conclude without mentioning the presence of the very same cultural stererotype, the love-lorn consumptive, in other medical traditions and especially in ancient Rome: See Amundsen (1974); Gourévitch (1984: 73-77, *le mal d'amour*).

## References


AMUNDSEN D.W. 1974: Romanticizing the ancient medical profession: The characterization of the physician in the Graeco-Roman novel. *Bulletin of the History of Medicine* 48: 320-337.

*Aṣṭāṅgahṛdayasaṃhitā of Vāhaṭa* (VIIth cent. A.D.). Bombay: Nirnaya Sagar 1939.

BOPP F. 1819: *Nalus, carmen sanscritum e Mahabharato, edidit, latine vertit et adnotationibus illustravit* Fr. Bopp. London, Paris, Strasbourg.

– 1832: *Nalus, Maha-Bharati episodium* (...). Berlin.

–1838: *Nalus und Damajanti, eine indische Dichtung aus dem Sanskrit*. Übersetzt von Fr. Bopp. Berlin.

BOTTERO A. (forthcoming): *La consomption par déperdition séminale en Inde et ailleurs.* Annales de la Fondation Fyssen.

*Carakasaṃhitā* (Beginning of our era). Bombay: Nirnaya Sagar, 1941.

CARSTAIRS G.M. 1957: *The Twice-Born. A Study of a Community of High-Caste Hindus.* London: Hogarth.

DUMEZIL G. 1969: Augur. Reprinted in *Idées Romaines*, pp.79-102. Paris: Gallimard.

– 1985: *Heur et Malheur du Guerrier.* Paris: Flammarion.

GOUREVITCH D: 1984: *Le Mal d'Etre Femme. La Femme et la Médecine dans la Rome Antique*. Paris: Belles

Lettres.

HABERMAS J. 1981: *Theorie des kommunikativen Handelns*. Frankfurt: Suhrkamp.

*Kāmasūtra* (VIIth cent. A.D.?). Varanasi: Chaukhambha, 3rd ed., 1982.

KLEINMAN A. 1981: On illness meanings and clinical interpretation: Not 'Rational Man', but a rational approach to man the sufferer / man the healer. *Culture, Medicine and Psychiatry* 5 (4): 373-377.

– 1988: *The Illness Narratives, Suffering, Healing and the Human Condition*. New York: Basic Books.

MALAMOUD C. 1989: Par coeur. Note sur le jeu de l'amour et de la mémoire dans la poésie de l'Inde ancienne. Reprinted in *Cuire le Monde, Rite et Pensée dans l'Inde Ancienne*. Paris: La Découverte.

MARRIOTT M. (ed.) 1990: *India Through Hindu Categories*. New Delhi: Sage.

*Naiṣadhacarita of Harṣa* (XIIth cent.) Bombay: Nirnaya Sagar. Reprint New Delhi: Meharchand Lachhmandas 1986.

OBEYESEKERE G. 1985: Depression, Buddhism, and the Work of Culture in Sri Lanka. In: KLEINMAN A. and GOOD B. (eds.) *Culture and Depression*. Berkeley: University of California Press.

REINICHE M.-L. 1971: *Les Erections de Pierres dans les Rites Funéraires et les Cultes 'Ancestraux' en Inde du Sud*. Paris, Thèse de Doctorat de 3e Cycle, Université de Paris III (inédit), 189 pp. et bibliogr.

SHWEDER R.A. 1984: Anthropology's romantic rebellion against the enlightenment, or there's more to thinking than reason and evidence. In: R.A. SHWEDER and R.A. LEVINE (eds.). *Culture Theory*. Cambridge: Cambridge University Press.

*Suśrutasaṃhitā* (Beginning of our era). Bombay: Nirnaya Sagar 1939.

TERRAY E. 1990: *La Politique dans la Caverne*. Paris: Seuil.

VAN BUITENEN J.A.B. 1975: *The Mahābhārata*, 2/ The Book of the Assembly Hall, 3/ The Book of the Forest. Translated and Edited by J.A.B. Van Buitenen. Chicago: University of Chicago Press.

WEINBERGER-THOMAS C. 1989: Centres d'immortalité. La crémation des veuves en Inde. *Archives de Sciences Sociales des Religions* 67 (1): 9-51.

ZIMMERMANN F. 1987: *The Jungle and the Aroma of Meats. An Ecological Theme in Hindu Medicine*. Berkeley: University of California Press.

– 1989: *Le Discours des Remèdes au Pays des Epices*. Paris: Payot.

# Traditional European and Chinese Definitions of Illness and Medical Practice

## A Reevaluation of Differences and Parallels on the Basis of the Work of Hsü Ta-ch'un, an 18th Century Intellectual, Physician, and Medical Author

Paul U. Unschuld

### Some Introductory Remarks

So-called Traditional Chinese Medicine has become, over the past years, a focus of interest of a few European and North American academicians pursuing historical, anthropological, or related research, and of a much larger number of health care providers searching for alternative or complementary perspectives to what they consider the inadequate or erraneous approaches of so-called Western medicine.

In this situation one of the most challenging issues is the extent to which traditional Chinese medical thought and conceptualization stand apart not only from the basic tenets of current Western medicine but also from pre-20th century European medicine. We have witnessed, especially during the past decade, the appearance of quite a few more or less scholarly publications emphasizing an alleged fundamental antagonism between Chinese and Western health care approaches. It has never become clear, though, whether this presumed antagonism is limited to surface phenomena or permeates all levels of Chinese and Western health care theorization, whether it is true only of comparisons of mid-twentieth century Chinese and Western medicine, or whether it holds for a comparison of former periods too. We have not seen any study so far, investigating the proximity or distance between, for instance, 18th century Chinese and European basic medical concepts, and we know of no attempts to pursue the question whether Chinese and European medicine followed separate conceptual paths right from their documented beginnings, or whether a divergence occurred only later, and if so, when.

With this paper I wish to draw attention to some of the ideas written down in the middle of the eighteenth century by Hsü Ta-ch'un 徐大椿 [1]. My discourse will be based on a translation and annotated edition of Hsü's book *I-hsüeh yüan liu lun* 醫學源流論 , a unique document revealing the world of thought of a Chinese intellectual and physician shortly prior to the massive encounter of Chinese health care culture with medical concepts and practices developed in the West [2].

### Biographical Data of Hsü Ta-ch'un

Hsü Ta-ch'un was born, in the city of Wu-chiang 吳江 , Kiangsu province, in 1693, into a family without riches but with some remarkable personalities in former generations. A paternal great-grandfather, Hsü Yün-ch'i 徐韞奇 , was a man of letters and author of several works, including at least one medical title. Hsü Ta-ch'un's grandfather Hsü Chiu 徐釚 (1636-1798) was noted as a poet, landscape painter, and participated in the compilation of the official history of the Ming dynasty. Hsü Yang-hao 徐養浩 (d.c. 1721), finally, Hsü Ta-ch'un's father, was an expert in the water control system of his native province [3].

Hsü Ta-ch'un himself never competed in any official examination for a civil service career, yet he became a famous scholar and physician during his lifetime. He published poems, essays, and books on topics including medicine, philosophy, and music among others. Continuing the expertise of his father, he was consulted at least twice on major water control projects, and likewise he was called twice to the imperial capital to offer his medical views on the illnesses of high ranking officials. Hsü Ta-ch'un died during the second of these visits to the capital in 1771; Yüan Mei 袁枚 (1716-1798), the famous poet, literary critic and essayist, who had been a patient and acquaintance of Hsü, wrote a posthumous biography of his [4].

### Medical and Political Ideology in Hsü' s work

At the age of 35, Hsü Ta-ch'un published a commentated edition of the *Nan-ching* 難經. Disregarding or simply misunderstanding the innovative nature of this first century classic, he criticized the unknown author for the work's many deviations from the contents of the 'real' classic, the *Huang-ti nei-ching* 黃帝內經 [5]. Obviously, Hsü Ta-ch'un must be counted among the 'conservatives' of his time; his medical work reflects the political tendencies of the "Han-hsüeh p'ai", i.e., the school of Han learning. Consequently, in the *I-hsüeh yüan liu lun*, written thirty years later than the *Nan-ching* commentary, Hsü Ta-ch'un responded with biting criticism to the impact of Sung-Chin-Yüan learning on medicine. Again and again he regretted what he considered the loss of the genuine foundations of medicine that had been developed by the Sages in antiquity but had been distorted or neglected especially under the influence of Sung-Chin-Yüan theorization later on.

Further studies of Hsü Ta-ch'un's medical and non-medical writings may reveal more clearly to what extent his medical perspective coincided with his political views. In the same way as Chang Yüan-su 張元素 (ca. 1180) had reflected, in his opposition to an unconditional continuation of ancient therapeutic patterns [6], the basic tenets of Sung Neo-Confucianism, Hsü Ta-ch'un's discourse appears to mirror the basic concerns of the 'back to the Han-sources movement' of his time. His numerous complaints about a lack of courage on the side of his contemporaries to attack problems directly, and about an unwillingness to fight rather than palliate, may have been focussed as much on the treatment of illness as on the handling of government affairs.

To point out these links between Hsü's political and medical thinking is not as far-fetched here as one might think; Hsü Ta-ch'un himself suggested such interpretations of his work by giving his essays titles such as "On parallels in the use of drugs and the use of soldiers", "On similarities between the ways of medicine and the ways of government", or "On parallels between illnesses and the changing (conditions) of states", and it may not be without political background that the last two of the three examples mentioned here were left out of the government sponsored *Ssu-k'u ch'üan-shu* 四庫全書 edition of the *I-hsüeh yüan liu lun* in 1778.

### Hsü Ta-ch'un and European Medicine

Of the many concepts delineated explicitly or implicitly by Hsü Ta-ch'un in the one hundred essays of his *I-hsüeh yüan liu lun*, and lending themselves to comparisons with concepts prevalent in Europe, I wish to concentrate here on merely two major issues, i.e., his understanding of the role of medicinal drugs in the treatment of illness, and, second, his notion of illness. Hsü Ta-ch'un's discourse allows - better than any other work of an individual author known to me - for insights into the thoughts of a highly educated pre-modern Chinese physician-intellectual, and his detailed argumentation provides valuable data for a reevaluation of parallels and divergencies in the history of European and Chinese medical thinking.

I shall draw attention to sufficient evidence for close parallels in the thinking of Hsü Ta-ch'un and contemporary as well as 19th/20th century concepts in European medicine, and I shall hypothesize that these parallels offer an explanation why Western medical thought, when it finally entered China on a broad scale in the second half of the 19th and in the first half of the 20th century, was not really alien to Chinese culture. Hsü Ta-ch'un's writings are representative of the fact that virtually all the conceptual images at the basis of modern Western medicine, as it developed in the late 19th and early 20th century, were present in Chinese thought too, and could have served as a starting point for similar developments in China.

*Hsü Ta-ch'un and medicinal drugs*

Hsü Ta-ch'un's statements on the meaning and usages of drugs may serve as a first example to support this argument. Hsü Ta-ch'un claimed to be "able to design (therapeutic) patterns in thousands of adaptations following the thousands of ends and tens of thousands of beginnings of the illnesses under heaven" [7], but he also acknowledged the value of "ready-made cure-all prescriptions that cured very many illnesses and that achieved miraculous successes in all cases" [8]. That is to say, some problems need individually formulated prescriptions, taking the very specific circumstances of an individual illness into regard, while other problems may be treated successfully with formulas that are effective irrespective of the idiosyncrasies of an individual patient. The former, one might also state, are directed at the individual patient, while the latter focus on a specific disease.

In many cases, the antagonism between drugs and illnesses is seen and explained, by Hsü Ta-ch'un, in militaristic terms. Drugs "attack", "encircle", and "eliminate" intruding evil, and to find the exact location of an evil in the body is as important for a physician, as to know the whereabouts of enemies in a country is essential for the commander of an army: "If one disregards the location of an illness", Hsü stated, "and employs drugs to attack, in contrast, a place that is absolutely free of illness, this is what the *Nei-ching* calls 'punishment without crime'." [9].

One interesting point should be added here. Hsü Ta-ch'un was aware of relationships between the places of growth of plants and their respective pharmaceutical potential; in his own words:

> There were specific areas where (specific) drugs grew. These were their native places. Hence (each drug) was rich in influences and full of strength. [10]

Elsewhere, Hsü Ta-ch'un wondered why specific drugs exert specific functions:

> When it comes to (the ability of) turtle shells to dissolve blockages, or to (the ability of) shih-chün tzu to kill the hui-worms, or to (the ability of) ch'ih-hsiao-tou to eliminate skin swellings, .. then (these are drug effects) that are even harder to explain. [11]

The mere fact of Hsü's curiosity may tell us, together with his perception of varying strengths of one and the same substance if grown in different places, how open-minded he might have been if he had ever had a chance to learn of such approaches as the chemical analysis of drugs and the search for active ingredients.

The step from Hsü Ta-ch'un's militaristic and ontological notions of drug effects to an understanding of the workings of chemotherapy is almost negligible. When bacteriology emerged in the late 19th century in central Europe, breaking after almost two millennia the domination of Galenic concepts of humoral pathology in the application of pharmaceutics, its basic concepts appear - in hindsight - to have been more alien to European culture - where it emerged - than to Chinese culture - where it was introduced soon afterwards. It fell on pre-conditioned grounds.

One might go a step farther and wonder whether Hsü Ta-ch'un might have had a conceptual potential for an appreciation of the even more recent concepts of immunology. Hsü stated:

(Superior practitioners) see to it that the original influences (of a person) are complete and remain in control so that they themselves may keep evil (influences) outside. If, however, evil influences gain the upper hand, and cause harm, the (superior practitioner) activates original influences that remained dormant so far. He has them fight with the back towards the wall and burst (into action) all at once. [12]

and elsewhere

If, however, those agents that are responsible for warding off (evil intruders) fail to do so, those (intruders) attacking them will gather in their places. [13]

All these metaphorical statements, it should be to emphasized, have, of course, primarily nothing in common with chemotherapy, bacteriology, or immunology. They are based on a transfer of environmental images - visualized originally, we may assume, during the period of the Warring States, and reinforced continuously by the reality and requirements of warfare and defense in later centuries - to the explanation of the invisible. However, to repeat my earlier argument, it may well be that the very existence of such metaphors and explanatory models facilitated, as long-standing conceptual antecedents, the acceptance of chemotherapy, bacteriology, and immunology in China, apart from the persuasive force of the results of the clinical application of the latter.

### *Notions of Illness*

In contrast to the *Huang-ti nei-ching* which attributed a normal life-span of about one hundred years to every person [14], Hsü Ta-ch'un saw man endowed with an individual share of resources lasting for individual periods of existence [15]. As interesting as this concept is with regard to a protection of medicine against accusations of failure in case of a patient's early death, it is even more important in that it demonstrates what little difficulties Hsü Ta-ch'un might have had in consenting to a concept of a genetically pre-determined, individually varying life-span.

Hsü Ta-ch'un attributed to nature a much more decisive role in the course of illnesses than is usually outlined by pre-modern or contemporary Chinese and Western writings on traditional Chinese medicine. Even though he had no specific term for it, he conveyed a concept equivalent to the *vis medicatrix naturae* of European medicine:

If an illness is not accompanied by fatal pathoconditions, any external affection will slowly retreat, and any internal damage will gradually recover. (That is, the illness) will heal by itself. [16]

The patho-conditions emerging as a result of affections from the outside, or of harm from the inside, they heal, for the most part, by themselves (even though) they receive treatment. Minor illnesses in particular may heal gradually even without any intake of drugs. Even in case the strength of the illness is quite dangerous, the evil influences will gradually recede, and a state of normalcy will return by itself, if the physician in charge does not commit too great a mistake. [17]

The 'success' of a treatment may, therefore, be an illusion if it was not to the physician's credit but to a trans-individual regularity of a disease that a given ailment reached its happy end.

Hsü Ta-ch'un recognized, in agreement with most authors of his own and earlier times, both external and internal factors, such as climatic and emotional irregularities, as causes of illness. Several passages in his essays emphasize his unambiguous notion of cause-effect relationships not only in the emergence of illness but also in therapy. His most clearcut statement reads:

Whenever an illness emerges, it must have a cause. [18]

Or:

If two (persons suffer from) such illnesses, it may be that the physical manifestations of their illnesses are identical while the causes of their (respective) illnesses are different. [19]

Most illnesses are to be interpreted, in Hsü's eyes, as ever varying combinations of pathological standard situations (such as "struck by wind" of "harm caused by cold") with individual circumstances. Identical causes may generate different illnesses in different patients, and apparently identical illnesses in two different patients may have resulted from different causes. Also, a patient may suffer from more than one single illness at the same time, with Hsü Ta-ch'un pointing out sophisticated differentiations among illnesses, so-called concomitant illnesses, and so-called separate illnesses affecting a single patient simultaneously [20].

An illness may be an abstract problem manifesting itself solely in functional disturbances, but it may also be a tangible evil moving through the body, from one well defined region, or location, to another.

In this regard it should be pointed out that the *I-hsüeh yüan liu lun* is a splendid example of the all too rarely recognized multi-faceted nature of learned traditional Chinese medicine. Hsü Ta-ch'un demonstrated a broadness of vision that is difficult to reconcile with the narrow conceptual framework attributed to Chinese medicine in most surveys of the past few years.

For instance, many illnesses are marked, Hsü Ta-ch'un pointed out, by "transmission" (*ch'uan* 傳 ) and "transformation" (*pien* 變 ) in the organism [21]. Transmission and transformation are processes marking the individual course an illness may take in an organism. Both these concepts should be of particular interest here since they demonstrate close links between a functional-individualistic view of illness and an ontological-localistic point of view. Hsü Ta-ch'un placed equal emphasis on both these perspectives; his discourses are a most convincing argument against an unnecessary antagonism created, in recent times, between Chinese medicine on the one side and European or modern Western medicine on the other side, in that the ontological-localistic approach is usually associated with Western medicine only.

Hsü Ta-ch'un recognized the tangible nature of various organs, and associated them not only with primary and secondary functions throughout the organism, but also with specific locations, as well as with concrete lesions. His essay "On intra-abdominal abscesses" [22] as well as many other of his statements might have enabled him to have an informed discussion with his European contemporary Giovanni Battista Morgagni (1682-1771), the author of "On the seats and causes of illnesses" and protagonist of European morphological pathology, if the two had ever had a chance to meet.

When Hsü stated "In case of jaundice, a bag containing yellow liquid is located in one's flanks. This bag has absolutely no opening. Drugs can reach only the outside of this bag, but they cannot enter it. ...One must make use of light and penetrating prescriptions which either break the bag, or drain the water from it" [23], then this is merely one example of a morphological pathology that could easily be reconciled with subsequent developments in Europe.

### Concluding remarks

This necessarily and intentionally highly selective presentation of some of the views written down by Hsü Ta-ch'un in his *I-hsüeh yüan liu lun* suggests a need for a reorientation in the Western interpretation of the basic tenets of traditional Chinese medicine. Over the past few years, those facets of traditional Chinese medicine appear to have been at the focus of attention in the West that seem to be antagonistic to Western medicine. The arguments in the present paper are not meant to obfuscate any real differences between traditional Chinese and Western medicine. Rather, they should demonstrate that it is equally possible to write lengthy discourses pointing out basic agreements between fundamental perspectives in Chinese and Western medicine. The point to be made here is rather simple but has not been acknowledged so far. Traditional Chinese medicine is a very heterogeneous and yet systematic construct combining a broad spectrum and many levels of different perspectives on health and illness. It appears, now, as if we approached the complex conceptual edifice of traditional Chinese medicine too often like a quarry from which we are prepared to take only those materials that are needed to construe a building considererd complementary to the edifice of so-called Western medicine. A close look, though, at the conceptual basis of tradtional Chinese medicine reveals that the depiction of Chinese medicine in Western secondary literature tends to follow the needs and desires of Western audiences, rather than the historical reality of traditional Chinese medicine.

When I have pointed out close parallels between Hsü Ta-ch'un's thought and the thinking characterizing European medicine, this is not to repeat a Needhamian argument, that is, I do neither wish to imply that all Western science is derived from Chinese stimuli, nor do I intend to imply that traditional Chinese medicine was 'scientific' in that many of the concepts of Western medical science appear to have conceptual antecedents in China. My point is that mankind has not been very creative in its cultural handling of illness, and that even in such geographically distant cultures as China, dominated by Confucian, Taoist, and Buddhist world views, and Europe, influenced by the Judaeo-Christian world view, only two basic approaches have been developed in reaction to illness.

The first notion on which both Chinese and European as well as modern cosmopolitan medicine are built is one of the normalcy of fight and attack, and of the necessity of defense. The second is a notion contradictory to the first, i.e., harmony or equilibrium is the normal or healthy condition of systems.

Both these notions find their justification in man's natural and social environment. The phenomena of harming, killing, and devouring among animals, plants, and humans, suggest a fundamental principle of attack and protection. On the other hand, the regular sequence of the four seasons, of day and night, of high tide and low tide, among many other regular phenomena, imply regularity, or harmony, as a basic principle. Given the fact that man has at all historical times resorted, consciously and/or subconsciously, to the images of his social and physical environment in order to understand the processes of his own organism, and in particular the origins and the nature of illness, it should be no surprise that both traditional Chinese medicine and European medicine are built around the two basic principles outlined above, and that both traditional Chinese medicine and European medicine contain, therefore, ontological *and* functional notions with all their logical consequences. If we penetrate the surface phenomena constructed around these basic notions, we shall realize that more bridges exist between Chinese and European or even modern medicine than are commonly imagined. Hsü Ta-ch'un' s *I-hsüeh yüan liu lun* may serve as evidence.

## Notes

[1] Style-name: Ling-t'ai 靈胎 ; sobriquet: Hui-hsi lao-jen 洄溪老人 . Cf. Hummel (1943: 323).

[2] Unschuld (1990). This book reproduces the entire Chinese text of the Ssu k'u ch'üan-shu 四庫全書 edition of the I-hsüeh yüan-liu lun, amended by those portions of the original text that had been left out in this edition. For a complete text see also Chiang Jen-an 江忍庵 and Lin Chih-ch'ing 林直清 (eds.) Hsü Ling-t'ai i-shu chu`üan-chi 徐靈胎醫書全集 Taipei: Wu-chou Publishing Co. 五洲出版社 (1969: chapter 1, 54-132).

[3] Sung Ta-jen, 1963, 30 宋大仁 清代伟大医学家徐灵胎的一生 江苏中医

[4] Hsieh Sung-mu, 1934, 12 謝誦穆 徐靈胎評傳 現代中医

[5] cf. Unschuld (1986b: 46-49 and passim)

[6] cf. Unschuld (1986a: 103)

[7] See the essay "Ch'u ch'i chih ping lun" 出奇制病論

[8] See the essay "Chih ping pu pi fen ching lo tsang fu lun" 治病不必分經絡論

[9] See the essay "Chih ping pi fen ching lo tsang fu lun" 治病必分經絡論

[10] See the essay "Yao-hsing pien ch'ien lun" 藥性變遷論

[11] See the essay "Yao-hsing chuan ch'ang lun" 藥性專長論

[12] See the essay "Yüan ch'i ts'un wang lun" 元氣存亡論

[13] See the essay "Ping you kuei shen lun" 病有鬼神論

[14] See Huang-ti nei ching, Su-wen 素問 , the first treatise: "Shang-ku t'ien-chen lun" 上古天真論

[15] See the essay "Yüan-ch'i ts'un wang lun" 元氣存亡論

[16] See the essay "Ch'ing yao yü ping lun" 輕藥愈病論

[17] See the essay "Ping you pu yü pu ssu sui yü pi ssu lun" 病有不愈不死雖愈必死論

[18] See the essay "Ch'ü-k'o ching-lo tsang fu lun" 軀殼經絡臟腑論

[19] See the essay "Nei shang wai kan lun" 內傷外感論

[20] See the essay "Ping cheng pu t'ung lun" 病症不同論

[21] See the essays "Ch'ü-k'o ching-lo tsang fu lun" 病情傳變論 , and "Ping-ch'ing ch'uan pien lun"

[22] "Fu nei yung lun" 腹內癰論

[23] See the essay "Ping you pu pi fu yao lun" 病有不必服藥論

## References

HUMMEL A.W. 1943: *Eminent Chinese of the Ch'ing Period (1644-1912)*. Washington.

CHIANG J. and LIN C. (eds.) 1969: *Hsü Ling-t'ai i-shu ch'üan-chi*. Taipei: Wu-chou Publishing Co.

UNSCHULD P.U. 1986a: *Nan'ching. The Classic of Difficult Issues (The Chinese Classics)*. Berkeley, Los Angeles, London: University of California Press.

— 1986b: *Medicine in China. A History of Pharmaceutics*. Berkeley, Los Angeles, London: University of California Press.

# Section 5

# Medical Systems and Cultural Change: An Analysis of Facts and Theories

# The Hierarchies of Medicines
## A Contextual Analysis of Schismogenic Processes

Gilles Bibeau

**"Ex divisione crescet divisio"**

It is generally assumed that protest leads to more protest: it would be in the nature of opposition to create more opposition and to engender division. Conservative Catholic historians who studied the Reformation have assumed that the rejection of a central Roman authority in Christianity led inevitably to a multiplicity of reformers and the formation of national churches. Specialists of political sciences have also often stressed the natural multivocality of all ideological movements when they oppose a dominant and well-organized ideology.

What I want to examine in this paper concerns the schismogenic process which seems to be at work in all kinds of alternative movements: is it in the nature of peripheral undertakings (groups, ideas, institutions) to be loosely organized, potentially antagonistic and anarchic, and unable to create unity? Two case studies of alternative therapeutic movements will be examined: the first concerns the movement promoting traditional medicine in Zaire between 1973 and 1988, while the second centers around the Québec political and legal context in which a large number of schools of Chinese medicine have emerged in Montréal during the last decade.

On the basis of those two cases, I will demonstrate the inadequacy of the prevalent conceptual positions which assume that division is naturally inscribed in alternative movements as soon as they oppose dominant systems. In my view, difficulty in maintaining consensus and unity is not an innate, inevitable feature of alternative or protest movements. The absence of a well-defined dogma and of a high-priest to maintain orthodoxy and to sanction heretic thinking should certainly be taken into account, but the endogenous dynamic that originates from these internal characteristics would be incomprehensible without a contextual analysis of the constraints imposed from outside. It is precisely those external and contextual constraints that I identify in the following analysis of two relevant cases.

Three contextual factors play a decisive role in the schismogenic process affecting the evolution and transformation of the two alternative therapeutic movements studied in this paper. The movements will first be studied against the background of the general cultural orientation in each country. The Zairian 'nationalist ideology of authenticity' clearly served the interests of traditional healers in the early 1970s, and its replacement by the economic rationality of the World Monetary Fund in the 1980s created a new context in which the policy of self-reliance in all domains was no longer dominant; likewise, the development of Chinese medical schools in Montréal must be examined against the background of Canada and Québec immigration and cultural policies.

Secondly, legal and policy provisions concerning the 'art of healing' should be examined. The examination of laws about the 'art of healing' in any country cannot be divorced from its more basic policies regarding cultural life: is cultural homogeneity promoted at the expense of heterogeneity? Is tolerance of diversity privileged over the search for unity?

Thirdly, the existential daily conditions of the various socio-economic strata in the national population should be known. The catastrophic deterioration of life conditions in less than a decade has led Zairians to survival strategies; in Québec, the search for other forms of therapy is today leading hundreds of persons to be trained in Montréal-based schools of Chinese medicine.

## Healer's Medicine in Zaire: from 1973 to 1988

### *Up to 1977: A nationalist ideology of cultural authenticity*

In order to understand why Zaire was the first country (in 1974) on the African continent to conduct a national survey on healers' medicine, one should examine closely the political imperatives which at that time permeated all aspects of national life. In 1973, President Mobutu Sese Seko put forward an original political ideology termed 'politique de l'authenticité' and 'politique de l'affirmation nationale', grounded in the conviction that African countries must frame original solutions to their current problems in all domains, using their own ancestral traditions as a basis. It was thought that the nationalist policy of cultural authenticity would lead to self-reliance and to development of a genuine African lifestyle. Although, everybody was aware at that time that this promotion of indigenous cultural values and behaviors constituted at least partially a way to hide the dependence on prevalent Western models, the nationalist ideology generally produced a wave of enthusiastic support in the country (Bibeau 1984).

Against this nationalistic background of 'self-reliance', Zairianization of the entire economy was decided in 1974; all trade houses, plantations and private enterprises owned by foreigners were expropriated and handed over to Zairian citizens. This sweeping reform proved to be economically unsound, and scarcely two years later Mobutu was inviting the former foreign owners to come back and participate in joint ventures with their Zairian counterparts.

University professors, researchers, journalists and artists were in this same period deeply involved in an intellectual movement known as the 'mental decolonization'. Without rejecting Western scientific paradigms, they became convinced that most of their efforts should be directed to the development of an intellectual and artistic life deeply rooted in their traditional heritage. This movement of 'mental decolonization' has remained powerful in Zaire over the years, but the very bad conditions in which intellectuals are obliged to work have made implementation of the program impossible. In late 1974, the Ministry of Health published a document proposing a radical move away from exclusively curative activities relying on sophisticated technology and based predominantly in urban hospitals. Instead, these services were complemented by preventive and educational activities, where there would be increased reliance on community resources, and services would be anchored in the cultural and sociological characteristics of these communities. To implement the Manifesto, a five-year National Sanitary Plan was elaborated in 1975 with the main objective of setting up a network of rural and urban health centers which would combine curative, preventive and educational activities, with professional practitioners as well as first-line personnel coming from the communities (Nguete, Bibeau and Corin 1979).

The necessity of adapting health services to the local cultures, and the effort to identify and to train individuals to help in delivering first-line services led directly to the survey on traditional medicine referred to earlier: following the popular Chinese model, many thought at the time that it should be possible to minimally retrain the healers already at work in rural and urban communities, with the aim of using them in a new reorganized and integrated health system (Bibeau 1985). The legal status of traditional healing practices at that time was ambiguous: although traditional medicine was never outlawed by legal codes, either during colonial times or since Independance, there were no clear provisions in the laws permitting regulation of ancestral medicine.

At government request, the Zairian National Institute for Scientific Research launched in 1974 a four-year project with the objective of assessing the therapeutic efficacy of the traditional healer's practices: botanists, pharmacologists, physicians, psychologists and anthropologists pooled their expertise in designing and carrying out this research which was done on a national basis. A total of some 500 healers, constituting a representative cross-section of the three major categories of traditional practitioners (herbalists, ritualists and spiritualist) active in all areas of Zairian society, were studied over a two-year period.

On the basis of this information, an overall picture emerged concerning the categories of health problems treated by healers, the socio-cultural characteristics of clients and health-seeking processes of recourse to modern and traditional therapists. In-depth case studies of treatments performed by traditional healers led to articulation of their implicit conceptual frame of reference: classification of diseases (nosology), etiology and diagnostic procedures, underlying conceptions of the body and the person, and therapeutic practices. Research also focused on the numerous and radical changes mode to traditional forms of treatment in order to adapt to modern society and to the increasingly depressed living conditions of the majority of the population.

The authors of the report concluded that reform of Zairian health services should be guided by four main principles. First, integration of healers would not be feasible without a general paradigm shift from a curative, technology-intensive, centralized system to a decentralized system primarily oriented towards prevention. Secondly, coordination and interfacing of traditional and modern therapeutic endeavors was doomed to failure unless Western-style practitioners were prepared to understand the conceptual framework underlying traditional therapies and to recognize their relevance as part of a more holistic system of healing. Thirdly, there must be formal legal recognition of traditional medical practices. Each of the three main categories of healers (herbalists, ritualists and spiritualists) would be constituted as a professional corporation that would assume full responsibility for regulation of its members' therapeutic activities. The healers associations were seen as a pivotal element in the new national health policy; it was felt that they would ensure maintenance of an authentic healing tradition that could evolve to meet new needs while remaining faithful to the fundamental cultural conception underlying the tradition.

Finally, it was proposed that healers be sensitized to priorities of the national health system without transforming their traditional conceptions and practices. It was also recommended that they be authorized to intervene at all stages of the therapeutic process rather than being confined to primary health care.

At the time, these proposals were made, in 1978, the nationalist policy of cultural authenticity seemed to prove an exceptional and unique context for promoting an original national health system, and there appeared to be a political will to implement this approach. But in less than five years, from 1973 to 1978, the whole situation in Zaire had radically been transformed.

### *After 1977: the economic bankruptcy of the country*

Zaire moved in a very short period of time from a promising nationalistic ideology of authenticity to the total collapse of the country. The low performance of the state in many sectors and the failure of economic authenticity led to a wave of popular discontent regarding the policy of authenticity. The breakdown of socio-economic institutions, unemployment and low wages for workers (the purchasing power of an average Zairian was approcimately - in 1986 - 10 percent of what it was in 1960) created for most people a catastrophic situation which forced them to develop survival strategies, including underground parallel trading activities in cities and return to a subsistence economy in rural areas.

Now faced with the total incompetence of the state, people not only avoid using official services, but have also built up parallel institutions and services. An increasing sphere of action outside of the state system has rapidly developed since 1978 in all sectors: the official health structure became less and less accessible for reasons of cost and non-availability of services, and the population started to resort more and more either to private and often unlicenced clinics established by unemployed physicians and nurses, or to healers' services. The state has become absent, non-existent, and in this vacuum its nationalistic ideology of authenticity has been swallowed up (Young and Turner 1985). In the face of this entire collapse, people were no longer interested in hearing about cultural authenticity and primarily stressed the necessity of restructuring the national economy.

Such a restructuring was imposed by the World Monetary Fund (WMF) in 1978 to avoid the bankruptcy of the country. The rule of budgetary equilibrium of state expenses has been drastically imposed on administrators in all ministries; a very large proportion of the funds provided by external trade has been applied to the payment of the interests of the $ 6 billion national debt; and a neo-liberal economic frame has been imposed (MacGaffey 1987).

The remedy administered to the quasi-bankruptcy of the country has produced dramatic side-effects in the general population of Zaire. In order to balance the expenses of the state, 52, 000 persons working in primary and secondary schools have been fired, and about 6, 000 persons of the 18, 000 who were employed by universities and research institutes have also been pushed out; the implementation of the health plan has never occurred because there were no funds to create rural and urban health centers. A deterioration of the situation in all other domains has also been observed: roads have been abandoned, public hygiene and sanitation totally neglected... In such a context, there was evidently no time and no interest for discussing the legal reform of the art of healing; and gradually, people no longer felt concerned by the project of cultural identity expressed by the policy of authenticity.

In making our propositions to the Zairian government in 1978, we were assuming that traditional medicine was legitimated on two grounds: on a socio-cultural ground through its historical existence as the medicine of the ancestors, on a political ground through the policy of cultural authenticity. We had not foreseen at that moment that its legitimation in the near future would come more and more from the inescapable economic conditions in which people would have to live. The collapse of the official health delivery system led rapidly to the multiplication of small dispensaries run by untrained nurses, absent physicians or herbalist healers. Spiritualist churches in which patients are treated by the force of the healing 'word' have also multiplied since 1978: a conservative estimate indicates that 40% of the Kinshasa population resort to one or more of the thousand known healing churches offering services in the capital city.

The utilization patterns of healers have radically changed in the context of survival strategies: ordinary people tend to use what is cheap and among the services given by healers they prefer spiritualist prophets who are cheaper than ritualists while they continue to rely on herbalists.

The general conditions in which the associations of healers had to function have likewise produced deleterious effects on leaders and members of these associations. Leaders of these associations have not always resisted the temptation of abusing in collecting fees from members and selling 'permits of practice' of any kind. The three associations, respectively of herbalists, ritualists and spiritualists, which were founded in 1978, have multiplied through competition for leadership because these associations were seen by many as a potential source of income. A survey which I carried out in 1986 of a sample group of twenty healers clearly shows that they have many complaints about the abuses of their leaders; healers also complained about the fact that the agents of the state were constantly after them requiring money for one thing or another.

Radical changes in the national situation of Zaire between 1973 and 1988 give us materials for examining the changing status of traditional medicine from the contextual analysis point of view. The acceleration of the schismogenic process and the increase of competition between healers after 1978 become understandable only when we analyse behaviors and conceptions in reference to the general situation of deep crisis that characterizes post-1978 Zaire.

## Competing Schools of Chinese Medicine in Montréal

Every year, a few hundred acupuncturists are trained in Québec. Up until 1987, students received a diploma from one or another school offering courses in Chinese medicine, but there was no official recognition by the Ministry of Education and no licence was legally delivered by the Ministry of Justice. Training in Chinese medicine was dispensed in Montréal by three main schools, each with its own program, its own conception of the theoretical background necessary to practice acupuncture and its original philosophy of medicine; very strong competition exists amongst the leaders of these schools not only in their effort to attract students but also in their conceptions of what is orthodox in Chinese medicine. Among them, there is a continuous schismogenic process of mutual exclusion and disparagement. At least four other smaller schools have been founded as alternatives to the three dominant schools (Bibeau 1986).

Over the years, competition amongst leaders of the various schools remained strong because they were convinced that the government would ultimately recognize only one school as being entitled to train all students. However, in 1987 the Québec government chose neither to grant hegemony to one of the existing schools nor to endorse the continued coexistence of a broad range of approaches to acupuncture taught by a variety of schools. Instead, hegemony was in practice granted to junior colleges (CEGEP's) which are an integral part of the public education system. Although existing Chinese medicine schools are allowed to continue to carry out research and teaching activities, all future acupuncturists will be required to follow the collegial program or to successfully pass official exams.

In designing the collegial curriculum for training in acupuncture, the various schools of Chinese medicine were consulted; indeed, several professors were hired to teach within the CEGEP system. The forthcoming junior college course will be a patchwork made of bits and pieces borrowed from the curricula of each of the existing schools, while its conceptual frame of reference is likely to be a mishmash of the diverse and sometimes incompatible philosophical approaches characteristic of each school. Acupuncture is reduced to a mere technique, divorced from the cultural heritage and the original world view which make it a meaningful alternative therapeutic system. In all likelihood, quality of training will decline. One may even suspect that disqualification of this profession was one of the unavowed objectives of the legislator.

Fully licenced acupuncturists (those who have succeeded their exams) are only allowed to treat diseased persons under supervision from a doctor. A client must get a medical certificate from a physician before consulting an acupuncturist; after 10 treatments of 7 weeks, the client must again visit the physician before being allowed to go on with acupuncture. In addition, clients have to pay the acupuncturists while the state health insurance system permits universal and free access to physicians. Yet, what makes a physician competent to determine whether acupuncture is an appropriate treatment for a particular patient? Acupuncture has been transformed by the 1987 law into a second-line resource with a low status that remains entirely under the control of Western medicine.

The law concerning acupuncture runs counter to the principles expressed in a number of other laws and official policies of the Québec government. Chiropractics, for example, is now a recognized and fully autonomous therapeutic alternative in Québec, although it is not covered by the state health insurance system.

More fundamentally, one of the basic premises governing consumer protection is that consumers are capable of making decisions, so long as they are adequately informed of risks and advantages inherent to various options. In the case of acupuncture, however, exclusive authority is vested in physicians to decide in lieu and place of the client whether acupuncture is an appropriate form of therapy.

Finally, the law on acupuncture contradicts one of the basic tenets of the official policy regarding ethnic communities, namely protection and promotion of the distinctive cultures and languages of the various communities. Respect for diversity is intended to complement integration into the majority francophone Québec culture.

When we look at the legislation regarding the practice of acupuncture against the background of the cultural policies, one can easily identify basic contradictions which create a favorable context for schismogenesis.

### *Margins and Periphery in Context: the Case of two Societies*

Linguistics has taught for many years that words carry meaning only when they are listened to or read in a specific context; the pragmatics of communication has also demonstrated how the same words or sentences produce different meanings according to the context in which they are uttered. Although they constitute the basic elements of any language, words are meaningless by themselves; their informative and communicative capacity emerges only when they are enacted in actual discourses uttered by living persons in concrete situations.

A similar theoretical trend dominates contemporary social sciences: more and more anthropologists define culture not only as a system of meaning and signs, but also as a system of action and practice; they gain access to the meaning system in looking at its actualization in practical contexts. Post-Saussurian semiologists see the system of signs that characterizes a culture not only as a blueprint for specific series of actions and behaviors; they also explicitly invert the direction in moving from life to thought, from action to meaning, and from actual praxis to the culture as design for action. This new frame of interpretation forces anthropologists to center their analysis on the praxeological dimensions of culture: praxis, action and context become the keywords in this renewed theoretical approach. Culture is seen as a series of relations that produce meaning when they are actualized in specific contexts by real human beings. In order to explain the two different kinds of schismogenic processes at work in the Zairian and Québec cases, I will refer to a semantico-pragmatic framework.

#### *Bateson's rules of progressive differentiation*

All his life, Bateson sought to understand why adverse and antagonistic elements do not produce de facto the breakdown of cultures in which there exists oppositions and a series of competing structures. This led him to develop the concept of schismogenesis, and later to elaborate on the cybernetic principle of negative feedback.

Bateson's structuralism, inherited from biology, integrates as key notions the ideas of symmetry, segmentation, serial division and above all the notion of pattern. According to Bateson, all segments in a given society, such as moieties and lineages, form an interrelated system with various possible states of equilibrium and disequilibrium. The relational dynamics is made possible because there exist latent contradictions and paradoxes that produce movement and activity within the system. Two fundamental processes organize the functioning of the system: *symmetric differentiation* which is based on competition: A responds to the boastfulness of B by boasting; and *complementary differentiation* which is built on cooperation: A responds to the domination of B by submission. These two processes can lead ultimately to destruction of the culture unless there are built-in corrective factors that counter the trend to schismogenesis.

Within Iatmul society, the Naven ceremony plays this role of auto-correction. The Naven ceremony is performed by the maternal uncle and his nephew. The nephew behaves symmetrically in an excessive and arrogant way towards his uncle; the uncle responds with an exaggerated and caricatural relation of a complementary nature, disguising himself as a woman and prostrating himself in front of his nephew in a position of extreme submission (he goes so far as to mimic sexual relations with his nephew). Through a double caricatural inversion (men behave like women; women like men), the Naven reaffirms the uncle's superior status. In face of his uncle's extreme submission, the nephew is forced to read in a complementary way by more submission; therefore in caricaturing submission, the uncle reaffirms his own domination.

The Naven combines in one dramatic event two opposite structural truths that coexist in Iatmul society; the uncle-nephew relation is both symmetrical and complementary, and the Naven ceremony constitutes a paradoxical resolution of the opposition between symmetry and complementarity. This double inversion plays on the following dominant Iatmul social rules: a) the relations between men and women are complementary (domination-submission); b) the relations between men are symmetrical (domination-domination); c) the relations beetween lineages are symmetrical (the maternal uncle and the son of his sister belong to two lineages within the patrilineal Iatmul society); and d) the relations between the wau (uncle) and the laua (nephew) are symmetrical. The Naven permits to respond to an excess of symmetry at a given moment (when the adolescent nephew reaches manhood through excessive complementary behaviors; the caricatural inversion which is then enacted protects the socio-cultural system from destruction: the nephew is paradoxically invited to occupy his rightful position in relation to the maternal uncle, a symmetrical one.

The Naven ritual reveals two essential mechanisms that explain the functioning of social life in most societies; as an auto-corrective structure, the ritual permits at a certain point to put a stop to the destructive character of the contradictory systems upon which all cultures are necessarily built, though the ritual reaffirms in a parallel and complementary way the structural value of these basic contradictions for the life of the group. The reversal of the schismogenic process contributes directly to a stronger maintenance and development of the basic dialectical order in the culture.

In his 1958 epilogue to *Naven*, Bateson declares that anthropological analysis should always focus on the interplay amongst three domains: the 'eidetic' cultural premises which organize the prescribed rules of behavior and the cognitive dimension of culture; the affective 'ethological' system which articulates a specific language of emotions associated with particular social positions and special events; and the pragmatic and sociological context in which the eidetic and ethological dimensions are de facto enacted.

Bateson insists on the fact that it is precisely the pragmatic context that creates a dynamic movement within the combined eidetic and ethological systems. In *Steps to An Ecology of Mind*, Bateson writes explicitly: "In 1935, I certainly had not clearly grasped the central importance of 'context'. I thought that the processes of schismogenesis were important and nontrivial because in them I seemed to see evolution at work; if interaction between persons could undergo progressive qualitative change as intensity increased, then surely this could be the very stuff of cultural evolution" (1972: 155). According to Bateson, the anthropological explanation implies that the anthropologist makes explicit the 'pattern which connects' the cognitive, emotional and pragmatic elements at work at a given moment in a cultural setting.

Following this line of thought, I feel that we cannot understand why the Zairian competing healers' associations so dramatically exploded in number after 1978, if we do not take into account the general socio-economic crisis of Zaire as well as the changes in the cultural policy and the ambiguity in the legal provisions regarding the 'art of healing'. Healers' associations then began to compete in a caricatural way to the point that they finally emerged as the dominant structures in the vacuum created by the absence of official health services; complementary relations were replaced by symmetrical attitudes and behaviors among healers who claimed to have the same rights and duties as the dominant licenced physicians and nurses whose social roles were then declining. This double inversion was made possible because of the general context of the country: the dramatic living conditions of the population favored resort to healers rather than to physicians, and the ambiguity created by contradictions within the laws accentuated the process of inversion.

The caricatural situation that characterizes post-1978 Zaire can probably be interpreted as a sort of auto-corrective process which reaffirms on an ironic mode the dominance of modern values and Western models. In order to understand the new roles played by healers' associations and more generally by traditional medicine in Zaire after 1978, we should not only look at the general political, economic and legal context of the country; we should also closely examine the place assigned to the margins and to the periphery in a developing society like Zaire. Such a society undoubtedly permits a stronger expression of differences and assigns larger space to marginal movements.

The radical difference between the fate of the schools of Chinese medicine in Montréal and healers' associations in Zaire probably lies in the differential status assigned to marginality in the two societies. Like most other Western societies, Québec society tends to reaffirm in all domains the centrality of its founding eidetic principles: even in cases where tolerance of cultural differences is legally and politically prescribed, the dominant bodies of the society accept the inversion with difficulty and reaffirm as often as possible the prescriptive order. Such a reaffirmation of the structural and eidetic values may even lead to a caricature as in the case of the new Québec legislation which requires the total submission of acupuncturists to physicians.

We find exaggerations and caricatural behaviors in both kinds of societies but the actors differ in the two societies: in the Québec society, physicians exaggerate while in Zaire, the exaggeration is on the side of healers. To better understand these patterns, we will examine closely the ways each society relates the dominant cultural codes to marginal values.

### *Hierarchy, centrality and marginality*

In his studies on the place of mysticism in Western society and in Christianity, Michel de Certeau uses a theoretical framework that may be useful for discussing the problem of relations between the central dominant codes in a culture and its marginal elements. De Certeau interprets mystical discourse and practice as a way of eroding the founding principles of the historical system in which mysticism functions; he sees in the mystics a rupture and a point of separation that delimits a new space for thinking and behaving. The dominant codes of normal spirituality are deconstructed by the practices of mystic men and women who cross the limits and establish new codes in the margins.

To better explain the relation of mysticism to the central spiritual codes of Christianity, de Certeau examines two kinds of rupture, schism and heresy. Heresy is an act of exclusion that opposes 'inside' and 'outside': a new space is created by heretic groups whose life after the rupture is organized around new beliefs and alternative values. Schism, on the other hand, implies inclusion within the same space: schismatic groups compete with the dominant group while remaining within the same political and judicial order. According to de Certeau, the 'schismatic mode' of opposition within the Christian church replaced the 'heretic mode' during the XVIth and XVIIth centuries: this period was marked by fundamental ruptures (urban versus rural; Old and New World; Reformation versus papal power) that led to the permanent disappearence of a single system of reference. The differences became radically inscribed within the system, making the loss of an unique central code to which, in preceding eras, all groups had been required to adhere.

Leaders in Post-Reformation societies were forced to deal with the political management of differences, which proved so difficult that most 'princes', like the theologians, preferred to rely on the principle of 'cujus regio, cujus religio'. The schismatic mode became the dominant form of protest in European societies during the post-Reformation period, but this mode was once again slowly replaced by the heretic mode in a context where many claimed to be sole possessors of true belief. The spreading of nationalism in the Europe of the XVIIIth and XIXth centuries and the emergence of the individual subject during the last century (see Dumont 1983 and Foucault 1966 for substantive historical analysis on this topic) also contributed to the consolidation of one central system of beliefs within each nation, a system which was slowly but progressively eroded by a countervailing belief in individual rights.

In the Zairian context, the schismatic mode dominates and I am tempted to draw an analogy between the present situation of this country and that of European countries when they were constructing their respective nations during the XVIIIth and XIXth centuries. The central code to be imposed upon everybody is not yet very well defined and therefore the social differences are accepted as a contribution to the building of such a code; diversity is played out within an all-inclusive framework. Such a schismatic mode is evidently also favored by the absence of a powerful state (the state is literally non-present in most sectors) and by the drastic living conditions that compel people to invent survival strategies.

Corin (1988) has also built up a useful conceptual framework in her analysis of the relations between centrality and marginality: she argues that the difference between the position assigned to deviant, insane people and marginal groups in Western societies and that which they hold in more traditional African societies is due to the fact that these societies are built around a different conception of cultural integration. Building on extensive ethnographical data, she has demonstrated that the construction of identity among the Zairian Yans follows multiple structural lines which are hierarchically organized: within the context of a bilinear system (matrilineage and patrilaterality), people also relate differently to two positional structures in the patrilineage, the paternal figure and the alternate generations. She concludes from her analysis: "Various indices point to the preeminance of a matrilineal ideology which gives a coherence to various levels of data, discourses and behaviors. However, in certain specific circumstances, we have observed the preeminence of other principles associated with different formal characters and with specific connotations. Thus, it seems evident that the preeminence of secondary axes in specific areas produces something new or different. [...] One could say that what seemed at first glance to be a 'left-over' or a 'margin' inserts creativity, fecundity and actuality into the central society" (Corin 1988: 13).

On the basis of this study of identity formation and some other cases borrowed from traditional societies, Corin has elaborated a model for rethinking the center-margins relations: the fact that a structural hierarchy is deeply inscribed in these societies within the dominant cultural codes allows subordinated structures to acquire a central status in certain contexts. Traditional societies like the Zairian one seem to do better than Western societies in developing a dialectical relationship between core and marginal elements in the culture: instead of being seen as purely marginal and peripheral, the value of subordinate elements is gauged on the basis of their potential relation to the center.

"Conversely", writes Corin, "the anti-hierarchic orientation of modern societies should handicap this process; the marginal status of deviance remains disconnected from the core society" (1988: 21). The absence of dialectical interplay between the center and its margins in modern societies comes from the fact that their dominant cultural codes are not dialectically structured. From her study on the status of insane and deviant people in North American society, Corin has concluded that their position is generally 'outside', at the periphery of the system, in a situation of exclusion. She writes: "North American societies let the margins exist as margins, without allowing the establishment of a link which could sustain either clear exclusion mechanisms or reintegrative processes. [...] In North America, the extremely individualistic ideology could aggravate the problem and the radically marginal status of the margins of our society would mirror the impossibility or the refusal to relate differences, as well as marginal people, into a unitary social and representational frame" (1988: 19-20). In de Certeau's words, this refers to the exclusive model (you are in or out), while Bateson would instead speak of the dominance of complemetary differentiation: the society functions on the basis of an oppositional structure (domination-submission) which is controlled by the preeminence of the dominating center.

Corin opens new avenues when she proposes that we systematically examine the internal ordering of cultural codes in our quest for understanding the status of marginality in various human societies. Margins gain the power of relativizing central ideology and of questioning dominant codes only in societies in which the structural hierarchy or normative codes is inscribed at the very center of the culture. And Corin concludes: "One gets the impression that traditional societies, more than modern societies, have retained the ability to really play and interact with their margins; the center can be questioned or even modified by what is occuring at the periphery of the system" (1988: 21). The process may go as far as inverting margins and centers: "The idea of a potential subversion created by the inversion is so great that the inversion transforms central values into contingent possibilities within various possibilities" (Corin 1986: 10). (See diagram on page 189)

### 'Primus inter pares'

A hierarchical ordering implies the existence of a point of reference. However, the adoption by a group of such a point of reference is in most cases pureley conventional: the Roman patriarch claims primacy over other patriarchs on the grounds that he is the successor of Peter who was the first apostle; the first born son of royal couples is automatically regarded as the successor of his father to the throne; in Tibet, the monk chosen by spirits through a complex ritual is acknowledged

**Diagram: A contextual analysis of schismogenic processes**

| **Context** | **Zaire** | **Québec** |
|---|---|---|
| *General cultural orientation of the country* | - 'from the nationalist ideology of authenticity (after 1973) to the WMF economic rationality (after 1978)'<br>- predominance of problems related to the building of the nation<br>- the cultural diversity (structural hierarchy inscribed within the dominant codes (Corin) | - 'from nationalist Francophone ideology (in the 70s) to an emerging policy of cultural heterogeneity'<br>- predominance of problems associated with individual rights<br>- the dominant cultural codes are not dialectically structured (Corin) |
| *Legal and political orientation* | - a minimal number of laws<br>- coexistence of different local systems (ambiguity)<br>- total absence of the state | - a maximum number of laws and rules<br>- one integrated system (over-systematization<br>- abusive presence of the state |
| *Economic situation* | - coexistence of economic orientations<br>- quasi-bankruptcy of the country<br>- survival strategies of individuals and families | - neo-liberalism<br>- sufficient wages for the majority<br>- consumer rights |

| **Schismogenic Processes** | **Competition between Zairian Healers' Association after 1978** | **Competition between Montréal-based schools of Chinese Medicine** |
|---|---|---|
| *Relations between center and margins* | - the subordinated structures may require a central status because there is a *center-margins dialectization* (Corin)<br>- coexistence of parallel normative codes with multiple margins<br>- dominance of the *inclusive mode*: the margins may become alternative centers (de Certeau) | - deviancy remains outside because there is *no dialectization between the center and the margins* (Corin)<br>- existence of one powerful and normative central code with restricted margins<br>- dominance of the *exclusive mode*: either in or out (de Certeau) |
| *Process of differenciation, competition, opposition* | - prevalence of the *schismatic model*: the competition takes place within the same legal-political space (de certeau)<br>- dominance of *symmetrical differentiation* (Bateson)<br>- *double inversion* (Bateson): the margins can be transformed into centers, and centers into margins<br>- caricatural-excessive behaviors on the part of the dominated and dominant groups (Bateson) | - theoretical prevalence of the *heretic model* (de Certeau): heretic groups are not de facto excluded but submitted<br>- dominance of *complementary differentiation* (Bateson)<br>- *no inversion*: the prescriptive group absorbs the marginal group<br>- *caricatural-excessive behaviors* on the part of the dominant group (Bateson) |

by all as the new leading Lama; in democracies, the egalitarian status of all citizens is purely theoretical and differentiation is built around money, education, name of families... Even in cases in which it is assumed that all people are equal (pares), social mechanisms have been invented for identifying who is the first (primus).

The problem of hierarchical ordering that I have expressed through the Latin motto: 'Primus inter pares' is homologous to the much more difficult question of the relation between centrality and marginality in human societies. In order to understand the specific schismogenic processes at work within various human societies, researchers must identify, as in the case of hierarchical ordering, the points of reference which serve as a basis for structuring the relations between cultural centers and their margins. In this paper, I have tried to elucidate the referential systems which explain why competition between healers' associations in Zaire is structured according to the inclusive mode, the schismatic model, symmetrical differentiation and double inversion, while competition among the Chinese schools of medicine in Montréal tends instead to be characterized by the exclusive mode, the heretic model, complementary differentiation and the absence of inversion.

The Québec and Zairian societies both possess their own privileged forms of schismogenic processes. To be able to understand the specificity of the processes at work in each country, I have proposed to anchor them in the general cultural orientation of the country, its political and legal characteristics, and its economic situation. Such a contextual approach assumes that semantic elements must be combined in the model with pragmatics; the cognitive eidos and the affective ethos must also be articulated against the background of the sociological context. It is a conjunctive-disjunctive model which sees culture as a co-existent double system of meaning and action.

## References


BATESON G. 1958: *Naven: A Survey of the Problems Suggested by a Composite Picture of the Culture of a New Guinea Tribe Drawn from Three Points of View.* Stanford: Standford University Press (Second Edition with added "Epilogue 1958"; first ed. 1936).

-- 1972: *Steps to an Ecology of Mind.* New York: Ballantine Books.

BIBEAU G., CORIN E., MULINDA H.B., MABIALA M., MATUMONA M.M., MUKUNA K.M., NSIALA M.M., and NGUETE K. 1978: *Traditional Medicine in Zaire. Present and Potential Contribution to the Health Services.* Ottawa: International Development Research Center.

BIBEAU G. 1983: *Le Zaire et son développement social.* Montréal: Groupe Culture, Santé et Développement (roneotyped).

— 1984: Authenticité et ambiguité d'une implication dans un institut africain de recherche. *Anthropologie et Sociétés* 8, 3: 95-115.

-- 1985: From China to Africa: The Same Impossible Synthesis Between Traditional and Western Medicines. *Social Science and Medicine* 21(8): 973-943.

-- 1986: Des pratiques différenciées de la santé. In: DUFRESNE M. and DUMONT F.. (eds.), *Traité d'anthropologie médicale.* Montréal: Institut québécois de recherche sur la culture.

-- 1987: *A la fois d'ici et d'ailleurs: le communautés culturelles du Québec dans leurs rapports aux services sociaux et aux services de santé.* Québec: Publications du Québec.

CORIN E. 1986: Centralité des marges et dynamique des centres. *Anthropologie et Sociétés* 10, 2: 1-21.

-- 1988: *Meaning games at the margins: The Cultural Centrality of Subordinated Structures.* (Manuscript: Paper presented at the 12th International Congress of Anthropological and Ethnological Sciences, Zagreb, July 1988; Symposium "Between Semantics and Rationality").

DE CERTEAU M. 1982: *La fable mystique*. Paris: Gallimard.

DUMONT L. 1983: *Essais sur l'individualisme*. Paris: Le Seuil.

FOUCAULT M. 1966: *Les mots et les choses. Une archéologie des sciences humaines*. Paris: Gallimard (above all: Chapter 10).

MACGAFFEY J. 1987: *Entrepreneurs and Parasites. The Struggle for Indigenous Capitalism in Zaire*. Cambridge: Cambridge University Press.

NGUETE K., BIBEAU G. and CORIN E. 1979: Steps Towards a New System of Public Health in Zaire. In: AYOADE J. et al. (eds.) *African Therapeutic Systems*. Waltman, Mass.: Crossroads Press.

YOUNG C. and TURNER T. 1985: *The Rise and Decline of the Zairian State*. Madison: University of Wisconsin Press.

curare Anthropologies of Medicine Special Issue 7/91, 221-258

# Cultural Constructivism:
## Sickness Histories and the Understanding of Ethnomedicines beyond Critical Medical Anthropologies

Atwood D. Gaines

### Introduction

The present paper has two goals. First it attempts to identify and outline some central assumptions of 'cultural constructivism' in order to unify and provide a common basis for a variety of interpretive approaches in medical anthropology. In line with this goal, an example of the cultural constructivist approach to sickness and biomedical knowledge is presented. This model of sickness, called a 'Sickness History', is offered to demonstrate the historical basis of the construction and meaning of contemporary sickness realities, realities found in both folk and professional Western medicines.

The second goal of this paper is to outline the views of a congeries of approaches in American medical anthropology currently vying for the label of 'critical medical anthropology'. Those appearing under this rubric claim a unique ability to truly explain and understand ethnomedical organizations and sickness realities, especially those of professional, Western ethnomedicines. A review assesses the validity and utility of the various critical medical anthropological approaches.

My reading of critical medical anthropological work is itself highly critical. However, the term 'critical' here refers to my use of interpretive anthropology to deconstruct forms of empiricist anthropology, i.e., the critical medical anthropologies, analogous to the "critique of culture" (Marcus and Fischer 1986), provided by anthropology with reference to our own culture. My notion of critique is not to be confused with cultural criticism, another version of so-called critical theory. The review develops a typology of critical medical anthropologies, highlighting the distinctive views and contributions of cultural constructivism and shows the numerous problems and limitations of the traditional, conservative ideology represented by critical medical anthropologies.

It is shown how the latter views are grounded in assumptions and ideology which are products of a particular European time and social space and social group, i.e., 'ideological group' (Geertz 1983: 148) which are no longer tenable in modern anthropology or social science. In this sense, the paper suggests, as earlier argued (Gaines and Hahn 1985), that interpretive or constructivist medical anthropology can provide a critique not only of medicine but of (medical) anthropology itself. This unveiling of the problematics of some forms of 'modern thought' being but an overture to an ethnography thereof, as Geertz has called for (1983).

The presentation is divided into two major sections. In the first section, the critique of critical medical anthropologies is advanced. After delineating the various positions self-labeled as critical medical anthropology, the critique focusses on the problems of Marxist and political economic critical medical anthropologies. Weaknesses in metatheoretical assumptions, theory and data are highlighted. Following this, I present cultural constructivism and outline its assumptions. Contrasts between cultural constructivism and social constructionism are presented. This is followed by the presentation of constructivist conception of sickness, the 'Sickness History', a notion of the cultural historical semantic origin and development of folk and professional nosological entities. It is advanced to demonstrate the highly complex cultural historical nature of sickness realities as a contrast to the functionalist empiricism of critical medical anthropologies.

## I. Critical Medical Anthropologies

### *Preliminary considerations*

At recent meetings of the American Anthropological Association (e.g., December 1986, November 1987), there has been some tumult and shouting, some sound and fury, apparently it seems, signifying perhaps less than meets the eye. The events relate to the development of a congeries of approaches working under the label of critical medical anthropology. Morgan (1987), in a timely and important review from which I will draw here, indicates that most of what goes under the rubric of critical medical anthropology is one form or another of the political economy of health. As Farmer (1988) notes, members of this group are without a single agenda. And, as we shall see, the same authors float among the various perspectives without recognizing shifts and contradictions in their own views. For example Singer, Frankenberg, Navarro, Scheper-Hughes and Baer, among others have done work classifiable under several different rubrics.

Each of the several variations of critical medical anthropology has vied for the label and articulated its position(s) usually in direct opposition to 'meanings and symbols' approaches here developed as 'cultural constructivism'. It is, incidently, relevant in terms of an anthropology of science that such opposition has been marked largely by ad hominem attacks. Interestingly, we find this same response in biomedical sciences to disagreement (see Good, Good and Moradi 1985).

### *The problems of the term 'critical'*

It should be noted at the outset that advocates of the various critical medical anthropologies take this label toward some end. The term is clearly a pleonasm since the critical character of their perspectives is not in fact distinguishable from any good medical anthropological accounts, especially of a constructivist turn [1]. The use of the term strikes one, at least this one, as a parallel of such terms as 'The Moral Majority' in American religious politics. Both are terms which, simultaneously, seek to validate the in-group by wrapping it in a mantle of respectability and to denigrate the chosen out-groups, i.e., everyone else.

The term recalls the critical theory of European Marxist sociology, as Morgan (1987) notes, but seems to have forgotten that the Frankfurt school's intention was to be *self critical* (in terms of Marxist theories and constructs), to avoid dogma and to reestablish a neglected aspect of early Marx's writings; that is, an *idealist* concern for human consciousness as the agent of construction of the world (Jay 1973). Critical medical anthropologies, as we shall see, seem more in line with critical sociology whose name serves, according to one wag, to denigrate the West and other theorists through several forms of uncritical Marxism.

Three general political economic views are outlined in Morgan's review, each of which is subsumed under the label of critical medical anthropology by its proponents. The first of these are said to be the more or less "orthodox" Marxist approaches. However, since followers may adhere to one or more of the various persons of Marx ("the young/early Marx", "the mature/late Marx", etc.) or to any of a number of interpretations of one or more of those 'Marxes', the "orthodoxies" can be and are quite numerous (e.g., we note even an idealist version of an 'early' Marx above, from the Frankfurt School).

*Marxist political economies in medical anthropology*

Marxists insist that medicine be analyzed within the context of "capitalism", a putative entity which may or may not include "the modern world system" ("capitalist world system") from dependency theory, discussed below. "They (Marxists) draw attention to the interplay between social class formation and the power of the postindustrial nation-states" (Morgan 1987: 133). For these authors explanations are phrased in terms defined and set forth by Marx well over a hundred years ago (e.g., "commodities, class, class struggle, infrastructure, superstructure, ideology, exploitation, oppression, mystification, contradiction and revolution"). Notions of unilineal evolution also appear but well concealed. Medical knowledge is, expectedly, seen as "elite class ideology" (Navarro 1986, cited in Morgan 1987). Its function is to obscure the "real" cause of mortality and morbidity (capitalism) (Singer 1986, Taussig 1980, Waitzkin 1979, 1981b). Medical knowledge is here represented as ahistorical and as a conscious creation the functions of which are control and mystification.

For Taussig, disease signifies societal 'contradictions', said to exist in reality, not as constructs of the analyst's mind, which are expressed through the medium of the body. For Taussig and Scheper-Hughes, illness is potentially "revolutionary" (though Scheper-Hughes [1988] has argued that Biomedicine makes a folk illness of a 'biological' condition while others argue that, as an instrument of social control, medicine responds to illness by biologizing it. Biomedicine serves the "ideological needs of the social orders to the detriment of healing and our understanding of the social causes of misfortune" (Taussig 1980: 3).

All medical knowledge is mere ideology,and health status and the organization of health care are seen as direct results of "the capitalist socioeconomic formation" (Morgan 1987: 133). Since medicine is seen as an instrument of control and domination of the capitalist class, Singer suggests it should be labeled "capitalist" or "bourgeois medicine" (1986). The label Biomedicine, proposed by the early constructivist works and intended to designate a particular culturally constructed professional ethnomedicine (Gaines and Hahn 1982; Hahn and Gaines 1985), however is the term now almost universally employed even by Marxists (Frankenberg 1988). Below, we consider a case study of a Marxist clinician and test the notion of an exclusively biomedical conspiracy to biologize sickness.

Medicine is also seen as a commodity which is poorly distributed to the detriment of most (Waitzkin 1981b, Navarro 1976). While the views expressed are clearly ahistorical, we also see a point of view uninformed by understandings of the social and 'natural' ecology of disease. Lacking is an understanding of the role of population density and sedentary settlement, patterns of social interaction, dwelling types, waste disposal, food and shelter supply problems, animal and human interactions and a host of other factors (e.g., Dunn 1977) which demonstrably vary with urban culture, not economy. Another group uses some of the notions of this group but differs from it in significant ways.

*Cultural criticism*

The Ehrenreichs (1971) are most closely identified with the second group of political economic viewpoints, the "cultural critique of medicine". This group criticizes medical practices and organizations as iatrogenic and communalistic but sees such practices as caused by group interests and a system geared for profit but which needs reforming (1971: 1-28). Illich's famous book *Medical Nemesis* (1976), as well as the work of anthropologists Stebbins (1986) on the maldistribution of Mexican medical services and Justice (1986) on health programs in Nepal are examples of this approach. Illich lays the blame for sickness on industrial society and medicine, not on capitalism per se, we note.

However, this group appears misnamed. The researchers in this field actually have a poor or nonexistent grasp of a modern concept of culture, nor is their evidence of an understanding of Biomedicine as a "cultural system" (Gaines and Hahn 1985, Hahn and Kleinman 1983). As such, the group would be more appropriately referred to as the "group interest approach". The term cultural critique is best reserved for those recognizing cultural realities as cultural constructions (Marcus and Fischer 1986). It is also true that this group may be seen as utilizing some version of the archaic economic man theory found also in Marxism and elsewhere.

These writers direct their criticism toward reform and rational organization as unsatisfied consumers, and/or advocates thereof, within a society (also see Knowles 1977). Their criticisms are thus part of intercultural processes rather than the illumination of cultural processes, structures and forms. As such, I shall have little to say further about this group of writers beyond noting that their work can produce clues useful in constructivist investigations of medical knowledge, production and expression.

### *Dependency*

The third form of political economy in medical anthropology is the group of explicit or implicit dependency theorists who see illness as caused by poverty, underdevelopment and maldistribution, themselves caused by the "penetration" of capitalism into the world's "peripheral" countries. This notion uses Wallerstein's model of the Modern World System, defined as a world economy in "which ... a single political system does not exist over all, or virtually all, of the space" (1976: 230). A unified political entity would be a "world empire". "Capitalism as an economic mode is based on the fact that economic factors operate within an area larger than that which any political entity can totally control" giving capitalism a freedom to maneuver. Marxists may or may not adhere to this theory.

The ideas are renderings of Baran's earlier work (1957), *The Political Economy of Growth*. The central dynamic idea is that for one area to develop, another must be "dedeveloped" or "underdeveloped" (used as a verb). "Under such conditions, national and international social relations are determined by the expansion of capitalism and the intrusion of the (capitalist) world market into the periphery" (Morgan 1987: 136).

Further, the approach sees classes world-wide as things in themselves (*an sich*) (Wallerstein 1976: 233), and classifies "core areas" and states as opposed to and exploitiung "peripheral" and "semiperipheral" areas. The latter two terms are applied according to a perceived relationship to what is perceived as a core area of production. Marxists, while at first taken with the world system idea, later felt it was not Marxist enough and so it was dropped, e.g. Navarro (1986) (Morgan 1987: 134). This group (Baer, Singer, Johnson 1986, Baer 1986, Singer 1986, Elling 1977, Maclean 1986) is also noteworthy for their avowed political stance and their calls for various reforms typical of Marxists. They seek, after Marx, to convince us that their mission is to both understand and to change the world in spite of their own assertions that all is determined from above.

### *Hegemony*

Most recently a new development has been a concern for hegemony, the notion being taken from the Italian Marxist, Antonio Gramsci. Similar ideas had been expressed earlier (Frankenberg 1980) but recently these ideas have been grafted onto the work of Gramsci, a figure whose work was popular in the 1970's and 1980's in social science areas outside of medical anthropology but whose currency has declined there. Frankenberg's (1988) and Gramsci's notion of hegemony of Biomedicine ultimately rests on faith, for the formulation asserts that it is present and if one is not aware of it, its unremarkable character only demonstrates the efficacy of its domination.

Morsey (1988) also argues for a hegemonic view of Biomedicine. But her article argues contradictorily, that local level voluntarisms' development, of clinics and hospitals and which have succeeded in "power sharing" with the Egyptian state, is an example of biomedical hegemony because such facilities do not revive and install traditional Islamic medicine. Morsey has elsewhere argued for a form of postmodern political economy which while emphasizing Marxist interests also recognizes the existence and influence of culture (1990).

### *Critical theory*

A final form of critical medical anthropology is defined by Lock. It is, "one which pays attention to macro-structural questions, the role of power in social life, and the way in which Biomedicine is culturally constructed, in addition to the more familiar and traditional ethnomedical approaches". She says that it, "is indeed a formidable task. It cannot be undertaken satisfactorily without recourse to theories and knowledge acquired within the disciplines of history, economics and political science, in addition to those of sociology and anthropology" (1986: 110). This view is not political economy and is articulated in opposition to it and is referred simply to as a "critical approach".

Morgan, in her review, incorrectly assigns Taussig (1980) to this group. He most decidedly belongs to the Marxist group, as Young (1982) notes. The critical group is clearly closest to those that I would term constructivist. However, this view mixes empiricist cause and effect with interpretive approaches without a synthesis. For example, considering "macro-structural" questions such as power is not different than considering how Biomedicine is culturally constructed, since power and medicine are both culturally constructed.

Power may be seen not as an empirical entity but as a dynamic cultural construct. After Weber, we recognize that power is the ability to realize one's will in an action in which others are participating. But the realization of one's will is made possible by cultural conventions. The conventional nature of power is apparent when one considers that those with power in one circumstance have none in another or that those in power one day may the next be 'out of power', whether through elections, coups, purges or 'retirement' and 'reassignment', or, as we have seen over the last year (1989) the collapse of dictatorial communist states and their ruling parties.

Likewise, contrary to Foucault's body of work, knowledge is not power unless it is culturally construed as such. In Foucault's French context, power is possessed and wielded by one group over others. However, research indicates that physicians are hardly all-powerful or autonomous or in league with the state (e.g., Rhodes, n.d.). What we see in Foucault's work may be more a reflection of power and authority in a particular place, France, not in Biomedicine in general or some generalized modern industrial society (Gaines 1989). As Goldstein's (1987) excellent historical study shows, French professional psychiatric medicine was a creature of French history and social organization, especially that of the French state. The notion of medical power and authority is thus shown to be culturally specific and grounded in culture history [2].

In the present critical perspective of Lock, the body is seen as the locus of sickness which reflects social and economic conflicts. Sickness is seen as a metaphor expressing the phenomenology of the 'indignities of postindustrial life', oppression and/or exploitation (Scheper-Hughes and Lock 1987).

Illness is seen not so much in terms of political economy as in terms of culturally constructed conflicts in social relations and roles (ethnic, gender).

A problem here is that the view of illness as metaphor for contemporary problems, or the 'indignities' of work in modern industrial countries, while on the right interpretive track, produces synchronic, functionalist explanations and makes it difficult to localize, and thereby situate, explanations in real human contexts. Also, it is noteworthy that there are clearly more illnesses than there are problematic social statuses; and each status does not have a 'type' illness corresponding to it as this view would suggest.

## The Critique of Political Economy, Dependency Theory, Hegemony and Marxisms in Medical Anthropology

My principal criticisms, then, are directed toward dependency theorists, hegemonists and the Marxists who argue that they seek "to understand health issues in light of the larger political and economic forces that pattern human relationships, shape social behavior, and condition collective experience, including forces of institutional, national and global scale" (Singer 1986: 128).

Although this statement clearly shows the empiricist bias, the simple determinism, the materialism and the deductive nature of critical medical perspectives, writers outline their views in several different ways. Since I will focus my critique on them, I shall present their views largely in their own words.

Singer (1986) argues that world system and dependency theory paradigms have, "had a telling impact on the political economy of health literature. It is the starting point of both these theories that we cannot comprehend the contemporary world unless we understand global economic processes and class relationships brought into existence by the growth and spread of capitalism".

"Consequently, it has been the tendency of world system and dependency theorists to focus their attention on the macrolevel, analyzing how nations at the core of the capitalist system have methodically extracted wealth from, and thereby caused the underdevelopment of, peripheral areas" (1986: 128). As Baer puts it, critical medical anthropology seeks to "understand health related issues within the context of the class and imperialist relations inherent in the capitalist world-system" (Baer 1982: 1). Navarro (1976) also argues that Biomedicine is a creature of capitalism; an economic regime creates a medical system which in turn is an agent in its spread and the domination of others (Singer 1986) and which thereby creates the predisposition to mortality and morbidity (Waitzkin 1981a).

Waitzkin argues that medicine and health are so such creatures of capitalism that, "the Marxist viewpoint questions whether major improvements in the health system can occur without fundamental changes in the broad social order" (1981b: 333). Elling (1977), for example sees diseases as generated within and by the "capitalist world system". This view is, of course, remarkable only in its lack of accounting or recognizing such other influences such as ecological, genetic or those having to do with primate evolution and culture. As such, the perspective suggests that human diseases have appeared only recently in human experience, after the advent of capitalism. Also ignored are the rather obvious and dramatic improvements in health in capitalist countries over the last century (McKeown 1979).

The "microlevel" is distinguished and said to include, "the particular configuration of class, gender, and ethnic relationships, the availability of resources and technology, demographic and ecological factors, and the historic and cultural patterns (which) contribute to the short- and long-term effects of capitalist penetrations of health care, as well as to any micropopulation's ability to resist the agents, agencies, and agendas of biomedicine". "In short, critical anthropology *struggles* to synthesize the macrolevel understandings of the political economy of health with the microlevel sensitivity and awareness of *conventional* medical anthropology" (Singer 1986: 128, emphasis added).

Essentially, the task appears to be the humanization of Marxist thought through a synthesis with the 'sensitivity and (human-centered) awareness' of what is misstated as "conventional" medical anthropology. An important point here is that proponents recognize the non-, even anti-humanist trend of Marxist thought, something that has been quite clear in China, Russia and Eastern Europe for some decades. However, the problem of mechanistic views remains despite Cooter's (1982) suggestion that attempts to "mechanically and causally to relate medical ideas directly to social, political and economic circumstances" are (i.e., must be considered) moribund (p. 87). In the following, we consider the rhetoric, metatheoretical problems, theories and ethnographic support and the stance vis-à-vis change of the critical medical anthropologies.

### *A. Problems of rhetoric*

As opposed to what is claimed as "conventional" medical anthropology, it is easily demonstrated that the theories of critical medical anthropologies are themselves conventional and traditional and, in fact were such even in the 19th century. However, their proponents obscure this by significant distortions in their representations of opposed viewpoints, especially those of constructivists. As an example, Singer (1986) refers to the sensitivity and understanding found in "traditional" medical anthropology. In point of fact, he refers to the rich (or "thick") descriptions and interpretations of things biomedical supplied by constructivist research (there called "interpretive" or "symbolic"). This research perspective represents the revolution in social science constituted by the turn toward an interpretive social science (Geertz 1973, 1983; Rabinow and Sullivan 1977, 1989). Clearly then, the interpretive work cannot be appropriately referred to as "traditional medical anthropology" nor traditional social science.

Constructivist approaches represent a genuine radical (instead of rhetorically radical) shift in the nature, form, method and goals of social science itself. The interpretive turn in social science rethinks the nature of the scientific enterprise and redirects us from things-in-themselves and cause-effect to meanings and understandings outside ourselves and which reflect back upon our own constructed realities and, in fact, demonstrate both their constructedness and their plurality.

Partly because of the traditional, 19th century scientific basis of the schools of thought under the label of critical medical anthropology, its members are led into making (false) rhetorical statements about regularities and patterns. A second reason is their tendency to convey a stance of certitude and great knowledge about their subject(s) (see Geertz 1988). That is, their terms have largely rhetorical functions. As examples, one can use the terms "advanced capitalist societies" (Baer 1982, Singer 1986, Waitzkin 1981b), often used in conjunction with the phrase, "typical of". Since the nations of the United States or, for that matter, Germany or France, are unique, one cannot assert that particular features are "typical" of anything. As well, changes over time do not necessarily imply a direction or stage (e.g. "advanced") even in evolutionary biology. The implicit, and quite archaic, evolutionary view of Marx remains thus encoded in the views of critical medical anthropology.

The use of rhetorical methods is 'typically' employed by critical medical anthropologists. Other examples are noted elsewhere in this paper (e.g., Marxist social science referred to as a "struggle"). In my review I shall reverse things and appropriately employ terms (to critique critical medical anthropologies) which are often used by critical medical anthropologists, or Marxists before them, to criticize others.

*B. Problems of metatheory*

Utopianism

Under this heading can be considered assumptions that form the background of political economists and Marxist thought in medical anthropology. First, it needs to be emphasized that these political economic views of health appear guilty of naive utopianism. Navarro, Waitzkin, Elling and others argue that capitalism is the sole cause of sickness in the world and that without it, human life would be sickness-free. Such an odd position denies, as Morgan notes, any role of biology or ecology in disease etiology. But also ignored are the health implications of other economic and technical regimes. That is, in order to argue that capitalism has dire health conseuqences, one must show the data on health consequences of other regimes, subtract these from figures for capitalist societies and have a positive difference. Such a process would yield the differences in health consequences. However, no writer has ever tried this. Rather, they use a form of the centuries old "argument from design" in a slightly different guise; i.e., if there is illness and capitalism at the same time, the illness must be caused by capitalism. Thus, the assertion that capitalism causes an increase in mortality and morbidity remains unproven and indeed, even from a common sense point of view, without merit.

Epistemology

The epistemology of this group is highly problematic. Like Hume and others, this group makes two epistemological assumptions which, in light of modern cultural anthropology, are quite untenable. First, they persist in asserting a reality of ultimate truth and true knowledge. Second, they believe that true knowledge has but one source, science. Of course, with regard to the latter, Marxist critical medical anthropologists have a much narrower view of science than did philosophers or do contemporary scientists; they believe with other Marxists that only their truth is true and their knowledge real, all else false, "being mere ideology and expressions of "bourgeois class interests" (Hawthorne 1987: 1).

And, we note that while all others' actions are "determined" by capitalism, the same appears not to be the case for critical medical anthropologists (or for Marxists in general). In fact, their views disprove their own assertions of economic determinism of local ideology.

Contradictions

Political economists, e.g., Navarro (1976), Waitzkin (1981b) often maintain a flatly contradictory position. On the one hand, they argue that medicine under capitalism is deleterious, even evil. On the other hand, they decry the commoditization, organization, ideology, etc., of medicine and the barriers to access encountered by "peripheral" groups to biomedical technology and services. If the initial characterization of medicine as fundamentally deleterious to health were correct, then marginal status and/or powerlessness would indeed be their own reward in capitalist society.

Obscurantism

Political economists typically argue that ideology, including disease theories, are designed to obscure and mystify class relations and obscure the means of oppression and, for cultural critics, for social control. However, it seems clear that political economic research is itself profoundly obscurantist. It conceals and/or avoids the very real thoughts, experience and actions of people at local levels. It conceals the cultural and social construction of reality, whether of illness or of social categories.

Political economy accomplishes this by avoiding examinations of local level thought and action by remaining solely at an experience- and culturally-distant, putatively, "macro" level [3]. Thus, political economic studies, such as those cited here and in Morgan's detailed review, tell us nothing about the actual thoughts and actions of healers, patients or their significant others. After asserting a problem is caused by capitalism or biomedical hegemony, sentiment is then *imputed* to others such that actors views are portrayed as justifying and corroborating the original analysis (e.g., Frankenberg 1988, Scheper-Hughes 1988) [4].

Political economic and Marxist perspectives take the traditional tact of viewing medicine and medical knowledge as distinct from other social knowledge (Wright and Treacher 1982a, b). Medicine is taken to be an independent, autonomous enterprise set apart and opposed to society and culture (Navarro 1976, Baer, Singer, Johnson 1986, Turner 1987). Since, political economists believe medicine is distinct from society, social forces in the form of recent or remote history and current events in society and culture can not be seen as playing significant parts in their analyses or their understandings. This leaves us with a view of professional ethnomedicines that is thin indeed.

As well, the distinctiveness accorded Biomedicine actually *supports* the claims of the medical profession to a singular knowledge base and unique sociolegal and sociological status. This implicit advocacy of this view assists in obscuring and mystifying the cultural and social historical origins of professional ethnomedical knowledge and practice and the human agents (not impersonal system) responsible for them. The view of "system as cause" precludes meaningful discussion and development of research on ethics, a topic that should soon be enormous in the study of both professional ethnomedicine (Lieban 1990, Weisz 1990) and ethnopsychiatry (Gaines n.d.,d, Young 1990).

## *C. Problems of theory*

### Unacknowledged Theoretical Shift

Theorists (Baer, Singer, Navarro, Johnson, cited in Morgan, Waitzkin) have adopted and apply a market definition of capitalism rather than utilizing Marx's notion of capitalism as a mode of production which creates relations of production. Thus, the nature of the central explanatory concept is dramatically changed without a working out of the vast number of consequences of such a change. For example, the notion of class derives from the mode of production definition, but is problematic in the market model, yet theorists use class as an explanatory concept despite dropping the theoretical basis for its use as such.

### Ahistoricity and functionalism

Political economic theories are either ahistorical or have only a very shallow historical depth. While Morgan notes this great deficiency, she does not indicate an awareness that this omission derives from the functionalist nature of such theories. Additional criticism may be placed at the doorstep of the various political economic and Marxist perspectives in several domains. As Morgan puts it, a notion of "social dynamism is curiously absent from works of political economists of health" (1987: 131). Actually, given the inherently functionalist nature of such formulations, the absence is not curious, as is noted elsewhere in the present paper.

Problematic and archaic key theoretical conceptions

Several key formulations used by political economists are archaic. One central notion is that of class. The concept of class (and class relations) appears to be an expression of the same empiricism and ethnocentrism (or Eurocentrism as Dumont [1980] called it) which reifies a world-wide prime mover, "capitalism". The notions of class and determinative class relations are artifacts of 19th century European, i.e., English social classification, not scientific truths [5]. Researchers, though they call themselves "critical", would do well to critically examine such seemingly unproblematic notions, taking note of their culture history, as is being done with other key realities of professional ethnomedicines, e.g. nature (Gordon 1988), "race" (Brandt 1978, 1985, Gaines 1985a, 1986a,b, 1987a), emotion (Lutz 1985, Obeyesekere 1985), mind (Gaines 1989, Kirmayer 1988), biology (Gaines 1986b, 1987a), medical competence (DelVecchio Good 1985), errors (Bosk 1977, Paget 1988), and gender in medicine (Barrett and Roberts 1980, Marantz-Sanchez 1985).

Many of these conceptions, such as "race" and "class" and notions of gender, were clearly imported into medicine and social science from popular society. In using popular social classifications, it may be argued that political economists support the continuation of culturally based invidious distinctions by reifying them and presenting them as natural elements and foci of 'scientific' study.

It should also be noted that the notion of 'the state' as a universal evolutionary development is also highly suspect. The 'state' appears to be a cultural creation or an "invention" (Badie and Birnbaum 1984), not a result of some asserted social evolutionary process. This accounts for the fact that the French state is, in structure, organization and recruitment of personnel, radically different from that developed (and developing) in the United States (see Birnbaum 1981).

To Morgan's criticism, one can add that critical medical anthropology also renders Marx ahistorically, failing to situate him, his ideas or his concepts in the context of European culture history. Related to this timeless view of Marx are the contemporary images of Marx or Marxism which are almost purely ahistorical, as Morgan notes, but also largely functionalist in nature. It might be useful to suggest a view of Marx as ethnographer, albeit not an extraordinary one, of 19th century English society. One then would regard his concepts as local cultural constructions and processes.

Another point may be added here. While critical theorists claim that their key insight is the social cause of illness, e.g. as mental illness, we find that this is *precisely* the official professional view of the etiology of mental illnesses held by American psychiatry in the 19th century (Grob 1987). Thus, what is heralded as a revolutionary insight today is the 'elite' (psychiatric) ideology of a century ago.

The problem of communist and socialist regimes

The final, and perhaps the most important point, is the fact that the key explanatory tool, the "modern world capitalist system" is most problematic if not entirely fictitious. The concept accepted and used willy nilly by political economists does not note, let alone explain, the central fact that most of the people in the world live under communist or socialist regimes. (This appears to be an area of change in the near future.) For some unexplained reason, political economic theory does not posit a "modern communist world system" or a "modern socialist world system" or a "communist-socialist world system".

The world's socialist and communist states appear, at least in Marxist-influenced political economic theory in medical anthropology, to have not the slightest power or influence in the world. The recognition that most of the world's people are not controlled by a capitalist system or systems starkly highlights the fact that "modern world capitalist system" is a construct of some social theorists, not a global socioeconomic reality. In fact, the very existence of non-capitalist states disproves the assertions of critical medical anthropologists that 1) all is determined by larger capitalist economic forces and 2) all contemporary human disease is caused by capitalism (though it is unlikely that any non-Marxist needed any proof of the latter).

### *D. Ethnography and the problems of evidence*

To this point we have considered rhetoric, metatheoretical and theoretical problems with critical medical views. Below, we consider a number of challenges to critical medical anthropological views related to problems of ethnographic evidence. Morgan considers nine major deficits of political economic work in the domain of evidence. Political economic works may be shown to:

1. grossly simplify the complex nature of social interaction and social action.
2. fail to predict local level realities, i.e., hypotheses are not borne out by local level ethnographic research.
3. minimalize the influence of local institutions and actions, including resistance, and local history, on developments and organizations in the local context.
4. overlook the central importance of local level social and class relations.
5. ignore extant and evident medical pluralism in favor of (inaccurate) characterizations of biomedical hegemony.
6. ignore abundant evidence demonstrating that economic formations do not correlate with specific forms of medicine or disease.
7. deemphasize, if not totally deny, biology and ecology in disease etiology, stressing capitalism as the primary cause.
8. ignore improvements found in health status in locales where Biomedicine has been introduced.
9. ignore data which shows even the development of dependence is not infrequently beneficial to health status.

Criticisms in this domain actually allow us to focus on a single fact; political economic theories are not in the least predictive and despite strident claims, do not address anything approaching "determinants" of health and sickness. Ethnography does not demonstrate uniformly negative consequences of capitalism or Biomedicine nor does it show a unique etiological agency of capitalism. The notion of "rules" by which capitalism purportedly works espoused by critical theorists and which are said to give rise to mortality and morbidity appear unrelated to reality. The notion of "rules" itself seems little more than a holdover from early twentieth century logigal positivism (Rabinow and Sullivan 1979). Thus we find the reason that critical theorists ignore ethnographic data or trivialize its meaning. Their theories find confirmation only in self reference, not in ethnography.

It is not difficult to see that for many political economists the real *bête noire*, is actually America (see also Geertz 1984) to which all manner of evil power and influence, past and present, is attributed, a conceptualization which, while more articulate in its presentation, is conceptually analogous to former President Reagan's simplistic cartoon quality notion of the USSR as an "evil empire". However, if this is the case, another major difficulty with the position arises and which is not confronted by theorists. That is, as Gilpin (1987) has recently detailed, America cannot be regarded as the hegemonic power in today's world because of declines at home and increases elsewhere in Japan and Europe, although Gilpin argues one such power will be needed to maintain the stability of a future world economy that is highly interdependent. We may note also that America could not be considered the world power even as recently as 50 years ago, making the case for American dominance in world events and its role as cause of all human mortality and morbidity since industrialization extremely dubious.

In general, political economists, far from being critical, accept uncritically concepts from other times, other places and other disciplines ("class", "political economy", "world system", "capitalist world system") and convert them into immutable timeless universal realities and truths. In doing so they act as intellectual colonists imposing the "truth" of their Eurocentric views of the world on others.

*E. Applicability, relevance and the status quo*

Given that all is determined at the macro-level "in the last instance", local initiatives can count for naught in the alleviation of human suffering. Thus, social science education, clinical anthropological work, patient advocacy, etc., of the constructivist bent should not have a salutory effect on patient care or health care delivery. Aside from the fact that a considerable amount of clinical and other medical anthropological work demonstrates the falsity of this prediction and assertion, the political economic viewpoints can offer no improvement at the local level and, indeed, thus argue implicitly for leaving things as they are.

The stance or pose of change agent asserted by advocates, then, is entirely illusory; the position in fact advocates inactivity unless or until 'the revolution comes', since change short of that is irrelevant (for Marxists and political economists but not for cultural critics). In addition, the position makes it unlikely that change can or will be made at the local level for another reason; advocates cannot enlist the aid of the medical people concerned since they are predefined as "oppressors". But, since all is determined from above, there would be no reason to enlist the aid of locals to make changes since only 'changes in the social order' are seen as significant and these must result from vast economic changes.

Political economists fail to note change instituted by constructivist approaches and which may be categorized as falling into two distinct types.

A. *Direct influence*: research, and the various aspects of clinical work may directly alter the ideology and practice of Biomedicine.
B. *Mirror for Physicians*: Work that is not applied may be used in medical education of students or practicing professionals. As much, even purely theoretical, medical anthropology can act as a "mirror for physicians" literature in a fashion analogous to the "mirror for magistrates" literature in Medieval Europe.

Both applied and theoretical medical anthropology can and does affect medical theory and practice at both local and national levels. In this regard, it is noteworthy that the influence of constructivist research and teaching (e.g., Kleinman and his mentors' (Eisenberg, Engel), associates' and followers' work) may be seen in the changes affecting American medical education. Consideration of psycho-social and cultural aspects of illness and treatment are being expanded at the expense of biological study. This is a major change which was initiated at local levels. Thus constructivists, accused of overlooking the "real" determinants of sickness, may claim some responsibility for minor and major changes in medical practice, education and even health status.

Since political economic theorists are materialists and empiricists, as are biomedical advocates, they indirectly support the ideology of science which both share. Political economy supports the status quo of medical and social scientific methods of discourse. One means of both is the eschewing of the interpretive turn in social science.

Since their conservative stances implicitly advocates maintenance in science and society, they are but *rhetorical* advocates of change and improvement in human affairs. As well, political economic theories conceive of others in different-and-therefore-unequal terms. Cultural differences are dismissed as reflections of economy and development/evolution. Differences in beliefs and behaviors are thus seen ultimately as illegitimate, as results of some form of deprivation, underdevelopment or dedevelopment, not as distinct, valid forms of being in the world. Local cultural forms, unless they are rebellions, are thus seen ahistorically as ephemeral, inept and misguided actions (e.g., see Baer 1981 on enthusiastic Black religious experiences). This perspective does little but trivialize and render inconsequential the very essence of meaningful human action. The moral high ground claimed by critical medical anthropological advocates thus becomes awash.

The problems with political economic views as seen by this reviewer render it far beyond saving. Morgan states political economists "have much to learn from anthropology", especially cultural anthropology. However, this reviewer does not agree that such a learning process should or can lead to a "new" synthesis of anthropological and political economic approaches to health and illness (see also Lazarus 1988). Rather, given the very poor track record of the paradigm, its 19th century, Eurocentric, empiricist and anti-relativist characteristics (Geertz 1984), the perspective is evidently more ideology than old or new science. Thus, the salutory change would not be a synthesis, but a shift to interpretive paradigms.

### *F. The illogic of attribution in clinical Marxism*

Typical of the work of those espousing putatively critical medical anthropological views is the assertion that capitalism (or "American capitalism" or "modern capitalism" or "the modern world capitalist system") is to blame for the incidence and prevalence of any form of morbidity and mortality, as if, depending on the writer, other modes of production produce no health consequences. Agriculture, or industrial work are seen as having negative consequences for health only in the context of capitalism (e.g., Susser 1988, Scheder 1988, Waitzkin 1981, Navarro 1976). Related to this issue, Waitzkin (1981b: 346) writes of maldistribution and says that "people in the U.S. do not have access to even the simplest forms of care. Because of maldistribution, there are still individuals who suffer death or permanent disability every year". Of course, Waitzkin's utopian implications set one to wonder about a society where death and disability would be entirely absent. But, relevant here is the means he uses to prove his point. He considers three cases. They are short and can be presented here almost in their entirety and afford us a chance to see clearly the logic of attribution in Marxism. What we will see is purely impersonal, ahistorical, economic and biological reductionism in Marxist clinical practice.

The cases are of farm workers in the state of California. The first case is of an "11 year-old Chicana girl" (sic-gender indicated by final vowel) who cuts her feet on a rock in a fall. Though the wounds fester, the parents wait two days more after her feet become "painful, hot, red, and swollen (symptoms of inflammation with infection)" (1981b: 342-343). The final diagnosis showed acute glomerulonephritis that was a complication from a streptococcal skin infection. "The patient's wounds were cleaned and dressed with an antibiotic ointment. A course of penicillin, rest, and a nutritive diet were (sic) started" and follow-up showed recovery.

The second case is that of a boy with congenital heart disease. "B.C. is a 5 year-old Chicano boy (sic) whose grandmother brought him to the UFW (United Farm Workers in California) clinic because of poor appetite. The child had been delivered at home and had not been seen by a doctor previously.

No one in the family speaks English. Physical exam revealed a very small boy, with height and weight below the 3rd percentile. The child's fingernail beds and lips were slightly cyanotic; the fingernails showed clubbing. Examination of the heart showed a harsh grade V/VI systolic ejection murmer ... tentative diagnosis (later confirmed) was pulmonic stenosis with possible septal defect and possible pulmonary hypertension". Waitzkin concludes with this comment: "This child faced the probability of permanent functional deficit from heart and lung disease. If he had had access to a health worker by age 3, the defect probably could have been corrected without permanent serious deficit" (1981b: 347).

The final case is that of a 14 year-old boy, a member of another California migrant farm worker family. The boy had no immunizations or contact with health workers until age 9, when he developed a fever and subsequent nearly complete paralysis of both arms and the right leg. Details of contacts with other ill individuals at the time are sketchy. He is followed at the clinic for vague abdominal symptoms that are thought to be psychosomatic, probably deriving from upset that he cannot follow his peers into farm labor. This is concluded with the statement, that the, "incidence of polio has dropped markedly since the introduction of vaccines but has not yet been eradicated from the U.S. because of incomplete immunization" (1981b: 347).

As is apparent, the cases are presented in biophysiological terms only. Each is followed up with comments indicating that the case would not have occurred if it were not for "maldistribution" of resources which in turn is caused by capitalism. In point of fact, all cases show failed responsibility of families and that late intervention of the biomedical system, in the first two cases, actually *saves* the lives of the patients.

Waitzkin, in his failure to contextualize the cases, leaves unexplained and unexamined the conceptions, beliefs, logic, understanding or anything else, about the patient or his or her significant others. What were the perceptions of parents to a listless child of abnormal stature and clubbed fingers? What did they think was happening? Why was assistance not sought for these problems? We note the child was presented for treatment because of poor appetite, not because of the preexisting obvious physical problems. We have a clue that there may be a strong peer pressure to perform and hence there may be attempts to mask symptoms. Though this is noted, it remains unexplored.

We also see in his account an ahistorical view typical of biomedical sorts, as McKeown has noted ("an absence of any real interest among clinical teachers in the origin of disease, apart from its pathological and clinical manifestations" (1979: xi)). Waitzkin is so concerned to blame an external system that he completely ignores the one in front of him and in which people live and construct their world, including the reality or importance of physical symptoms. He does not mention, for example, the very well known fact that migrant workers are often illegal and or mistrustful of authorities. As a consequence, they *purposely* keep their children out of school and away from other institutional contexts in which their poor health otherwise would be noted.

Problems of seeking help late can and do occur in cities near medical facilities and among the poor *and* the well-to-do as in cases of drinking, STDs, as well as when one overlooks potentially serious symptoms. These are dependent upon local definitions and local knowledge of the situations, not global forces. If the illnesses above were caused by capitalism, how did the clinical Marxist intervene and blunt these all-powerful forces? The presentations in these cases are in biomedical and biomedical compatible voice only and are models of biomedical reductionism, consonant with the author's material/economic reductionism. There appears no place for a human story or human understanding as the scientistic stance dehumanizes the account in two ways, through the biomedical and Marxist gazes.

Waitzkin's approach shows us several of the many central failures of Marxist medical anthropology. First, because something untoward occurs in America or some other society labeled "capitalist" does not mean the event is caused or even related to economic organization. Second, and most critical, putatively critical medical anthropologists fail to make valid comparisons if they make them at all. Generally, the comparisons are made with implicit utopian standards of perfection, not to any known human reality. And finally, we see how the human dimension, in phenomenal terms and in terms of agency, is effaced in a quest for empirical, extra-human causes.

### *G. Problems of interpretation in critical medical anthropologies*

Singer, Baer, Navarro and others have noted, after McKeown (1979), that medicine cannot be credited with the great improvements in health over the last century or so. While they are correct, they fail to recognize that the decline in infectious diseases and the great increase in life expectancy and the decline of suffering from dental disease (McKeown 1979) must be attributed, according to their usual causal chain, to capitalism. It is in these countries so labeled by these writers where the greatest strides have been made. Instead, these authors point to capitalism as the cause of disease without taking into account the relative impact of different forms of disease in different social constellations. We may note here that ideas of the English hygienists and clean air groups of the 1800's were not based upon any notion of infection or contagion or bacteriologic theories, in fact many were anti-contagionists (see Cooter 1982) and so were outside of the medical establishment. However, they were yet most influential in the development of parks and sanitary conditions and alleviation of problems later recognized by Biomedicine as contributing to increased mortality and morbidity in urban populations.

Three points are important here. First, ideas such as these led to new social and political actions (i.e., local initiative, local changes). Second, these novel ideas, which improved urban living conditions, must be seen as creatures of capitalism. Third, the hygienists' ideas originated outside of medicine and later became a part of it. This suggests that a less than hegemonic reality of medicine, and a relationship of medicine to society (i.e., as part of it) not seen in Marxist works. Another example of this relationship appears in Johnson's work on physician impairment. He shows that this medical interest is a reflection of a society-wide interest originating in the human potential movement (1988).

In terms of the putative hegemony of Biomedicine in other contexts (Frankenberg 1980, 1988), it is clear that the "penetration of capitalism", does not eliminate other forms of medicine. It should also be stressed that the spread of aspects of the West's versions of capitalism and medicine are often cases of *borrowing* rather than imposition, as accounts of Biomedicine in foreign cultural contexts demonstrates this (see Janzen 1978, Lock 1981, 1985, 1988, Ohnuki-Tierney 1984, Weisberg and Long 1984), as Leslie had pointed out earlier (1980).

In such contexts, Biomedicine or capitalist medicine is in fact made into another option for health care, as in China (Kleinman 1980) and Africa (Janzen 1978) and or restructured according to local values of great historical depth, as in Japan (Ohnuki-Tierney 1984, Lock 1981). As Lock (1986: 99) says, "Medical practice and health seeking behavior in Japan, while superficially very similar to corresponding activities in societies which embrace Western-style capitalism, are, on closer inspection, grounded in an unfamiliar ideology such that the paradoxes, ambiguities, and fears associated with illness are cast in a rather different light." And Ohnuki-Tierney (1984) shows that while conceptions of health, illness, and cleanliness are couched in biomedical terms, they in fact refer to conceptions of purity and pollution that even predate the ancient established religions of Japan.

It can be argued that capitalism, like Biomedicine or Catholicism, is a set of historically generated cultural ideas which create social and economic forms. As it, or they, diffuse from one area to another, changes necessarily occur in local contexts. I would suggest that capitalism is not just an idea of private enterprise, but has a myriad of related cultural notions including those of self, time, affiliation, world mutability, locus of and self control, perfectibility, truth, efficacy, novelty and more. Hence, only aspects of it translate to other cultures. For this reason, we find *forms* of capitalistic activity, e.g., Confucian (Lock 1986), Burmese (Spiro 1970) or Gallic (Gaines, n.d. c) capitalism, not a hegemonic "world capitalism".

We see in this the primacy of local knowledge over putative world forces in the shaping of local realities. Evidence of this with respect to local constructions of Biomedicine is likewise quite overwhelming (e.g., Brandt 1985, Gaines and Hahn 1985, Gaines and Farmer 1986, Gaines 1979, 1982a, 1985b, 1986b, 1989, Gilman 1988, Good and DelVecchio Good 1981, Goldstein 1987, Lock 1981, Maretzki 1988, Maretzki and Seidler 1985, Norbeck and Lock 1988, Ohnuki-Tierney 1984, Pernick 1985, Weisberg and Long 1984, Young 1988). Given the emerging recognition of quite distinctive features or ideology, practice and the social organization of medicine in modern countries such as Germany, Canada, America, Japan, Mexico and France, there is little evidence for authors' insistence on a unitary world hegemonic Biomedicine (Frankenberg 1988, Morsey 1988).

In the next section, I will present cultural constructivism. The constructivist perspective takes culture history, meaning, human agency, human experience and responsibility as focal, not ephemeral, concerns. It seeks to locate contemporary illness experience in continuous cultural historical processes which serve to frame, interpret and give meaning to experience.

## II. Cultural Constructivism

### *On terminology*

The term I employ here, "cultural constructivism", has an analogue in sociology, i.e., "social constructionism" (Wright and Treacher 1982a). I suggest the alternate term here for several reasons. First, I am concerned to distinguish an anthropological enterprise from a sociological one, and to stress the important differences in key concepts in each discipline. For sociology, the key concept is society, seen generally, synchronically, while for anthropology, it is culture, necessarily viewed, I would argue, diachronically. The implications of this difference are worth emphasizing and suggest the appropriateness of a distinct term. This anthropological enterprise takes as its task the 'deconstruction' of social and cultural realities.

In the last decades, anthropological research has developed in ways which strikingly distinguish it from its 19th century empiricist, and culturally untutored, ancestors (Elkana 1981, Geertz 1973, 1984, Marcus and Fischer 1986). The distinctiveness has been generated by a concern for indepth, "thick" descriptions, and attempts to approximate "'the native's point of view' (Geertz 1983, after Malinowski) through the use of the ethnography of other people in other places. "... (I)f the locus of order and the source of modern anthropology's major intellectual contribution to scholarship were to be identified, it would be the ethnographic research process itself, bracketed by its two justifications. One is the capturing of cultural diversity ... and the other is a cultural critique of ourselves, often underplayed in the past, but having today a renewed potential for development" (Marcus and Fischer 1986: 20).

Such research replaced the grand, data- and "*verstehen*"-poor theorizing of 19th century anthropologists, sociologists and political economists of the chair, thus confronting us with the Other in detailed, rather than vague and confused, portraits. Implicit themes have also changed. Today, we are less enthralled with the idea of progress, "a synthesis of the past and a prophecy of the future" in

J.M. Bury's apt phrase, than were our intellectual ancestors. Inseparable from a unilineal notion of time (Nisbet 1980: 5), the idea was confounded with that of evolution (an idea which, in its social form, is dubious at best). Authors looked into the future to see what great stage was to come, or was already here, each brought into existence by sweeping natural forces beyond the will of humankind.

Anthropological enlightenment has come in two forms. First, social cultural anthropology furnished detailed accounts of the culturally different. Second, such accounts allowed reflection on ourselves and our own world. "In using portraits of other cultural patterns to reflect self-critically on our own ways, anthropology disrupts common sense and makes us reexamine our taken-for-granted assumptions" (Marcus and Fischer 1986: vii).

In this way, cultural constructivism (and much social constructionism) provides the basis for important critiques of ourselves and our culture, including its ethnomedical beliefs and practices. However, as I argue below, cultural critiques from a cultural constructivist perspective are quite distinct from critical medical anthropologies which, alas, are still true to their 19th century intellectual ancestors (see Geertz 1973, 1983, 1984, Marcus and Fischer 1986, Mendelsohn and Elkana 1981, Rabinow and Sullivan 1979, Sahlins 1976, Wright and Treacher 1982a, b).

What is required is a perspective on ethnomedicines cognizant of their expressive, cultural nature, for such systems are not autonomous, isolable cultural units (Elkana 1981). Rather, they are phases or expressive moments of historically derived cultural systems. They are but aspects of their respective, larger cultures-in-the-making; one window among many we may use to gaze into the house of culture.

I use the variations of the root term constructive (e.g., constructivist, constructivism) in the stead of those of construction (e.g., constructionist, constructionism) used in sociology to distinguish the anthropological emphases in the study of medical cultures. Anthropological accounts should be seen as distinct from sociology's social histories of (professional) ethnomedicine (e.g., Starr 1982; Brown 1979) or of professional medical specialties (e.g., Baer 1987), for such histories are frequently quite acultural (e.g., Starr 1982, Brown 1979, Wright and Reacher 1982, but see Fox 1989).

In its initial dictionary meaning, the term constructivism conveys the notion that such an approach, "serves to advance a good purpose and is helpful". The good purpose for which the constructivist approach is helpful is the pursuit of cultural understanding of uniquely human beliefs, experiences and practices in their cultural context. However, the term construct*ionism* in legal circles connotes an approach which is retrogressive and opposed to civil freedoms (in the guise of intellectual rigor). The legal constructionists ("strict constructionists") bring preconceived ideas and prejudices to a text and present these as "interpretations" said to be textually derived. Cultural constructivism, in contrast, takes a perspective which does not prejudge or assume an end result. Constructionists thus find only their preconceptions in new texts, a problem, that is found not only in the contemporary American juropolitical circles.

Another resonance in my usage derives by analogy from contemporary art, where constructivism refers to a movement utilizing (industrial) elements (glass, metal, etc.) "to create nonrepresentational ... objects". Cultural constructivists would use animate (people) and inanimate (archives, film, plastic arts) cultural material to deconstruct and construct cultural meaning systems, replete with inconsistencies, multiple values, and beliefs and assumptions sometimes in conflict and sometimes in concert, but always cognizant that what they interpret (aspects of cultural systems) and what they produce (accounts, ethnographies) are not things in themselves.

Finally, I find the term cultural constructivism suitable because it conveys the notion that human conceptions, experiences, as well as endeavors, are constructed by and through cultures, seen as historical processes, including histories of contact with other cultures. They are constructed and maintained in social processes and relations by means of negotiated definitions and redefinitions of life situations and events. Reality, the important, the good and the bad are those life aspects defined as such and maintained as such through human interactions wherein actors bring and exchange meanings and derive consensus, conflict or 'new' (variations of old) meanings.

The term constructivism allows the use of deconstruction as a definition of its enterprise which seeks to decode and reveal the underlying, fundamental constituent historical semantic elements of cultural processes including medical systems. I do not see this in Derridian terms wherein anything can mean anything, but rather as suggesting a notion of cultural systems with finite underlying folk theories (see M. Turner 1987) and assumptions, including ethnopsychologies, which generate surface symbols of words (e.g., metaphors, metonymies), acts, gestures and or events as people try to make sense of experience and to construct it. Cultural meanings are thus seen as finite and constrained, not infinite and indeterminate.

### *Constructivism and the summation of extant perspectives*

At the outset, it should be noted that my term "cultural constructivism" groups perspectives particularly relevant to medical anthropological research that are extant, but are now without a common designation. In this paper, I suggest a name and attempt to lay a coherent foundation for such research as well as to show some of its current limitations. I refer to work labeled "interpretive", "hermeneutic", "cultural/symbolic studies", "Kleinman's school", "the explanatory model (EM) approach", "meaning-centered", "semantic approach", "The Anthropology of Biomedicine".

Most recently, such perspectives have been referred to as "microlevel approaches" by some critical medical anthropologists (e.g., Singer 1986, Baer, Singer and Johnson 1986). These may all be subsumed under the cultural constructivist rubric, though some limitations are noted with respect to these approaches (or approaches called by these various terms) and we note that those who condemn so-called microlevel approaches sometimes engage in microlevel research (e.g., Waitzkin 1984) calling such a focus "contextual". I employ the term cultural constructivism more as a summarizing rubric for I intend it to group all of the other of these terms except for the last ("microlevel").

My primary objection to the term "microlevel" is its inapplicability. Culture, history, social classificatory systems, status systems, systems of purity and pollution, conceptions of person, theories of the body, of experience and suffering, ethnophysiology and ethnoanatomy, local health care systems, and systems of power are all key conceptions and foci for understanding and interpreting medical experiences. Such foci simply cannot be appropriately characterized as "EM" or as "microlevel" approaches (Kleinman 1981, 1986)..

While apprehended at the level of the clinic, the home, or other local area, which, one notes, are the only areas in which humans appear, individuals are always seen as representatives of extra-personal dimensions (e.g., culture, history and society). They are seen as isolates in the biomedical view and, apparently, the critical medical anthropological view. The outside is always and everywhere also inside. The distinction between the two amplified by critical medical anthropologists in fact suggests a particular, Western, view of self and a distinct, extra-human social order with its own 'laws'.

My second objection is, of course, the pejorative intent of the label. Critical medical anthropologists believe they are concerned with "broader", "larger" issues, for which read "more important". We note that they deal with them from local levels, usually precisely the same (the clinic, the library) as those they criticize. As Kleinman wrote some years ago, in what is still the most essential medical anthropological study, his model of health care systems emphasizes, "microscopic, internal clinical view, but the model I employ does not ignore the large-scale external factors that other models emphasize" (1980: 27). Since the intent is to provide a semiotic portrait of human sickness and healing, systems of meaning and knowledge are central, not ephemeral. The interpretive approach necessarily considers extraclinical material as a matter of course.

In this vein, I note the use of the term EM (Explanatory Model) theorist is also incorrect (Baer, Singer and Johnson 1986, Young 1982, Lazarus 1987). And, Lazarus' view that the term (from Hahn and Gaines [1985], "The Anthropology of Biomedicine" is "another term for an Explanatory Model approach" is simply wrong.

The authors referred to by the "EM" label employ forms of interpretive anthropology (semantic, symbolic, hermeneutic, cultural) to medical issues. However, as is, or should be widely known, Kleinman's (1980) Explanatory Model is but one instance of the interpretive approach; there it is employed with respect to particular illness episodes of individuals in a cultural context. It highlights the nature of symbolic interaction (i.e., the meanings brought to and exchanged in social interactions as per Mead, Blumer and others) in clinical domains to understand participants' thoughts and actions. The related notion of Good's Semantic Illness Network (Good 1977) specifies the meaning context of particular symptoms, the surrounding cultural ideology that is the source of the salience of sensations (which thereby become "symptoms") and their personal and group meaning(s).

The use of the appellation 'EM theorist' distorts the nature of the interpretive enterprise and makes its foci appear small and trivial and its concerns parochial. In fact the opposite is true. As interpretive social science, the approaches of constructivists call into question not only biomedical knowledge, practice and organization, but all the givens including "laws" and seemingly irreducible realities such as "biology", "disease", capitalism", "class" or "domination", "power", "economy", and other conceptions and constructs of traditional, 19th century medical and social sciences.

## Assumptions in Medicine and Medical Science

### *Assumptions of social sciences of medicine*

The traditional approaches to Western professional medicine, Biomedicine, including various Marxist and other sociological forms of inquiry, all see medicine as autonomous, as an independent, sovereign institution set apart from and, especially for Marxists, opposed to society (e.g., Baer 1986; Baer, Singer and Johnson 1986, Navarro 1975a, b; 1976, Taussig 1980, Waitzkin 1981a). Biomedicine in the view of many is seen as "having a unique scientific status which separates it from the influence of social forces" (Wright and Treacher 1982b: 5). Wright and Treacher enumerated four traditional perspectives or assumptions which they have identified in the sociology of Western professional medicine. These are:

1. Medical knowledge and or medicine is seen as unproblematic, as given in nature.
2. (Related to the first assumption) Medical knowledge is distinct from other social knowledge because it is built upon modern science and is, therefore, efficacious.
3. Diseases are naturally existing entities: they exist independent of their isolation and designation by medical science.
4. External social forces (since they are seen as distinct from medicine), may be ignored in the study of medicine or its development.

*Assumptions of Biomedicine*

Of primary interest to me is that Wright and Treacher are in fact indicating that sociological (and political economic) approaches employ some of the very same assumptions held by Biomedicine itself. AMishler, et al. (1981), Engel (1977) as well as others, have enumerated specific assumptions held by Biomedicine. These are:

1. Diseases are deviations from measurable biophysiological norms.
2. Diseases are universal.
3. Diseases are generic.
4. Medicine (practice and research) is an acultural, neutral scientific enterprise.

Although Mishler points to these as "silent assumptions" in Biomedicine, such a description is suitable for only three of the assumptions. The definition of disease (1) is actually explicit and articulated by members of Western medicine. The idea explicitly serves to organize their medical actions. At a deeper level, like that of purity and pollution in Japan, Gordon (1988) shows there is an unarticulated philosophy of "Naturalism". This philosophy underlies and forms the "background" for medical thought. As a consequence, "biomedical practitioners approach sickness as a natural phenomenon, legitimize and develop their knowledge using a naturalist method (scientific rationality) and see themselves as practicing on nature's human representative - the human body" (Gordon 1988: 24).

Fundamental assumptions of constructivist approaches need to be explicated here. These key assumptions are my first attempt at outlining a common basis for the several constructivist approaches and I present the list as somewhat preliminary. There are four key assumptions, as I see them, which researchers should keep in mind in the conduct of medical anthropological research in complex or simple societies.

## Assumptions of Cultural Constructivism

### *1. Ethnomedical Knowledge is Problematic*

The first assumption needing explication is that medical knowledge is to be seen as problematic. While perhaps assumed early on for folk medicines, research has recently clearly indicated the cultural bases of knowledge and practice in Western Biomedicines (note plural) as well. Therefore, these now may be termed (professional) ethnomedicines of Asia, Europe or North America (e.g., Brandt 1978, 1985, Kleinman and Good 1985, Freidson 1975, Gaines 1979, 1982a, b, 1985b, 1986b, 1987a, n.d., Gaines and Hahn 1982, Hahn and Gaines 1985, Good and Good 1981, Hahn and Kleinman 1983, Katz 1985, Lock 1981, 1985, Lock and Gordon 1988, Maretzki and Seidler 1985, Mendelsohn and Elkana 1981, Norbeck and Lock 1988, Ohnuki-Tierney 1984, Townsend 1978, Weisberg and Long 1984, Young 1977).

Though axiomatic for interpretive sorts, it should be noted that many medical anthropologists, such as those working in international health, in the ecological and epidemiological fields and in physical medical anthropology, maintain views of biomedical knowledge as unproblematic (see Greenwood et al. 1988b). And, not all ethnomedical researchers see Biomedicine as a "culturally constructed professional ethnomedicine" (Gaines 1982b, Hahn and Gaines 1985), for example, see Hughes (1968), and Foster and Anderson (1978).

However, cultural constructivism and social constructionism alike take as a central hypothesis that medical knowledge, including nosologies, theories, therapies, research practices, etc., is problematic and not given in nature, an entity which is itself a cultural construct (Sahlins 1976). No aspect of medical knowledge or practice should be seen as unproblematic or as a natural object existing independent of an ethnomedicine's isolation, description and labeling.

Ethnomedicines are, to praphrase Evans-Pritchard's post-functionalist view of society, 'moral, not natural systems' and thus are human creations. Aspects of medical enterprises are to be seen as culturally constructed and not necessitated in form or function by biology, nature, society or economy.

In addition to the problematics of professional biomedical theory, the role of extramedical cultural beliefs, such as various forms of communalism, for example, American forms of agism, sexism, racism, elitism, are to be expected within the theory and practice of ethnomedicine (see Barrett and Roberts 1980, Brandt 1978, 1985, DelVecchio Good 1985, Hahn and Gaines 1985, Harding and O'Barr 1987, Lock and Gordon 1988, Marantz-Sanchez 1985). Also central to understanding medical cultures are definitions of self and of medical practice (Cassedy 1984, Gaines 1979, 1985, Johnson 1985, Williams and Boulton 1988), pretraining assumptions (Gaines 1979, Lock 1985), distinct definitions of the situation occasioned by differences in cultural and or social affiliation (Gaines 1982a, 1985a, 1986b, Good et al. 1985; Gordon 1988), and the nature and character of disciplinary training (Gaines 1982b, 1985a, Kleinman and Good 1985).

Related to this notion of the problematics of medical knowledge, is the problematics of all knowledge including anthropological and other social scientific knowledge. As Young succinctly puts it, "an empiricist (empirical medical sociology) works in an epistemology-free social science. He supposes that his language and techniques, once they have been suitably refined, uncover facts about the world rather than produce them." "Social anthropology (and its application in medical studies) is a science in continuous pursuit of a satisfactory epistemology. What separates the anthropologist from the empiricist is that he regards his own concepts and ideas as simultaneously privileged *and* a part of a cultural system" (1982: 259-261).

With this in mind, medical anthropologists legitimately seek the forms and sources of knowledge in medical cultures as well as their historical roots and development, distribution, legitimation, validation and accountability. In short, constructivism is an anthropology of medical knowledge seen as a dimension and a moment of culture.

### *2. An ethnomedical system is an unfinished product of its culture's history*

A second central notion of the cultural constructivist approach holds that medicine, professional or lay, can neither be fully understood nor explained without reference to history. Specifically, medical systems are seen as the never-finished, historically derived products-under-construction. I suggest here a processual view of medical systems, including their roles, nosologies, therapies and the like, in the stead of synchronic, empiricist views.

The constructive process is to be found in the historically based interactions of members of a culture and of those who come into contact with them. Constructivism thus agrees with Evans-Pritchard in recognizing the kinship of anthropological and historical research. Most important here are historical studies of components of medical education, research and practice as well as the development of conceptions of particular diseases such as obesity (Ritenbaugh 1982), depression (Jackson 1985), and of particular medical specialties (Hahn 1982, 1985, 1987), nosologies (Menninger et al. 1963) as well as categories of thought operative in medical domains, as studied by Foucault (also see B. Turner 1987).

Ethnomedicines and sickness experiences remain opaque and incomprehensible to synchronic, empiricist approaches, especially approaches which efface the human element in search of 'larger' systems. How apparent changes occur and how manifest local variations exist, as well as the historicity of medical realities, are problematic for macrolevel approaches.

*3. Ethnomedical systems are cultural expressions*

The cultural constructivist position views medicine, or any other identifiable human enterprise, as an expression of the culture wherein it is located. This is, of course a view argued long ago by W.H.R. Rivers, though he was referring to isolated, autonomous, indigenous cultures.

Today, one sees massive culture contact making autonomous local groups a rarity. In the context of plural medical situations, one recognizes that imported elements do not obliterate local medical constructions. This would produce the improbable situation in which people thought and behaved in ways completely foreign and incomprehensible to themselves. As long ago as Spengler, researchers have noted that even imposed practices are transformed in their new locales. Such is clearly the case with the importation of ethnomedicines, including Biomedicines, with which one finds an integration with indigenous systems, some of which are themselves imports, such as in India (Leslie 1977a, b), China (Kleinman 1980, Leslie 1977b), Japan (Lock 1981, 1985, Ohnuki-Tierney 1984) and elsewhere.

Also noteworthy is the influence of European medicine on American medical organization and practice, for example, the influence of German Biomedicine on American internal medicine (Hahn 1985) and American psychiatry (Gaines 1989, Gilman 1988, Young 1988). One needs to be cognizant also of the adoption of a range of foreign health beliefs and practices in America in both professional and lay sectors (e.g., acupuncture, aspects of Chinese and Japanese herbal medicine, forms of meditation and foreign health diets and heterodox medicines). The latter instances suggest that the modern world is witness to considerable borrowing, i.e., diffusion, of beliefs and practices rather than a one-way expansion of a single acultural European/ American system, as some authors argue.

Although I do not see cultural constructivism as Kantian, it does recognize practical understanding in context and cannot be reduced to a closed system of categories such as "capitalist" or "bourgeois medicine", "ruling/oppressed classes", "modern capitalist world system", "core", "periphery" and "semiperiphery". These terms are used as empiricist explanatory realities especially in the works of Singer, Navarro, Waitzkin, Taussig and Baer (although the latter gives us a nice history of manipulative medicine in the US which pays attention to local cultures and actors' situational definitions [1986]). These terms, however, are defined only in terms of their asserted relationship(s) to one another and thus show the complete relativity of what is said to be real, objective differences. This problem is found with Marxian analyses outside of medical anthropology as well (see Rabinow and Sullivan 1979).

I might suggest also that the distinction between culture and society (Leslie 1977b) is problematic. Society (i.e., roles, social relationships, institutions) is but an expression of culture on the social plane. For this reason, different cultures, including "Western industrial capitalist" countries, in fact have very different social arrangements in their medical systems. For example, there are sharp differences in structure and organization of French Biomedicine as compared to American, including differing medical specialties (Letourmy 1985), and a different status system involving the hospital, which itself plays a distinctive role in French medicine (Pouvourville and Renaud 1985). As noted here and elsewhere, there are also different sickness categories (Dodier 1985, Gaines and Farmer 1986, Gaines 1986b) and relations among physicians (Baszanger 1985, Gaines n.d., a). German medicine also shows distinctions (Maretzki and Seidler 1985, Maretzki n.d.). And, of course, these Biomedicines as well as all others in the West, except American, are socialized medical systems. Thus, the asserted key relationship seen by Marxists and political economists between capitalism and a particular form and organization of medicine has no cross-cultural reality.

As Ohnuki-Terney pointed out, a discussion of early worldview and the system of social classification of people was important to include in her study of illness and professional medicine in Japan. "It is vital to understand that categories of thought operative in the medical domain are related to thought governing other domains of Japanese culture, and that these categories show historical continuity" (Ohnuki-Tierney 1984: 3). The import of local popular medical knowledge is clearly seen too in works focusing on Biomedicine in America (Hahn and Gaines 1985, Lock and Gordon 1988). And Worsley (1982) argues that medical conceptions are framed in and by metamedical, cultural concepts, a point made well by Gordon in her recent analysis of the philosophy of Naturalism in American biomedicine (1988) and noted in other early (Gaines 1979, 1982a, Hahn 1982) and recent interpretive work on Biomedicine (Gaines, n.d., d,e, Kirmayer 1988, Nuckolls, n.d.). The notion of the sickness history, presented below, builds upon this central insight.

### *4. Ethnomedicines concern human experience-near realities*

Analyses of lay and professional ethnomedical systems should see them in human terms; in terms of real human experiences such as pain, relief, frustration, loss, joy, anger, fear and the like. Analyses which seek explanations in experience- and culture-distant terms omit the critical factors in health and illness, the phenomenal persons and groups wherein human experience aand intersubjective (not subjective) realities are constructed. While medicine is thought to be objective and social sciences subjective, cultural constructivism seeks to find the intersubjective reality underlying both. Thus the perspective is not the antithesis of 'objective', empirical views; rather it takes them as its subjects of research as well. The constructivist approach does not lose sight of or efface the voices of patients, healers and sickness.

In light of this point it is appropriate to focus attention not only on patients, their experiences and understandings but those of healers. This focus does not predefine healers as instruments or agents of abstract systems whose views are mere ideology. Rather, a view of medicine must include human understanding and experience, and include "ethnographies of emotion" (Gaines 1987b) in professional, heterodox and folk medicine and of physicians, nurses and attendants alike (see Dwyer 1987, Paget 1988, Hahn and Gaines 1985, Lock and Gordon 1988).

Medical systems schould be seen as novel recombinations of existing ideological and practical elements of the local culture and, in some cases reworked elements from other cultures. Folk or professional medicines are not bodies of instrumental, acultural scientific knowledge with praxis based upon such knowledge. Rather, medicines are systems of knowledge and understanding, criteria for interpreting and constructing culture members' sickness experiences (see also Wright and Treacher 1982b, Hahn and Gaines 1985, Kleinman 1980, Young 1977, 1982, 1988). The study of medical systems becomes a study of knowledge-in-formulation and knowledge-in-use. Knowledge-in-formulation refers to the processes whereby knowledge is created, historically altered, distributed and disseminated. Knowledge-in-use refers to knowledge as it is expressed in lay and professional ethnomedical ideology and praxis.

Having briefly outlined what I see are key assumptions in cultural constructivism, I now turn to a consideration the notion of Sickness History as an application of cultural constructivism. It shows that sickness invokes issues, themes, cultural history and practices far beyond the micro-clinical context.

## Sickness History

The notion of "Sickness History" might be applied profitably in medical anthropological research of the constructivist turn. As I see it, the construct is related to Kleinman's EM (1980) and Good's Semantic Illness Network (1977; Good and Good 1982) and Young's (1982) notion of prototypes and chain complexes but is also distinct from all three. These formulations tend to place the definitional apparatus of illness episodes in the hands of an individual and within the temporal boundaries of a given illness episode or an individual's illness experiences. I argue that illness is an expression of conceptual categories which are not individual products nor are they products of contemporary interactions. Rather, illness realities are historical products and historically meaningful categories and frames of and for experience. Thus, knowledge and thinking about illness may be seen largely as cultural historical productions rather than as individual, contemporary ones.

When presented with a sickness entity, whether from a professional or folk ethnomedicine, one might profitably assume that it is not a naturally occurring entity, but rather one which has been created, and is still in the process of creation and recreation. That is, the presence of a labeled, or even unlabeled, sickness gives one a glimpse at a moment in a cultural historical process. However, some writers explain the presence of illness in terms of putative world economies, or simply capitalism (Singer 1986; Navarro 1976; Taussig 1980; Waitzkin 1986). They thus avoid addressing the meaning of illness to human sufferers, as Scheper-Hughes and Lock note (1986).

However, their own approach suggests illnesses are forms of everyday rebellions against the "indignities of life and work under conditions of advanced industrial capitalism" (p. 139) which Marx and Parsons are supposed to have "understood well". (We will leave aside the vexing question of Marx's knowledge of something called "advanced industrial capitalist life"). This approach, however, reduces the meaning of illness to complaints about current life situations, and views people as Durkheimian mirrors of social structure or structural conflicts. Though attention is directly paid to the meaning and experience of illness in this view, as in the case of Kleinman, Young and to a certain extent Good, the perspective is too synchronic and, hence, too 'thin'.

A different approach might be suggested wherein sicknesses are viewed as historically created vessels of meaning. An understanding of a sickness designated in a culture, whether by folk or professional medicine, would need to attend to the cultural history of that sickness. Sicknesses should be seen as culturally constituted, historically derived modes of expression. Thus, while I agree that sickness episodes express something, I would argue that they symbolize important cultural historical meaning rather than a social structural or economic present.

Such a recognition allows us to understand the very different sickness episodes of people in different modern industrial societies, e.g., Japan, Taiwan, France, Germany, Canada and the like, for while they are all modern, capitalist, industrial, hierarchical, etc., they are also quite different from one another and exhibit distinct illness realities, beliefs, healing strategies and medical organizations. In what follows, I will look at just two forms of sickness known in the professional, popular and folk sectors of the French health care system.

### *Fatigué*

One disorder is labeled simply *fatigué*, used as an adjective and expressed as, "je suis fatigué" ("I am tired" or "fatigued") or "je me sens fatigué" ("I feel tired"). It is quite widespread in France. The expression refers to a state of being, an illness state, and may be contrasted with another experience of la fatigue, which expresses a sense of lassitude caused by work that is too difficult or too long (e.g., "Malgré la fatigue j' écris" - I write despite fatigue).

In a recent article on sickness and its moral justification in the French workplace (Dodier 1985), 'being tired/fatigued' was shown to be a common sickness. In fact, it is one of the most common disorders experiences by both blue- and white-collar workers, a distribution contradicting the illness-equals-structural-position notion.

Diagnosis, treatment and management of this condition by physicians in France is usual, expected and appropriate. Leaves of absence for "rest" ranging from a matter of days to a month or more are prescribed by physicians. The etiology of the disorder is not overwork or lack of a few nights' sleep, both causes of la fatigue (noun). Rather, the etiology of fatigué (adjective) is a perceived burdensome life situation. This condition may appear at any time and does not necessarily derive from an unusually stressful of difficult period in one's life.

The reason for this is that all aspects of phases of life are seen in France as difficult, burdensome and taxing. One's day to day life, seen as a constant "struggle", periodically simply "weighs one down". In France, it is deemed appropriate to have a view of life as tragic or sad. To view life otherwise is an indication of shallowness and a lack of "seriousness", (a Franch notion approximating, but not the same as, that of 'maturity' in America) (Gaines 1982a, Gaines and Farmer 1986; Wylie 1957). The same notions of life as tragic and maturity are to be found in other cultural groups of the Mediterranean tradition (for which see Peristiany 1966; Gaines and Farmer 1986; Gilmore 1980). These include Iran (see Good 1977; Good, Good and Moradi 1985), Italy (Cornelisen 1971) and Greece (Campbell 1964; Lee 1959) and in the New World, Latin ethnic groups which are daughters of the Mediterranean tradition. All have named illnesses marked by sadness (e.g., susto, pena, saladera, tristitia).

Medical leaves for 'being tired' routinely are viewed by employers, colleagues and physicians as appropriate for this recognized, morally sanctioned sickness. Illness and disease thus coincide, though in my perspective, one is the other (also see Obeyesekere [1985] on this point). We note that the notion of tiredness as a common, morally justified, medically recognized and treated diesease as well as a folk illness entity was entirely unproblematic to the French author (Dodier 1985), though such would clearly be problematic to both American clinicians, researchers and lay readers. For them, tiredness is a *symptom* of a number of disorders; it does not itself constitute a disorder.

The disorder appears in the form of a continuum. At one end is "fatigué" (tired), "un peu fatigué" (a bit tired) or "bien fatigué" (very tired). The phrases, which indicate severity of problems on a physical plane also suggest a low emotional tide. It is important to note that the sickness presents with particular behavioral signs such as listlessness, incautiousness and some physical signs, a drawn look, even "pimples" (Dodier 1985: 123), recognized by others as indicative of "being tired". Those who exhibit the signs are seen as a bit brave, even courageous, if they continue to work. Empathy and concern are voiced and, as Dodier also found, the recipients feel these are justified and appropriate. Our second illness entity, presented below, is a bit more complicated.

### *Triste/Fatigué Tout le Temps*

The second disorder I want to discuss is "triste (or fatigué) tout le temps" ('sad (or tired) all the time') (see Gaines and Farmer (1986) and Gaines (1986a) for fuller descriptions and a case study). This disorder is a chronic version of 'being tired' but differs in that it generally appears to have a precipitating event. As well, the condition has two names which are used quite interchangeably. Informants queried in Paris and Strasbourg (1982, 1983) all knew of the disorder, could describe it and its various possible symptoms and could relate a tale of one or more people personally known to them who had been afflicted by the disease/ illness.

Most interesting are the two faces of the sickness. On the one hand, one sees a physical problem, tiredness. On the other, one sees emotional difficulty, especially sadness. The symptoms of this French disorder are many and various and include chronic disturbances of vegetative signs, such as sleep and appetite, as well as problems of concentration and memory, sad affect, often psychomotor retardation, chronic tiredness and weakness, distractability, emotional lability, and other, but again nonspecific, multiple symptoms.

"Sad (or tired) all the time" is a syndrome with a widely varying symptom picture. Its origins are a reaction to some form of perceived traumatic event such as a miscarriage, a divorce, or loss due to the death of achild, spouse or close friend (Gaines 1986a, Gaines and Farmer 1986) or the termination of an important relationship. Informants in Paris and Strasbourg also suggested that the precipitating event may not be a loss but a failure to achieve some important personal goal, such as school entrance or a loss in a personally important sporting endeavor. The disorder is commonly depicted in films and novels and is seen as quite normal, i.e., not unusual or remarkable enough to explain or dwell upon.

An attempt to apply American psychiatric nomenclature to this condition would meet with failure, though it is in this domain of medicine that American diagnosticians would be led immediately by the presentation of the symptoms often associated with this disorder. Taking the descriptive approach championed by the 3rd Diagnostic and Statistical Manual of the American Psychiatric Association (1980), or its revision DSM III-R (1987), one could construct a label for the disorder, i.e., "chronic, reactive depression".

However, in American psychiatric theory and practice such an entity is both unkown and nonsensical as psychiatrist colleagues (at Duke Medical Center in 1983) informed me, "assuming a normal premorbid personality". As well, individuals afflicted by the illness were not unable to care for self and others. Though psychodiagnostic instruments indicated that several persons, locally referred to as "saints", with *triste tout le temps* suffered from severe clinical depression, ethnographic research and the saints' activity levels easily disconfirmed the diagnosis (see Gaines and Farmer 1986, for a detailed case history and psychodiagnostic test results). It is very important to note the clear relationship of suffering and a semi-sacred status in this disorder but it is also present in the case of fatigué.

Though this problem is recognized, active interventions to alleviate the condition appear to informants as inappropriate. Informants see the condition as understandable and as a reasonable response to loss or misfortune. In this, the French show the cultural basis of the logic that sickness conditions necessitate actions to eliminate them (Parsons 1951). This sociological perspective misses the cultural validity and value of illness in some cultures, such as those of the Mediterranean and its daughter cultures in the New World.

The primary response to the condition, as informants stated, is to allow rest and a reduction of expectations for the afflicted for months, years or even a lifetime. An attitudinal response of commiseration, support and indulgence from significant others is also appropriate. No cure is expected, just a modicum of relief from episodic exacerbations of the basic chronic picture. Thus, a reaction to a single loss or frustration perceived as traumatic produces a chronic sickness affecting mood, affect and cognition as well as behavior.

While manifested in the soma, these two disorders have neither a somatic seat nor etiology in American biomedical terms. Each sickness in its own way reflects culturally specific French views of the self (Gaines 1982a), the nature and the positive meaning of suffering in life (Cornelisen 1971, Gaines and Farmer 1986, Good, Good and Moradi 1985). The latter is commonly manifested as a culturally constituted "rhetoric of complaint", an aspect of local discourse which communicates the suffering and misfortune and, therefore, the worthiness of the speaker (Gaines 1982a).

The illnesses also demonstrate the patterning of dysphoric affect in France and in the Mediterranean tradition more generally (see Gaines 1982a, Gaines and Farmer 1986, and Good, Good and Moradi 1985, for fuller accounts of the patterning of dysphoric affect in this tradition). But how did these two forms of sickness appear? Why are these forms used in France today? The sickness history helps explain these questions which are not well answered by other notions of sickness including EMs and SINs, or by synchronic functionalist approaches.

*The past in the present*

Could the two disorders be related and share a common origin? A look at the history of the notions of "acedia", "tristitia" and "melancholie", forms of what Jackson (1985) calls "dejected states", may allow an understanding of them. And, we note the other writers have begun to show the role of theology and philosophy in the shaping of the experience of dysphoric affect and depressive illness across cultures (Kleinman and Good 1985, especially see Lutz 1985, Obeyesekere 1985).

In Jackson's work, he finds three forms of dejected states in the Latin tradition. These are "acedia", "melancholia", and "tristitia", all of which "belong to the history of religion, in particular to the religious scheme of the cardinal sins in Christianity" (1985: 44). The condition of acedia, a sin, developed out of the examination of the particular experiences of members of the desert anchorite community near Alexandria in the fourth century A.D. The isolation and inactivity produced mental states which detracted from contemplation. The unusual mental states of acedia soon acquired behavioral referents. Shifts in emphasis from one pole (mental state) to the other (behavioral modes) have occurred through the centuries. On the one hand, acedia was described as "carelessness", "weariness", "exhaustion", "negligence" (in contemplative duties), and on the other as "apathy", "anguish", "sadness" and "low spirits" (Jackson 1985: 45).

The concern for acedia, originally seen only negatively as a sin, changed and it acquired a positive meaning as well. This derived from the view that the condition of sadness and weariness were a result of one's "penitence for sin" and or one's "desire for perfection" (Jackson 1985: 48). These notions spread widely in Christendom beyond the clergy to the laity between 1200 and 1450, especially after the Lateran Council (1215-16) decided on the need to extensively disseminate penitential literature, clerical manuals (for the newly required sermonizing) and catechetical handbooks (Jackson 1985: 48, Wenzel 1960).

The penitential literature spread the benevolent view of acedia and "tristitia", a term used interchangeably with acedia. In this view acedia, since it was a condition which afflicted only the devout penitent, seems to have become thought of as penance itself (Jackson 1985, Wenzel 1960) but also perhaps for itself; dejection as its own reward. Penance was seen "as medicine for the soul" (McNeill and Gamer 1938: 44). The most common governing medical principle of the time held that contraries cured contraries (Jackson 1985, McNeill and Gamer 1938) (one is reminded of Homeopathy's dictum, "like cures like").

Jackson points out how acedia and tristitia gradually lost their theological affinities and the positive and negative connotations became part of the culture area's secular/philosophical theories of emotions. They later appeared in the context of medicine. The component of positive sorrow "became more identified with Christian tradition of the sufferer as an object for care, concern and cure" (Jackson 1985: 53-54). The conception developed further into "melancholie" as the Church lost its explanatory powers over human behavior and there developed a renewed interest in classical writers. This renaissance brought forth melancholie as both a medical term for an illness (Menninger 1963) and as a popular term referring to sadness (Jackson 1985). The popular term referring to sadness also developed an association (from Aristotle) of the later term "melancholia" with the very bright and the gifted; the two were seen as dispositional companions. It is possible to see this association in the last century in France relative to the experience of tuberculosis and the romantic vision of the pale, wan tubercular intellectual (see Sontag 1977) (see Gaines and Farmer 1986, for the relevance of hagiography).

This history, brief as it is, provides us with some understanding of the meaning of dejected states and dysphoria in French culture and history. The meaning of such states, their positive value and their cultural significance make us understand better the existence and elaboration of related sickness conditions in French popular culture and in that culture's professional ethnomedicine which is its expression.

It seems evident that acedia and tristitia are the source of fatigué and triste/fatigué tout le temps and probably "spasmophilie", a disorder of concern in popular and professional sectors since the late 1970's. The two faces of acedia are well preserved; the mental face (triste) and the behavioral/physical face (fatigué) in the contemporary disorder. Spasmophilie and fatigué may also represent the Janus nature of the ancient sickness of acedia. And it is noteworthy that elsewhere in the Mediterranean and Latin traditions one finds tristitia as well as other disorders indicating inactivity and sadness, e.g., pena (Tousignant 1984) and susto (Rubel 1977) among others, as well as named modes of expression thereof (e.g. corredos, ghinnáwa.

Here we see the historical development of folk theories which appear in a Biomedicine. Other, less historical works have also shown the presence of folk theories in professional ethnomedicine in America (e.g., Brandt 1978, 1985, Cassedy 1984, Gaines 1979, 1982a, b, 1985a, 1986b, 1987a; Gaines and Hahn 1982, Hahn and Gaines 1985), Germany (Maretzki n.d., 1988; Townsend 1978), England (Helman 1985), Canada (Lock 1985, Katz 1985, n.d.) and Africa (Swartz n.d.).

As well, the impress of culture on Biomedicine also has been noted in Asia (Lock 1981, Reynolds 1979, Weisberg and Long 1984). It is suggested that such differences have to do with culture history in the form of sickness histories. As a culture maintains continuity with its past, it would 'import', so to speak, its past into its present Biomedicine and thereby make it vary from another's (unless it was completely superimposed thereupon (see Kornfield 1986)). Diseases, as elements of medical knowledge, may be seen thus as products of sickness history, not as products of current hypothesized material forces.

## *Sickness as charter*

A culture's history contains various sickness histories which serve as charters for subsequent personal and group interpretations of and responses to experience. Their historical meaning makes them uniquely realistic and significant. The Sickness History, then, is a culture's historical experience with a particular form of distress which frames both folk and lay understandings and responses to contemporary experiences and concerns. I intend the term to encompass sicknesses included in epidemiology as well, such as the plague in Europe. The experience of the plague still colors contemporary perceptions of epidemics. Likewise, the American response to AIDS is grounded in its cultural historical experience with Sexually Transmitted Diseases and how, then and now, STDs are interpreted in the context of peculiar American forms of communalism (see Brandt 1978, 1985; Gaines 1986a). Thus contemporary sickness episodes may be seen to have deeper historical roots than synchronic approaches can appreciate and bear meanings and significance which critical medical approaches cannot approach.

### *Other sickness histories*

In terms of other sickness histories, one might inquire into the meaning of "crise de foie" the most well-known of French disorders. And here I note that some 15% of all biomedical diagnoses made in France are not shared with other Western countries. Likewise, one might inquire into "dropped stomach" in Asian cultures which is diagnosed and treated in Asian Biomedicines as was the condition of "floating kidneys" in America until the late 1960's. Shinkeishitsu's history could also be unraveled although Reynolds (1976) has provided us with excellent contextual material. But one should not omit the history of seemingly unproblematic diseases, i.e., those said to be natural and universal, such as diabetes, asthma (Wright and Treacher 1982) and 'genetic' disorders which American medicine tries to construct as "racial" (e.g., sickle cell anemia, cystic fibrosis, Tay-Sachs, etc.). This is a part of an unconscious cultural (and racist) effort to (re)construct "race" as a biological reality instead of the nonbiological, cultural system of classification that it is (Brandt 1985; Gaines 1985a, 1986b, 1987a, n.d., b). This notion appears derived from German culture, medicine and science (Gaines n.d., c, 1989; Gilman 1988).

Another problem for researchers was hinted at above. That is, the conception of illness as negative. Sicknesses can be assertions of the positive. Seeing sickness as always the result of contemporary class relations may simply express Western views of normality and abnormality in implicit forms of deprivation theory (e.g., Navarro 1976; Waitzkin 1981, 1986). Current research indicates the positive nature and dimensions of sicknesses and dysphoric states in at least the Mediterranean tradition (see for example, Gaines and Farmer 1986; Gaines 1987a; Good, Good and Moradi 1985) and the Buddhist (Obeyesekere 1985).

## Conclusions

Marxists and other political economists of health unwittingly seek to place the study of folk and professional ethnomedicines within a 19th century natural scientific paradigm, one which is a twin of that which so severely limits American professional Biomedicine itself. In their view, social action in historical and cultural contexts, is reduced to the operation of a purely epistemic subject (Rabinow and Sullivan 1979: 4).

Critical medical anthropologies are, if I may adapt Navarro's own description (1975: 361) of Illich's *Medical Nemesis*, "radical only in style and rhetoric but intrinsically conservative" in meta-theoretical assumptions, theory, methodology and interpretation. Like the anti- or critical psychiatry of the 1960's and 70's, political economists give only the appearance of being voices from a change-oriented, humanitarian left.

*Other Directions*

A host of new avenues of research present themselves. Myriad others will doubtless occur to the reader. One such might be the cultural historical study of medical social sciences, such as the political economy of health. The self critical study of anthropology has been and is being done and is the cause for development of improved interpretive approaches in social cultural anthropology in general and constructivist approaches in medical anthropology in particular. Constructivist approaches have a singular ability to reveal to us things we do not know. Other theoretical traditions, it may be suggested, tend to tell us things we already know, believe we know or could have guessed without, because, often like medicine itself, they reproduce conventional lay wisdom.

Since even constructivists sometimes may be seen to rely on an external, implicit natural "biology" (e.g., Kleinman and Good 1985), it appears that work needs to be done to examine the cultural construction of biology in different Biomedicines, i.e. biology as a cultural construction. The meaning of biology is by no means everywhere constant across time or within or across cultures and appears a worthy topic for cultural constructivist deconstruction (Gaines 1986b, 1987a).

Another area of importance would seem to be the study of medical practice from a more experiental viewpoint; and "ethnography of emotion in medicine" (Gaines 1987b). This would seem especially timely given the care demand of the AIDS epidemic on health care professionals. I would like to see more on background philosophies in culture in general and in medicine in particular, such as provided by Gordon (1988) and Kirmayer (1988). The construction of the notion of scientism, for example, and its association with numbers (statistics) in Western Biomedicine (see Cassedy 1984) seems a worthy topic of further study. Finally, the concept of Sickness History needs further development but may contribute to our understanding of the historical construction of illness realities; the distribution, the waxing and waning of same and of healing strategies and medical organization.

This paper has advocated the use of cultural constructivist perspectives in medical anthropology and beyond. Medical cultural constructivists see professional and lay ethnomedical realities as constructed through time in social interactions to which meanings are brought, exchanged and created and realities defined and actions charted. Medical knowledge and practice are held to be expressive moments of the cultures in which they are found and in the context of which they are and were created, recreated and transformed. Constructivist approaches allow for expanding our knowledge and understanding of this important domain of human experience, allowing its reconstruction and deconstruction and furthering our understanding of this vital domain and its webs of significance spun over and in time. These webs produce local versions of theories of coherence and give meaning to human health, life and suffering.

## Acknowledgements

My thanks to conference participants for their comments and support. Special thanks to Dr. Thomas Maretzki for initiating and co-organizing the conference.

## Notes

[1] e.g., Marcus and Fischer (1986) who cite and refer to works by constructivists, examples of cultural critiques of our own society and medicine.

[2] As well, it may be argued that Foucault's concern with the medical gaze replicates the clinical emphasis of French medicine, an emphasis not shared by other Biomedicines, e.g., German and American which because of historical

relationships, emphasize biological sciences and see clinical work as an applications thereof (Gaines 1989, Gilman 1988, Goldstein 1987).

[3] See Smith (1984) who notes the same tactic is used by political economists in social anthropology more generally.

[4] We should also here note that thesewriters also regularly impute motives, feelings and "needs" to their inanimate ideological constructs, e.g., to capitalism, production, industrialization, world system, etc.

[5] See Sahlins (1976), Smith (1984) for more on the limitations of these formulations.

[6] See Farmer (1989) on changing conceptions of AIDS in Haiti, and their integration with traditional etiological theories.

## References

AMERICAN PSYCHIATRIC ASSOCIATION 1980: *Diagnostic and Statistical Manual III (DSM III).* Washington, D.C.: American Psychiatric Association

-- 1987: *Diagnostic and Statistical Manual III-Revised (DSM III-R).* Washington, D.C.: American Psychiatric Association.

BADIE B. and BIRNBAUM P. 1984: *The Sociology of the State.* Chicago: University of Chicago Press.

BAER H. 1981: Prophets and Advisers in Black Spiritual Churches: Therapy, Palliative, or Opiate? *Culture, Medicine, and Psychiatry* 5 (2): 145-170.

-- 1982: On the Political Economy of Health. *Medical Anthropology Newsletter* 14 (1): 1-17.

-- 1986: Sociological Contributions to the Political Economy of Health. *Medical Anthropology Quarterly* 17 (5): 129-131.

-- 1987: Divergence and Convergence in Two Systems of Manual Medicine. *Medical Anthropology Quarterly* (n. s.) 1 (2): 176-193).

BAER H., SINGER M. and JOHNSON J. (eds.) 1986: Toward a Critical Medical Anthropology. *Social Science and Medicine* 23: 95-98.

BARRETT M. and ROBERTS H. (eds.) 1980. Doctors and Their Patients: The Social Control of Women in General Practice. In: SMART C. and SMART B. (eds.) *Women, Sexuality, and Social Control.* London: Routledge and Kegan Paul.

BASZANGER I. 1985: Professional Socialization and Social Control: From Medical Students to General Practitioners. Noal Mellott, trans. *Social Science and Medicine* 20 (2): 133-143.

BIRNBAUM P. 1981: *The Heights of Power: An Essay on the Power Elite in France.* A. Goldhammer, trans. Chicago: University of Chicago Press.

BOSK C. 1977: *Forgive and Remember.* Chicago: University of Chicago Press.

BRANDT A. 1978: Racism and Research: The Case of the Tuskegee Syphilis Study. *The Hastings Center Report.* December: 21-29.

— 1985: *No Magic Bullet.* Cambridge: Harvard University Press.

BROWN E.R. 1979: *Rockefeller Medicine Men.* Berkeley: University of California Press.

CAMPBELL J. 1964: *Honour, Family and Patronage.* Oxford: Oxford University Press.

CASSEDY J. 1984: *American Medicine and Statistical Thinking 1800-1860.* Cambridge: Harvard University Press.

COOTER R. 1982: Anticontagionism and History's Medical Record. In: WRIGHT P. and TREACHER A. (eds.). *The Problem of Medical Knowledge: Examining the Social Construction of Medicine.* Edinburgh: University of Edinburgh Press.

CORNELISEN A. 1971: *Women of the Shadows.* Boston: Little, Brown and Company.

DELVECCHIO-GOOD M.J. 1985: Discourses on Physician Competence. In: HAHN R. and GAINES A. (eds.) *Physicians of Western Medicine.* Dordrecht: D. Reidel.

DODIER N. 1985: Social Uses of Illness at the Workplace: Sick Care and Moral Evaluation. Noal Mellot, trans. *Social Science and Medicine* 20 (2): 123-128.

DUMONT L. 1980: *Homo Hierarchicus.* Chicago: University of Chicago Press.

DUNN F. 1977: Health and Disease in Hunter-Gatherers. In: LANDY D. (ed.) *Culture, Disease and Healing.* New York: Macmillan.

DWYER E. 1987: *Homes for the Mad.* New Brunswick, NJ: Rutgers University Press.

EHRENREICH B. and J. 1971: *The American Health Empire*. New York: Vintage Books.

EISENBERG L. 1977: Disease and Illness: Distinctions Between Professional and Popular Ideas of Sickness. *Culture, Medicine and Psychiatry* 1 (1): 9-24.

ELKANA Y. 1981: A Programmatic Attempt at an Anthropology of Knowledge. In: MENDELSOHN E. and ELKANA Y. (eds.) *Sciences and Cultures.* Dordrecht: D. Reidel.

ELLING R. 1977: Industrialization and Occupational Health in Underdeveloped Countries. *International Journal of Health Services* 7: 209-235.

ENGEL G. 1977: The Need for a New Medical Model: A Challenge for Biomedicine. *Science* 196: 129-136.

EVANS-PRITCHARD E.E. 1964: *Social Anthropology and Other Essays.* New York: Free Press.

FARMER P. 1988: Bad Blood, Spoiled Milk: Bodily Fluids as Moral Barometers in Rural Haiti. *American Ethnologist* 15 (1): 62-83.

— 1989: Sending Sickness. *Medical Anthropology Quarterly* 4 (1): 6-27.

FOX R. 1989: *The Sociology of Medicine*. Englewood Cliffs, N.J.: Prentice-Hall.

FOSTER G. and ANDERSON B. 1978: *Medical Anthropology*. New York: Wiley.

FRANKENBERG R. 1980: Medical Anthropology and Development. *Social Science and Medicine* 14B (4): 197-207.

-- (ed.) 1988: Gramsci, Marxism and Phenomenology: Essays for the Development of Critical Medical Antropology. *Medical Antropology Quarterly* (n.s.) Special Issue 2 (4).

FREIDSON E.1975: *Profession of Medicine*. New York: Dodd Mead.

GAINES A.D. 1979: Definitions and Diagnoses. *Culture, Medicine and Psychiatry* 3 (4): 381-418.

-- 1982a: Cultural Definitions, Behavior and the Person in American Psychiatry. In: MARSELLA A. and WHITE G. (eds.). *Cultural Conceptions of Mental Health and Therapy.* Dordrecht: D.Reidel.

-- 1982b: Knowledge and Practice: Anthropological Ideas and Psychiatric Practice. In: CHRISMAN N. and MARETZKI T. (eds.) *Clinically Applied Anthropology.* Dordrecht: D. Reidel.

-- 1985a. Alcohol: Cultural Conceptions and Social Behavior Among Urban 'Blacks'. In: BENNETT L. and AMES G. (eds.) *The American Experience with Alcohol: Contrasting Cultural Perspectives.* New York: Plenum.

-- 1985b: The Once- and the Twice-Born: Self and Practice Among Psychiatrists and Christian Psychiatrists. In: HAHN R. and GAINES A. *Physicians of Western Medicines.* Dordrecht: D. Reidel.

-- 1986a: Disease, Communalism and Medicine. *Culture, Medicine and Psychiatry* 10 (4): 297-303.

-- 1986b: Culture and Medical Knowledge in France and America. (Paper presented at the *American Anthropological Meeting.* Washington, D.C.) (Also in: GAINES A. (ed.) *Ethnopsychiatry.* Albany: SUNY Press. (in press).

-- 1987a: Cultures, Biologies and Dysphorias. *Transcultural Psychiatric Research Review* 24 (1): 31-57.

-- 1987b: Comments. *"Psychiatric Institutions". Symposium American Anthropological Association Meetings.* Chicago.

-- 1989: From DSM I to III-R, the Voice of Self and Mastery: A Cultural Constructivist Reading of Psychiatric Classification. Paper presented at the *American Anthropological Association Meetings.* Washington, D.C.

-- n.d., a: *Physician and Function: The Study of the French Pharmacy.* ms.

-- n.d., b: Ethnicity as Cultural System: 'Race' and Culture in Strasbourg. ms.

-- n.d., c: The Word and the Cross: Identity and Paradox in Alsace. ms.

-- n.d., d: Culture and Psychiatry: On the Cultural Construction of Ethnopsychiatries. In: GAINES A. (ed.) *Ethnopsychiatry.* Albany: SUNY Press. (in press).

GAINES A.D. and FARMER P. 1986: Visible Saints: Social Cynosures and Dysphoria in the Mediterranean Tradition. *Culture, Medicine and Psychiatry* 10 (4): 295-330.

GAINES A.D. and HAHN R. (eds.) 1982: Physicians of Western Medicine: Five Cultural Studies. *Culture, Medicine, and Psychiatry* Special Issue 6 (3).

GAINES A.D. and HAHN R. (eds.)1985: Among the Physicians: Encounter, Exchange and Transformation. In: HAHN R. and GAINES A. (eds.) *Physicians of Western Medicine.* Dordrecht: D. Reidel.

GEERTZ C. 1973: *The Interpretation of Cultures.* New York: Basic Books.

-- 1983: *Local Knowledge.* New York: Basic Books.

-- 1984: Anti-anti-relativism. *American Anthropologist* 86 (2): 263-278.

-- 1988: *Work and Lives.* Stanford: Stanford University Press.

GILMAN S. 1988: *Disease and Representation.* Ithaca: Cornel University Press.

GILPIN R. 1987: *The Political Economy of International Relations.* New Jersey: Princeton University Press.

GOLDSTEIN J. 1987: *Console and Classify: The French Psychiatric Profession in the 19th Century.* Cambridge: Cambridge University Press.

GOOD B. 1977: The Heart of What's the Matter: the Semantics of Illness in Iran. *Culture, Medicine and Psychiatry* 1 (1): 25-58.

GOOD B. and DELVECCHIO-GOOD M.J. 1981: The Semantics of Medical Discourse. In: MENDELSOHN E. and ELKANA Y. (eds.) *Sciences and Cultures.* Dordrecht: D. Reidel.

-- 1982: Toward a Meaning-Centered Analysis of Popular Illness Categories: "Fright Illness" and "Heart Distress" in Iran. In: MARSELLA A. and WHITE G. (eds.) *Cultural Conceptions of Mental Health and Therapy.* Dordrecht: D. Reidel.

GOOD B., DELVECCHIO-GOOD M.J. and MORADI R. 1985: The Interpretation of Iranian Depressive Illness and Dysphoric Affect. In: KLEINMAN A. and GOOD B. (eds.) *Culture and Depression.* Berkeley: University of

California Press.

GORDON D. 1988: Tenacious Assumptions in Western Medicine. In: LOCK M. and GORDON D. (eds.) *Biomedicine Examined.* Dordrecht: Kluwer Academic Publishers.

GREENWOOD D., LINDENBAUM S., LOCK M. and YOUNG A. 1988: Introduction. In: GREENWOOD D. et al. (eds.) Medical Anthropology Theme Issue. *American Ethnologist* 15 (1): 1-3.

GROB G. 1987: American Psychiatry: An Ambivalent Specialty. In: SALZMAN J. (ed.) *Prospects: Annual of American Cultural Studies.* Volume 12. Cambridge: Cambridge University Press.

HAHN R. 1982: "Treat the Patient, Not the Lab": Internal Medicine and the Concept of 'Person'. *Culture, Medicine and Psychiatry* 6 (3): 219-236.

-- 1985: A World of Internal Medicine: Portrait of an Internist. In: HAHN R. and GAINES A. (eds.) *Physicians of Western Medicine.* Dordrecht: D. Reidel.

-- (ed.) 1987: Obstetrics in the United States *Medical Anthropology Quarterly* (n.s.) *Special Issue* 1 (3).

HAHN R. and GAINES A. (eds.) 1985: *Physicians of Western Medicine:Anthropological. Approaches to Theory and Practice.* Dordrecht: D. Reidel.

HAHN R. and KLEINMAN A. 1983: Biomedical Practice and Anthropological Theory. *Annual Review of Anthropology*. Palo Alto, CA: Annual Reviews, Inc.

-- 1984: Biomedicine as a Cultural System. In: PIATELLI-PALMARINI (ed.) *Encyclopedia of the Social History of the Biomedical Sciences*. Massimo. Milano: Franco Maria Ricci Publishing Company.

HARDING S. and O'BARR J. (eds.) 1987: *Sex and Scientific Inquiry.* Chicago: University of Chicago Press.

HAWTHORN G. 1987: *Enlightenment and Despair.* Cambridge: Cambridge University Press. (2nd ed.)

HELMAN C. 1985: Disease and Pseudo-Disease: A Case History of Pseudo-Angina. In: HAHN R.and GAINES A. *Physicians of Western Medicine: Anthropological Approaches to Theory and Practice.* Dordrecht: D. Reidel.

HUGHES C. 1968: Ethnomedicine. In: *Encyclopedia of the Social Sciences*. New York: Free Press.

ILLICH I. 1976: *Medical Nemesis.* New York: Bantam Books.

JACKSON S. 1985: Acedia the Sin and Its Relationship to Sorrow and Melancholia. In: KLEINMAN A. and GOOD B. (eds.) *Culture and Depression.* Berkeley: University of California Press.

JANZEN J. 1978: *The Quest for Therapy in Lower Zaire.* Berkeley: University of California Press.

JAY M. 1973: *The Dialectical Imagination.* London: Routledge.

JOHNSON T. 1985: Consultation-Liaison Psychiatry: Medicine as Patient, Marginality as Practice. In: HAHN R. and GAINES A. (eds.) *Physicians of Western Medicine: Anthropological Approaches to Theory and Practice.* Dordrecht: D. Reidel.

-- 1988: Physician Impairment: Social Origins of a Medical Concern. *Medical Anthropology Quarterly* (n.s.) 2 (1): 17-33.

JUSTICE J. 1986: *Policies, Plans and People.* Berkeley: University of California Press.

KATZ P. 1985: How Surgeons Make Decisions. In: HAHN R. and GAINES A. (eds.) *Physicians of Western Medicine: Anthropological Approaches to Theory and Practice.* Dordrecht: D. Reidel.

-- n.d.: The Active Posture of Surgeons. ms.

KIRMAYER L. 1988: Mind and Body as Metaphors. In: LOCK M. and GORDON D. (eds.) *Biomedicine Examined.* Dordrecht: D. Reidel.

KLEINMAN A. 1980: *Patients and Healers in the Context of Culture.* Berkeley: University of California Press.

-- 1981: The Meaning Context of Illness and Care. In: MENDELSOHN E. and ELKANA Y. (eds.) *Sciences and Cultures.* Dordrecht: D. Reidel.

-- 1986: *The Social Origin of Distress and Disease.* New York: Yale University Press.

KLEINMAN A. and B. GOOD (eds.) 1985: *Culture and Depression.* Berkeley: University of California Press.

KNOWLES J.H. (ed.) 1977: *Doing Better and Feeling Worse: Health in the United States.* New York: W.W. Norton.

KORNFIELD R. 1986: Dr., Teacher or Comforter? Medical Consultation in a Zairian Pediatric Clinic. *Culture, Medicine and Psychiatry* 10 (4): 367-387.

LAZARUS E. 1988: Theoretical Considerations for the Study of the Doctor-Patient Relationship: Implications of a Perinatal Study. *Medical Anthropology Quarterly* (n.s.) 2 (1): 34-58.

LEE D. 1959: View of the Self in Greek Culture. In: LEE D. (ed.) *Freedom and Culture.* Englewood Cliffs, NJ: Prentice-Hall.

LESLIE C. (ed.) 1977a: *Asian Medical Systems.* Berkeley: University of California Press.

— 1977b: Pluralism and Integration in the Indian and Chinese Medical Systems. In: LANDY D. (ed.) *Culture, Disease and Healing.* New York: Macmillan.

— 1980: Medical Pluralism in World Perspective. *Social Science and Medicine* 14B (4): 191-195.

LETOURMY A. 1985: An Economic Approach to the Daily Activities of Private General Practitioners.Noal Mellott, trans. *Social Science and Medicine* 20 (2): 173-180.

LIEBAN R. 1990: Medical Anthropology and the Comparative Study of Medical Ethics. In: WEISZ G. *Social Science Perspectives on Medical Ethics*. Dordrecht: Kluwer Academic.

LOCK M. 1981: *East Asian Medicine in Urban Japan.* Berkeley: University of California Press.

-- 1985: Models and Practice in Medicine: Menopause as Syndrome or Life Transition? In: HAHN R. and GAINES A. (eds.) *Physicians of Western Medicine: Anthropological Approaches to Theory and Practice.* Dordrecht: D. Reidel.

-- 1986: Plea for Acceptance. *Social Science and Medicine* 23 (2): 99-112.

-- 1988: Introduction: Health and Medical Care as Cultural and Social Phenomena. In: NORBECK E. and LOCK M: (eds.) *Health, Illness, and Medical Care in Japan.* Honolulu: University Press of Hawaii.

LOCK M. and D. GORDON (eds.) 1988: *Biomedicine Examined.* Dordrecht: Kluwer Academic Publishers.

LUTZ C. 1985: Depression and the Translation of Emotional Worlds. In: KLEINMAN A. and GOOD B. *Culture and Depression.* Berkeley: University of California Press.

MCKEOWN T. 1979: *The Role of Medicine: Dream, Mirage, or Nemesis?.* Princeton: Princeton University Press.

MCNEILL J. and H. GAMER 1938: *Medieval Handbooks of Penance.* New York: Columbia University Press.

MACLEAN A. 1986: Family Therapy Workshops in the United States. *Social Science and Medicine* 23 (2): 179-189.

MARCUS G. and M. FISCHER 1986: *Anthropology as Cultural Critique.* Chicago: University of Chicago Press.

MARETZKI T. 1985: Including the Physician in Healer Centered Research. In: HAHN R: and GAINES A. (eds.) *Physicians of Western Medicine*. Dordrecht: D. Reidel.

-- n.d.: Georg Groddeck's Integrative Massage and Psychotherapy Treatment. In: GAINES A. (ed.) *Ethnopsychiatry: The Cultural Construction of Professional and Folk Psychiatry*. Albany: SUNY Press. (in press)

MARETZKI T. and E. SEIDLER 1985: Biomedicine and Naturopathic Healing in West Germany: A History of a Stormy Relationship. *Culture, Medicine and Psychiatry* 9 (4): 383-421.

MENDELSOHN E. and ELKANA Y. (eds.) 1981: *Sciences and Cultures*. Dordrecht: D. Reidel.

MENNINGER K. with MAYMAN M. and P. PRUYSER 1963: *The Vital Balance*. New York: Viking.

MISHLER E. et al. 1981: *Social Contexts of Health, Illness and Patient Care*. Cambridge: Cambridge University Press.

MORANTZ-SANCHEZ R.M. 1985: *Sympathy and Science: Women Physicians in American Medicine*. Oxford: Oxford University Press.

MORGAN L. 1987: Dependency Theory in the Political Economy of Health: An Anthropological Critique. *Medical Anthropology Quarterly* (n.s.) 1 (2): 131-154.

MORSEY S. 1988: Islamic Clinics in Egypt: The Cultural Elaboration of Biomedical Hegemony. In: FRANKENBERG R: (ed.) *Gramsci, Marxism and Phenomenology. Medical Anthropology Quarterly (n.s.) Special Issue* 2 (4): 355-369.

-- 1990: personal communication.

NAVARRO V. 1975a: The Underdevelopment of Working America. *American Journal of Public Health* 66: 538-547.

— 1975b: The Industrialization of Fetishism or the Fetishism of Industrialization?: A Critique of Ivan Illich. *Social Science and Medicine* 9: 351-363.

— 1976: *Medicine Under Capitalism*. New York: Prodist.

— 1986: *Crisis, Health and Medicine*. London: Methune.

NISBET R. 1980: *History of the Idea of Progress*. New York: Basic Books.

NORBECK E. and LOCK M. (eds.) 1988: *Health, Illness and Medical Care in Japan*. Honolulu: University Press of Hawaii .

NUCKOLLS C. n.d.: Notes on a Defrocked Priest. In: GAINES A: (ed.) *Ethnopsychiatry*. Albany: SUNY Press. (in press)

OHNUKI-TIERNEY E. 1984: *Health and Illness in Contemporary Japan*. Cambridge: Cambridge University Press.

OBEYESEKERE G. 1985: Depression, Buddhism and the Work of Culture in Sri Lanka. In: KLEINMAN A. and GOOD B. (eds.) *Culture and Depression*. Berkeley: University of California Press.

PAGET M.A. 1988: *The Unity of Mistakes: A Phenomenological Interpretation of Medical Work*. Philadelphia: Temple University Press.

PARSONS T. 1951: *The Social System*. New York: Free Press.

PERISTIANY J. (ed.) 1966: *Honour and Shame: The Values of Mediterranean Society*. London: Wiedenfeld.

PERNICK M. 1985: *A Calculus of Suffering*. New York: Columbia University Press.

RABINOW P. and SULLIVAN W. 1979: The Interpretive Turn: Emergence of an Approach. In: RABINOW P. and SULLIVAN W. (eds.) *Interpretive Social Science*. Berkeley: University of California Press.

-- 1987: *Interpretive Social Science: A Second Look.* Berkeley: University of California Press.

REYNOLDS D. K. 1976: *Morita Psychotherapy.* Berkeley: University of California Press.

RHODES L. n.d.: The Subject of Power in Medical/Psychiatric Anthropology. In: GAINES A. (ed.) *Ethnopsychiatry.* Albany: SUNY Press. (in press)

RITENBAUGH C.1982: Obesity as a Culture-Bound Syndrome. *Culture, Medicine and Psychiatry* 6 (3): 347-364.

RUBEL A: 1977: The Epidemiology of a Folk Illness: Susto in Hispanic America. In: LANDY D. (ed.) *Culture, Disease and Healing.* New York: Macmillan.

SAHLINS M. 1976: *Culture and Practical Reason.* Chicago: University of Chicago Press.

SCHEDER J.C. 1988: A Sickly-Sweet Harvest. *Medical Anthropology Quarterly* (n.s.) 2 (1): 251-277.

SCHEPER-HUGHES N. and LOCK M. 1986: Speaking "Truth" to Illness: Metaphors, Reification, and a Pedagogy for Patients. *Medical Anthropology Quarterly* 17 (5): 137-140.

-- 1987: The Mindful Body. *Medical Anthropological Quarterly* (n.s.) 1 (1): 6- 41.

SCHEPER-HUGHES N. 1988: The Madness of Hunger: Sickness, Delirium, and Human Needs. *Culture, Medicine and Psychiatry* 12 (4): 429-458.

SINGER M. 1986: Developing a Critical Perspective in Medical Anthropology. *Medical Anthropology Quarterly* 17 (5): 128-129.

SMITH R. 1984: Social Class. *Annual Review of Anthropology.* Palo Alto: Annual Review, Inc.

SONTAG S. 1979: *Illness as Metaphor.* New York: Vintage Books.

SPIRO M. 1970: *Buddhism and Society.* New York: Harper and Row.

STARR P. 1982: *The Social Transformation of American Medicine.* New York: Basic Books.

STEBBINS K. 1986: Curative Medicine, Preventive Medicine and Health Status. *Social Science and Medicine* 23 (2): 139-148.

SUSSER I. (ed.) 1988: *Health and Industry. Medical Anthropology Quarterly (n.s.)* Special Issue 2 (1).

SWARTZ L. n.d.: Ethnopsychiatry in South Africa: the Question of Relativism. In: GAINES A. (ed.) *Ethnopsychiatry.* Albany: SUNY Press. (in press)

TAUSSIG M. 1980. Reification and the Consciousness of the Patient. *Social Science and Medicine* 14B: 3-13.

TOUSIGNANT M. 1984: Pena in the Ecuadorian Sierra. *Culture, Medicine and Psychiatry* 8 (4): 381-398.

TOWNSEND M.1978: *Cultural Conceptions and Mental Illness.* Chicago: Chicago University Press.

TURNER B. 1987: *Medical Power and Social Knowledge.* London: Sage.

TURNER M. 1987: *Death is the Mother of Beauty.* Chicago: Chicago University Press.

WAITZKIN H. 1979: Medicine: Superstructure and Micropolitics. *Social Science and Medicine* 13a: 601-609.

-- 1981a: The Social Origins of Illness: A Neglected History. *International Journal of Health Services* 11 (1): 77-103.

-- 1981b: A Marxist Analysis of the Health Care System of Advanced Capitalist Societies. In: EISENBERG L. and KLEINMAN A. (eds.) *The Relevance of Social Science for Medicine.* Dordrecht: D. Reidel.

-- 1986: The Micropolitics of Medicine: Theoretical Issues. *Medical Anthropology Quarterly* 17 (5): 134-136.

WEISBERG D. and LONG S.O. 1984: *Biomedicine in Asia: Transformations and Variations. Culture, Medicine and Psychiatry Special Issue 8 (2).*

WEISZ G. (ed.) 1990: *Social Science Perspectives on Medical Ethics.* Dordrecht: Kluwer Academic.

WENZEL S. 1960: *The Sin of Sloth: Acedia in Medieval Thought and Literature.* Chapel Hill: University of North Carolina Press.

WILLIAMS A. and BOULTON M. 1988: Thinking Prevention: Concepts and Constructs in General Practice. In: LOCK M. and GORDON D. (eds.) *Biomedicine Examined.* Dordrecht: D. Reidel.

WORSLEY P. 1982: Non-Western Medical Systems. *Annual Review of Anthropology*. Palo Alto, CA: Annual Review.

WRIGHT P. and TREACHER A. (eds.) 1982a: *The Problem of Medical Knowledge: Examining the Social Construction of Medicine.* Edinburgh: University of Edinburgh Press.

-- 1982b: Introduction. In: WRIGHT P. and TREACHER A. (eds.) *The Problem of Medical Knowledge: Examining the Social Construction of Medicine.* Edinburgh: University of Edinburgh Press.

WYLIE L.1957: *Village in the Vaucluse.* Cambridge: Harvard University Press.

YOUNG A: 1977: Internalizing and Externalizing Medical Belief Systems. *Social Science and Medicine* 10 (1): 147-156.

-- 1982: The Anthropologies of Illness and Sickness *Annual Review of Anthropology*. Palo Alto, CA: Annual Review, Inc.

-- 1988: 'Unpacking' the Demoralization Hypothesis. *Medical Anthropology Quarterly* 2 (1): 3-16.

-- 1989: A Case Study Describing How an Institutional Ideology Shapes Knowledge of Mental Disorder (PTSD). In: LOCK M. and LINDENBAUM S. (eds.) *Analyses in Medical Anthropology*. ms.

-- 1990: Moral Conflicts in a Psychiatric Hospital Treating Combat-Related Posttraumatic Stress Disorder (PTSD). In: WEISZ G. (ed.) *Social Science Perspectives on Medical Ethics.* Doredrecht: Kluwer Academic.

# Holistic Health and a Changing Western World View

Carol P. MacCormack

## Introduction

When one listens to the messages of the holistic health, ecology, peace, feminist and creation spirituality movements one becomes aware of a congruence. Being within these movements as a participant observer, or reading a great cross-section of popular magazines from these movements, one realizes there is a great deal of overlap; they are all saying the same things. If we take as a speculative thesis that European society may be reaching the limits of dualistic thinking, and the scientific materialism and individualism it engenders, then this congruence may constitute a coherent 'code' in the sense that Anthony Wallace intended when writing about societies that become 'revitalized' through a transformation process. The code is a coherent message that leads people toward a realistic alternative future, and in the process makes them and the society feel whole again. But there are other messages about. It is quite possible that we could move toward another wave of fascism rather than a holistic future. The light and the dark are always with us; the healing creative possibility, and the destructive possibility that arises from shadowy personal and collective egoistic urgings are always with a population. Either way, this is a theoretical exploration of cultural change, focused upon the holistic health movement in Britain.

## Theoretical Considerations

This paper explores the search for holistic health and the broader Western hope for inner and outward peace. It is about millenarian movements, and the revitalization of societies. It draws on theory from theoretical physics and radical biology as well as anthropology. A quick survey of recent medical anthropology literature suggests that implicit functional theory remains most popular. It is based on an analogy with whole organisms, and any change in a part of the society will result in changes in other parts, thus altering the nature of the whole society. In Radcliffe-Brown's classic formulation of functionalism, societies that cope successfully with internal change and external threat succeed in maintaining themselves over time - that is their function. Systems theory augments functionalism, describing with more clarity the concept of a society with institutional sub-systems linked in feed-back relationships that facilitate mutual adjustments (see for example Bateson 1972).

Structuralism was a revolt against the shallow empiricism of many functional studies. When we read health care project reports we usually find our minds turning to what is left out; to the underlying cultural template that gives coherence to a society, but is not amenable to empirical measurements. Thus we are interested in systems of classification and symbolic analysis. Some legitimacy for an intuitive 'knowing' of the structure underlying surface manifestations is acknowledged in a structuralist approach. There is also an interest in exchange and reciprocity that binds individuals into groups, and groups into alliances, solidarity being a question functional theory did not deal with very well.

Neither functional/systems theory nor structural theory deal very well with change. Functional analysis is based on a slice of time much as the pathologist's knife slices the liver to know the state of its internal structures. Functionalists and systems analysts seek to see how the parts of a society are constituted and interact with each other at the analytical moment. With structuralism, the putative deep structure simply *is*. It may transform within the parameters of its fundamental rules, but is rather tethered by them.

Structural theory, in common with systems theory, is interested in cybernetics and information theory, the 'algebra of the brain' for Lévi-Strauss. But the underlying structure is rather like Plato's Ideas, or else it is explained by an extreme materialistic reductionism, being inherent in the mechanism of the physical brain (MacCormack 1980: 3ff.). Non-material ideas, and the physical brain, are the two horns of the dualistic dilemma European culture has struggled with since the time of Descartes. We are constantly being called upon to choose the mental/spiritual or the material alternative, and have perhaps grown tired and fragmented with the effort.

Marx chose the materialistic option, and Marxism takes as its central concern the process of historical change. The social relations of production allow one group to draw surplus value off the labour of another until the situation becomes too exploitive for the population to bear. Then the system progresses to a new mode. As with other theoretical perspectives, this approach explains certain kinds of phenomena very well, such as the relationship between apartheid and health in South Africa, for example (WHO 1983). This article, however, is concerned to understand the process of change in contemporary Western society and Marxist theory, with its intense focus on economic and political relationships, does not explain very satisfactorily the phenomenon of people seeking holistic health and ecological harmony. It especially is not an adequate theoretical framework to explain why, in the United Kingdom with a comprehensive national health service since 1948, in which anyone can see a real doctor without paying a fee, so many people are now paying fees for services in the alternative sector.

Part of the perceived sickness of our times is a world view based on Cartesian metaphysical dualism which splits off material bodies from their minds and souls. The holistic movement seeks an alternative view of reality, drawing insights from theoretical physics, chemistry and biology. For example, David Bohm, Professor of Theoretical Physics in the University of London, Einstein's pupil, has put forward a theory of wholeness and implicate order. Some of the basic concepts of relativity theory and quantum theory contradict each other, but what they share is a concept of undivided wholeness in the universe. Bohm (1980, 1985) encourages us to consider the universe not as empty space but the plenum, a vast sea of energy which is the ground of the existence of everything, including ourselves. The explicate things that appear to our senses unfold from this multidimensional implicate order. Their true meaning can be seen only when we consider the plenum. Bohm takes his particular meaning of implicate from the verb *to implicate* meaning *to fold inward*. Each region of time and space has, in some sense, the total universe of unbroken wholeness enfolded within it. In our universe of unbroken wholeness, space and time are no longer the dominant factors determining relationships. Rather, time, space and material particles are abstracted from the whole. They are explicate or unfolded forms, derived from the enfolded order as waves appear on the sea, projected in explicate form from the whole sea, then dip back into the whole. The whole is in every explicate part as the whole picture is in every part of a hologram when we shine light on a small part of it. But Bohm writes not of the hologram. Rather, he gives us the image of a holomovement, for the universe is dynamic. Our dynamic perceptions, attention, thought, feelings, desire, will and other manifestations of consciousness similarly unfold out of the same implicate order, the ground of all, transcending mind-matter dualism (Bohm 1980: xv, 149ff., 192-212).

Field theory, too, is beginning to help us develop new non-dualistic insights into the nature of human thought and action. Gravitational, electro-magnetic and quantum fields are related to matter.

They interact with matter and account for its patterns. This is not a Cartesian restatement of fields versus matter because matter is no longer thought of as hard billard balls. Rather, matter is thought to consist of rhythmic processes of activity; of energy patterns within a field. In Einstein's special theory of relativity the electro-magnetic field permeates space. It has energy and momentum. But it has no mechanical basis for interacting with matter. Rather, the field exchanges energy and momentum with matter. Later, in his general theory of relativity, Einstein described quanta – or units – of energy. Atoms absorb and emit light in these 'packets'. In 1924 de Broglie suggested that just as light waves have particle properties, so also particles of matter display wavelike properties. Electrons, in an electron microscope, for example, behave like waves, not little billard balls. De Broglie's theory has been extended beyond subatomic particles to all matter. Whole atoms and molecules have wavelike aspects. In the idea of quantum matter fields, particles are quanta of excitation. Different kinds of matter fields interact with each other. They can also interact with electro-magnetic fields. All interactions are mediated by quanta. There is no dualism of field and particle since fields are the basic reality and specify the probability of finding quanta at particular points in space (Zukav 1979: 218ff.).

Rupert Sheldrake (1988, 1990) has applied field theory to biology, and the implications for anthropology are important. All activity, whether thought or action, might be considered as quanta of excitation that constitute what he called morphic fields. These fields are non-material regions of influence extending into space and continuing in time. Morphic fields are fields of information shaped by influences from previous generations of an organism. Each type of matter and each species of organism has its field. We might think of a water molecule as a quantum, or unit, in a water field. We might think of a person as a quantum, or unit, in a human field. All awareness and activity of previous generations constitutes a cumulative memory for the species system. The memory, in fields, is accessible to living things and helps to explain the pattern of form and activity in the present. Individuals 'tune in' on the cumulative influence of past activity. Sheldrake calls the process by which the past influences the present *morphic resonance.* When events have occurred again and again over billions of years, as is the case with some kinds of atoms, molecules and crystals, the cumulative fields transmit causal influence through space and time.

Sheldrake suggests that the same process is at work in the history of crystals, amoebae and even human societies. With more complex entities, and specifically with self-aware human beings, we might think afresh about the nature of culture. Many people appear to 'know' the existence of the collective unconscious posited by Durkheim, the archetypes posited by Jung, or the deep structure of Lévi-Strauss. But heretofore no theorist has been very good at explaining the nature of those deep levels of culture, or how they might be linked with our everyday awareness. With the concepts of morphic fields and morphic resonance, Sheldrake has offered a theoretical explanation of such linkages in human history. Morphic fields may be compared to quantum matter fields, Sheldrake hypothesizes. If atoms can be said to have morphic fields, then they may be already described within quantum field theory. The morphic fields of molecules may already be partially described by quantum chemistry. But the morphic fields of cells, tissues, organs and living organisms need the kind of further systematic descriptions Sheldrake is attempting.

### Cultural Change

As with fields in quantum physics, morphic fields are not deterministic, but probabilistic. For example, even cells, tissues, organs or organisms which are genetically identical and develop under practically identical conditions are never exactly the same, and the influence of many past people, all slightly different, results in some central tendencies within the limits of probability. Furthermore, different types of fields influence each other, accounting for another aspect of variability within the limits of probability (Sheldrake 1988: 120).

We might think of societies as probability structures influenced by morphic fields, our 'history', our collective unconscious, or the deep structure of our culture. Morphic fields may have some role in structuring human language, thought and activity. They have an organizing influence on the interrelations of the component parts of societies. Societies are stabilized by self-resonance from their past and by morphic resonance from previous similar societies. Since morphic fields are probability structures, social and cultural regularities are also probabilities rather than determined (Sheldrake 1988: 246). *Homo sapiens* as an organism is subsumed within social and ecological systems. All the separate types of morphic fields, at different levels of complexity, are organized into nested hierarchies of fields within fields. Higher-level wholes (e.g., ecosystems) are made up of parts (e.g., organisms) that are themselves wholes at a lower level of organization. Culture and nature are unitary when we think of a sweet pudding as a whole dish, or cultural artifact, made up of whole sugar crystals, which in turn are made up of sugar molecules. The molecules are smaller wholes made up of carbon, hydrogen and oxygen atoms, and so on to ever smaller wholes of sub-atomic particles. The morphic fields at higher levels of complexity have a role in organizing lower-level entities within them (Sheldrake 1988: 94, 120). This helps us think of the influence of society, ecosystems, or the entire cosmos on individuals. This view, when applied to human society, gives us an alternative to the current prevailing philosophy of individualism which arose in the 17th century simultaneously with mechanistic science. It may help us move beyond Platonistic, dualistic, and reductionistic thinking about culture and society. Indeed, we might think of myths and scientific paradigms themselves as probability structures shaped by morphic fields and maintained by morphic resonance (Sheldrake 1988: 256).

The concept of probability in morphic fields is complemented by the work of Prigogine and Jantsch on self-organizing systems. Prigogine was awarded the Nobel Prize in chemistry for his work on the thermodynamics of nonequilibrium systems. A system is defined by interactions within it, and by interactions between it and the larger environment. Far from being at rest, the stability and duration of systems seems to depend on their capacity for self-organizing activity. In Prigogine's principle of order through fluctuation he describes systems as constantly adjusting to internal and external stimuli, achieving ever-changing homeostasis through feedback and adjustment. When there is non-equilibrium there is a flow of matter and energy which may be a source of order. However, extreme changes within a system, or from its larger environment, may push it beyond a threshold of adjustment causing it to transform into something else. Thus genes mutate, plants and animals evolve, and societies go through periods of destructiveness and revitalization (Bateson 1972, Prigogine and Stengers 1984, Jantsch 1980).

Mind is an aspect of all self-organizing systems. It is manifest as systems are 'aware' of stimuli and make organizational adjustments. The most simple molecules have this awareness and capacity to adjust. The concept of morphic resonance gives new insight into the meaning of mind. We think by combining memory and perception. But memory may not be entirely locked up in physical brain cells. To use an analogy, the brain may not be like a magnetic tape that contains all the information, but like a television set - an organ which is sensitive to information in external fields. But even without resources to field theory, we might have a concept of mind that is not bound by a material organ. In social and ecological systems, for example, it is located not only in individual bodies but also in messages, and the pathways of messages, outside of bodies. There may be larger manifestations of mind of which our inividual minds are only subsystems, as for example with all who sit in a room listening intently to a lecture. Our minds may also be integrated into layers of ecosystems and may participate with some kind of universal mind.

### The Contemporary Holistic Perspective

Marilyn Ferguson, in her popular book *The Aquarian Conspiracy* (1980) wrote of small groups and networks springing up everywhere, each in its way having the common theme of personal growth amidst cultural transformation, or cultural 'awakening'. In her chapter on health she is sensitive to consumer demand that is causing some American medical practitioners to once again acknowledge psychological, social and spiritual dimensions in people. Some are re-learning the healer's art of listening to people in a way that leads to intuitive diagnosis - at least as a complement to impersonal technological diagnostic methods (1980: 264ff.).

Fritjof Capra's critique of the biomedical model (chapter 5) and Newtonian psychology (chapter 6) in *The Turning Point* (1982) works out holistic and ecological concepts of health. The whole book is based on systems theory, showing how many institutions, including health and economics, might move toward congruence with physics and cosmology - how they might be conceptually integrated into a single world view or ethos which is greater than the sum of the parts. He defined, in layman's language, contemporary physicists' descriptions of matter, energy and all 'reality'. He has done much to consolidate and make coherent the conceptual and emotional shift from a mechanistic to a holistic vision of reality that may be taking place.

Capra's intention in writing *The Turning Point* was to help all the informal networks and 'alternative' organizations that have sprung up to see that they are all going in the same direction. They are nurturing only different aspects of a single new vision of reality. He states clearly that "the purpose of this book is to provide a coherent conceptual framework that will help them to recognize the communality of their aims" (1982: xviii). Are we observing the formation of a mass movement as surely as Ted Schwartz (1962) did when he documented a cult arising and transforming the life of Manus Islanders into the *newfela fashion*? On Manus Island, in the 1950s, the re-thinking of the culture, the *thinkthink* led by Paliau, made Manus Islanders whole again: the religious and the secular were only different aspects of single reality (Schwartz 1962: 261).

In medical anthropology, Nancy Scheper-Hughes and Margaret Lock wrote an important paper in 1987 on *The Mindful Body*, charting the extent to which our Western cultural tradition has encouraged science and clinical medicine to oppose spirit to matter, and mind to body. This conceptual fragmentation was itself the result of 17th century transformation in world view. Descartes was especially persuasive in defining people as being made up of palpable body on the one hand, and mind or soul on the other. By splitting us in two, Descartes could remain within Christian orthodoxy by preserving the soul as the domain of theology and the body as the domain of science. Thus biology and medicine were set on the radical materialist path that has brought contemporary crisis to the medical profession and a flourishing of alternative holistic therapies (Scheper-Hughes and Lock 1987: 8ff.). Christian orthodoxy was at the same time transforming. At least since it gave up attempting to control materialistic science, religious thinkers have turned away from concern about cosmic, earthly and even cultural matters. Instead, they have mostly concentrated on (1) the personality of Jesus, (2) the process of redemption and salvation, (3) the spiritual disciplines needed to guide a person's soul along the path of salvation, and (4) enforcement of judicial procedures to protect the path of salvation (McDonagh 1986: 62).

The contemporary holistic perspective arises from a congruence of theory in physics and biology with ancient Hindu, Buddhist, Taoist and Christian thought. Capra harmonized, in *The Tao of Physics*, a view of the world from theoretical physics with the Dance of Shiva and the Tao (Capra 1975; see also Zukav 1979). The Tao teaches the enquirer, whether physicist or mystic, to trust flashes of intuition in which oneness of being is experienced in a moment of heightened awareness. When qualities of *yin* and *yang* are harmonized thought is heightened, energies flow freely within a framework of simplicity, tranquillity and non-violence (Lao Tzu 1963). The greatest of theoretical physicists seem to have had experiences of unity consciousness and written about them (Wilber 1984). Bohm's theory of wholeness and the implicate order, exemplified by the holomovement, was described poetically in the *Upanishads*, written in Sanskrit between 800 and 400 B.C. For example, the Chandogya Upanishad instructs that "the little space within the heart is as great as the vast universe. The heavens and earth are there, and the sun, and the moon, and the stars; fire and lightning and winds are there; and all that now is and all that is not ..." (Upanishads 1965: 120).

Eastern thought is entering the mainstream of popular Western culture. It is there, unmasked as exotic or intellectual, much as a popular version of Freudian thought pervaded Western culture in the 1950s. In Britain, for example, many people initially attracted to yoga exercises or meditation classes for individualistic reasons began to see themselves and their everyday activity in a different cognitive framework.

Buddhism is the fastest-growing religion in Britain, and Buddhist concepts are beginning to appear in popular British culture. Buddhism reassures us that change is normal, but goes on to the more radical insight that all things are impermanent, especially the separate ego. All things are one and have no life apart from oneness. The one is all things, and is incomplete without the least of them. Yet the parts retain their full identity. They are interfused with the reality of oneness but not merged into it. Once this concept, so congruent with Bohm's theory of wholeness and the implicate order, is grasped it frees us from Cartesian dualism. It also transcends Marxist conflict theory and shows the way to inner and outward peace. Buddhism suggests that when we feel stressed, incomplete and unhappy the cause is the illusion of self; the belief that the part can pit its separate self against the nature and the welfare of the whole. Egoistic desire causes us to suffer. We give way to egoistic desire out of ignorance of the *holistic* nature of things. What is, is a wheel of becoming, and the law of the wheel is love. When we know all life is one, then when any other being suffers we suffer. Being kindred, we feel compassion and seek harmony (see for example Humphreys 1951).

The search for a holistic world view also looks back into our own cultural history, before metaphysical dualism changed our view of the world. For example, paperback books of the teachings of Hildegard of Bingen, a 12th century Benedictine nun born in Germany where the Nahe River flows into the Rhine, are readily available in London book shops. She wrote music, some of it recently recorded by the well-known soprano Emma Kirkby. She also painted pictures, was a herbalist and healer, and was consecrated abbess of the large convent she founded. "She castigated the pope for his timidity and an emperor for moral blindness. She taught scholars and preached to clergy and laity." The pope encouraged her to write more (Fox, in Hildegard of Bingen 1985: 14). She painted her insights or *illuminations* as mandalas, each a holistic symbolic expression. One illumination is the *egg of the universe* which unites biology and cosmology. Another pictures a child on the earth, attached to a kite in the sky by a string (or an umbilical cord). The kite string unites within one cosmology the sky and earth, bringing down the fire that possesses the heart of the child, its soul burning with the fire of deep understanding. "It pours itself through all the limbs of the person and gives the greenness of the heart and veins and all the organs - to the entire body as a tree gives sap and greenness to all the branches from its root" (Hildegard of Bingen 1985: 55). Matthew Fox's commentary on her work is woven harmoniously together with current discourse in physics, cosmology, the ecology movement, feminism and Buddhism. Body and spirit are united; people and cosmos are kindred.

## Alternative Therapies and Holistic Health

The holistic health movement has grown organically out of widespread understanding of at least some aspects of quantum and field theory, cybernetics and ecological theory, pre- and post-Freudian psychology, and cross-cultural theology. People in increasing numbers appear to be viewing themselves and all being within a framework of wholeness and implicate order. This is a world view in direct contrast to that of allopathic medicine which is based on a diagnostic process which splits people down to a diseased organ, and uses drugs which induce an opposite condition as part of the cure - to correct the malfunctioning part.

In contemporary European society these two world views now exist side by side, usually within individuals who switch from one medical system to another, depending on past experience and the current context. Although the Conservative government in Britain would prefer to privatize the National Health Service, it is too popular with the electorate for the government to risk more than cosmetic fiscal surgery. In 1986 the National Health Service treated over 6 million in-patients and 37 million out-patients in a population of 54 million.

At the same time the popularity of alternative therapies is growing rapidly. In 1983 there were five million treatments in the five major alternatives: homeopathy, osteopathy, chiropractic, herbal medicine and acupuncture. The number of treatments is estimated by the Institute for Complementary Medicine to be increasing at 15% per year, or 20% per year by a WHO estimate (Institute for Complementary Medicine 1984). The numbers are at least doubling every five years. In Britain there are 3000 practitioners in the five major therapies and a further 12,000 in other therapies, many training and practising part time (Endacott 1988). A 1981 survey asking about resort to all types of therapies estimated that ten million people had consulted an alternative practitioner, and the number would be much greater now (Price 1986: 16). The Healing Arts Exhibition, held in London from 2-6 November 1988, with an admission charge of £3.00, had 15,000 visitors who browsed among 145 stalls (Wilson 1988). There were lectures on such topics as near-death experiences, Buddhist and American Indian approaches to health. Groups such as Friends of the Earth, as well as explicitly health-oriented groups had stalls. A recent guide to alternative therapy centres and training centres in the London area alone runs to over 200 pages (Bradley and Considine 1988). Whole food and vegetarian diet appear to be becoming a mass phenomenon in Britain. The motives are concern about pest and preservative poisons, allergy-causing additives, but also a compassion for other animal species linked with us in a common environment.

The alternative sector is becoming 'established'. There is now an Institute for Complementary Medicine on an important site in central London. It is a meeting place for therapy organizations, whose representatives form an advisory and educational committee. It provides information to the medical and general public and holds classes, lectures, and other scholarly and self-help activities (Endacott 1988). Two years ago the Marylebone Health Centre, in the crypt of a fashionable London church, began offering a range of allopathic and alternative therapies (counselling, massage, osteopathy, acupuncture, homeopathy, music therapy, meditation, yoga, etc.). At the end of the first year of offering this range of therapies for all kinds of sickness patients presented, including the pain, anxiety, stress and other problems of our times, the prescribing costs for all drugs was 48% below the average for other National Health Service practices in the same area of London (Marylebone Health Centre 1989: 21). In December 1989 the Marylebone Health Centre expanded within the framework of a Trust under the royal patronage of the Prince of Wales, augmenting its research and teaching role, and thus strengthening its attempt to affect policy within the National Health Service (Pietroni 1990).

There are transformational stirrings within the orthodox medical profession. In 1983 a group of physicians and medical students founded the British Holistic Medical Association with the aim of introducing holistic approaches into medical practice. Most of its members are general practitioners rather than doctors doing hospital medicine. Originally members were all qualified physicians, but the distinction between allopathic and alternative medicine has become so blurred that the membership is now open to both types of practitioners. In 1988 there were 1470 members, about 600 of them physicians (BHMA 1988). That is not a large number compared with 89,641 physicians (about 34,000 general practitioners) in the United Kingdom (BMA 1988). Members of the breakaway group stress the uniqueness of the patient, and recognize people's ability, with help, to treat themselves. They encourage such self-help skills as breathing and relaxing routines, meditation, visualisation, exercise and diet, and have begun publishing monographs much sought out by medical students.

The British Holistic Medical Association publishes a well-written and widely-read magazine, *Holistic Medicine*. Two physician anthropologists are on the editorial board and other anthropologists have contributed articles. Recently, another well-written popular magazine has appeared entitled *Caduceus: Health, Healing and Holistic Medicine*, edited by an ethnically British woman who is a practising Sufi, in the Islamic mystical tradition. Serious articles by physicians with a holistic perspective are appearing in the established medical journals as well. For example, in 1987 Aldridge and Pietroni, writing in *Family Practice* raised several methodological issues in clinical research, asking for a broader science of human behaviour. They concluded that "the clinician is required to act as a clinical anthropologist as well as a clinical epidemiologist" (see also, for example, Chase 1988, Pietroni 1986, 1987a, b).

The scientific establishment is showing the straints of challenges to comfortable medical orthodoxy. The British Medical Association held an investigation of holistic medicine and found it lacking. But of course the theoretical assumptions and analytical techniques used in the study biased its findings, which failed to understand medicine in relationship to society or to historical context. The report was about diseases, and it ignored social facts, psychological correlates and illness behaviour. The prestigious journal *Nature* accepted a paper describing a phenomenon in which biological information appeared to remain in the molecular organization of water after so many homeopathic-type dilutions (and shaking) that none of the starting molecules of antibody substance was left. The work was done independently in laboratories in France, Israel, Italy and Canada, and all showed similar results. In France one in four French doctors prescribe homeopathic medicine, and the French researchers were sympathetic to homeopathic medicine. Within a fortnight of publication the editor of *Nature* (a physicist) went off to the French laboratory with a professional magician and a scientific fraud finder. None were biological nor medical researchers. They did not find fraud, and some experiments were successfully repeated in their presence (Nature 334: 287). This is remarkable considering that the editor of *Nature* and his two colleagues were a theatrical and upsetting intrusion into the laboratory, and it is therefore understandable that repetitive, rushed, complex experiments did not show consistent results. That, however, did not keep the editor of *Nature* from writing an article in which he called the original experiments a delusion (Nature 334: 287-290). Whether dilute homeopathic medicines are 'active' remains an interesting question, but of even equal interest are other kinds of research question, such as, does the sufferer feel better following a healing consultation?

## Ethos

This confused picture, and a certain amount of antagonistic opposition, is just what anthropologists expect when observing societies undergoing accelerated cultural change and perhaps transformation. The literature on millenarian movements, for example, documents the politics of change, whether among colonialized people, or the British Medical Association under challenge from patients and alternative practitioners, or even the editor of *Nature*, guarding the gates of established science against challenges to the 'law and order' of orthodox scientific theory.

When discussing these signs of cultural change, people often use the 'scientific' language of linguistics, writing of a 'paradigm shift' in Western culture (see, for example Kuhn 1962). But paradigm, merely pattern, does not seem a rich enough concept for the shift we are trying to understand. Therefore, I prefer to work with the concept of ethos. The term was used by Aristotle to refer to the characteristic spirit of prevalent tone or sentiment of a people. He also used it with regard to the 'genius' of an institution or system (O.E.D. 1971: 314). At the turn of this century, William Graham Sumner developed the concept to cover "the totality of characteristic traits by which a group is individualized and differentiated from others" (1960 [1906]: 76). In *Naven*, Gregory Bateson used ethos to mean "the system of emotional attitudes which governs what value a community shall set upon the various satisfactions which the contexts of life may offer" (1956: 220). Kroeber also extended the concept beyond the Greek meaning in order to link anthropology and psychology more closely. Ethos referred "not so much to the specific ethics or moral code of the culture as to its total quality, to what would constitute disposition or character in an individual; to the system of ideals and values that dominate the culture and so tend to control the type of behaviour of its members (1948: 294).

Ethos is of course a concept linked to the German concept *Weltanschauung*: a world view, ideology, philosophy of life and concepts of the way the world really is. Toward the end of his life, Kroeber, writing on world-view and historical change, concluded that the collective image of one's own culture is always idealized. This idealized view is then "imposed on history, law, religion and every other intellectual product of the culture". It is an organized and coherent picture which is at the same time a product of culture but feeds back and influences cultural products. Kroeber speculated that a society with a feeble or blurred image of its culture could not be a vigorous culture (1963: 91).

## Analytical Discussion

Numbers alone suggest we are not on the threshold of a sweeping change in world view. The British National Health Service saw 37 million patients in 1986 and the five major alternative therapies saw about 7 million people in the same year. However, we do not know how many additional people sought other kinds of alternative therapies. If only 600 out of almost 90,000 doctors are members of the British Holistic Medical Association, the profession is not being too seriously undermined from within. High circulation newspapers and magazines are still characterized by articles such as 'Britain's Highest Paid Man' (Sun, p.1, 4 Nov. 1988), or titillating descriptions of rape and other personal violence. But world views do not change by democratic vote. We do not know how many people - or which people - have to begin to think and live in another mode before a transformation occurs.

In my very random, but not scientifically-sampled conversations with people, I had become interested as an anthropologist when I realized that some people appear to be living within the materialistic, individualistic and technological ethos, but have begun to think a different way under the surface. This in not extreme cognitive dissonance as Festinger (1957) described it, but a good-humoured tolerance of simultaneous alternatives. As we understand more about the nature of self-organizing systems and their transformations, these subtle pre-disposing alternative undercurrents are of particular interest in understanding the process of cultural change. Prigogine's concept of 'order through fluctuation' helps us to see how normal it is to live in an unstable world where small causes can have large effects; in all kinds of systems from chemical compounds to whole societies (Prigogine and Stengers 1984: 206ff.).

In a historical perspective, for at least a century philosophers and social scientists have commented that we now see reality through a scientific world view. Indeed, they have suggested we have 'progressed' from magic through religion to science (see for example Taylor 1871, Frazer 1890, Bury 1932). They wrote that our mythic past is left behind and we live in a world which has been so accurately described that we can manipulate it with greater success than the shaman or priest could. But theoreticians such as Bohm are suggesting that Newtonian mechanics, and the manipulations they have facilitated, are merely the explicate order, and true reality is in the enfolded totality of the implicate order. Furthermore, some people are asking if being able to see the world with the 'inner eye' of love, conscious of a compassionate unity with it, might be of greater value than manipulating it to our individualistic purposes. Some are articulating the desire to live in a culture which is *simultaneously* enriched by magic, religion and science, harmonized within a unified world view. They realize that scientific theories are simply myths in the sense that they are mental constructs which attempt to make sense of the world. Also, science itself does not have a unified view. Some fields of science, such as theoretical physics, are leading a revolt against overly reductionistic technological definitions of science, while the extreme reductionist approach is still paying off in such fields as immunology. But even within theoretical physics, visions of 'reality' move in and out of vogue. At the moment a very reductionistic model is on offer in contrast, for example, to Bohm's view that all being is a web of unbroken wholeness (Isham 1988).

Scientific exploration, because of its nature, throws up novel observations, and theories to accomodate the observations. When novelty occurs it causes excitement for a few and discomfort for many. As further instances of novelty are observed, and further radical explanations offered, resistance may cause science to become increasingly rigid. Significant scientific novelty will often occur simultaneously in several laboratories. Kuhn (1962) has suggested that if a major shift in the dominant scientific paradigm actually occurs there will be a concomitant change in world view. This is a functional or systems theory description of the historical observation that all aspects of the society will adjust. The behaviour of the editor of *Nature* in challenging a possible scientific basis for homeopathic medicine, or in describing Sheldrake's application of field theory to biology as "infuriating ... the best candidate for burning there has been for years" are examples of resistance to novelty in biology and medicine. Widespread religious fundamentalism may be a manifestation of resistance to incipient transformation in the wider culture. However, in physics and cosmology the scientific revolutions of quantum theory and relativity theory has already occurred. It now appear to be spilling over into other disciplines such as biology, and even psychology (e.g. Wilber 1979, 1984).

Let us look into the future and suppose that David Bohm is correct. Our dualistic way of thinking is becoming increasingly dysfunctional. It fragments us and keeps us from feeling an integral part of all existence (Bohm 1980: xi, 2, 1985: 1, 2, 23). We long for holistic health and harmony, but are exhorted to seek material success individually. Should these mutually contradictory perceptions, knowledge, motives, values and hopes become too dissonant, we can act to reduce the dissonance by either changing the real world in some way so as to modify the information coming in, or modify our perceptions of self and the real world in such a way that one type of dissonance is no longer recognized (Festinger 1957). In the first approach we may throw our energies into alternative therapies, peace activism, or the ecology movement. On the other hand, we might look with Buddhist insight on egoistic striving as illusion that leads to suffering, or we may see in a flash of insight the Tao of physics in all reality. We might even take both approaches simultaneously.

Anthony Wallace has tried to analyze how a change in world view 'feels' to people and institutions undergoing extreme change. His analysis is shaped by familiarity with Weberian and psychoanalytic theory. He recognizes the role religion plays in helping people rationalize cultural contradictions. Lévi-Strauss (1978) has further explicated the way in which myth and folk tales are symbolic formulae for resolving conflicts engendered by a society with structural contradictions. The contradictions may be so great as to place ordinary people in such a moral quandary that religion may no longer serve as a guide. When social and cultural stress become too great, a new configuration of myth and ritual arises. The current European holistic movement may be such an incipient configuration. If this or any other new ethos is internally coherent, and also a workable description of the world, it may serve as a guide for efficient action in a transforming society (Wallace 1966: 29-30).

Wallace looked at a variety of millenarian movements in the anthropological literature. When such movements succeeded to the stage of efficient action in a transformed society he designated them revitalization movements. They are incipient when the society begins to move in and out of equilibrium with increasing change, but the levels of disorganization and stress in individuals and institutions is tolerable. As people experience more stress they begin to perceive that the institutions of the society are not dealing well with increasing social and political hierarchy, war, disease, or aspects of ecology. With another ratchet turn, serious cultural disorientation sets in. People attempt to restore personal equilibrium by adopting socially dysfunctional expedients such as alcoholism, drug-taking, scapegoating and similar alienating behaviour. If cultural disorientation becomes even greater the societies he reviewed either splintered into smaller autonomous groups or were absorbed into another more stable society. Another option was that the society quietly died off, an interesting observation when we realize that most European nations have such low birth rates that they are not replacing their population except through immigration. Another option in our nuclear age is to end with world-destroying bang, not a whimper. Or, a society may successfully transform through a revitalization movement (Wallace 1966: 158-159). We might speculate that European society is at the stage of 'serious cultural disorientation'. Alcoholism and drug-taking are endemic, and the scapegoating whiff of fascism is in the air. Is the holistic movement a viable alternative to the psychic numbing engendered by the enormities of our times (see for example Lifton 1967, Thompson 1986).

According to Wallace, societies that successfully transform and are revitalized have common features. There must be a code, model, vision or myth of a healthy future. The code, such as Capra's *Turning Point*, serves as a guide not only to thought and sentiment, but also to behaviour carried out in an organized way in order to reach the desired 'goal culture'. Often the code will be crystallized by a charismatic 'prophetic' figure. The new code must then be communicated, often in an evangelistic spirit since the aim is to convert people to the new vision. The benefits promised need not be immediate nor materialistic, but must hold out the hope of self-respect within a stable social system. Increasingly the evangelists become executives, responsible for administering new organizations as they arise. Most people, however, are followers. They give some time, money and allegiance to the movement. In a period of adaptation, resistance from those who had benefited from the previous *status quo* must be dealt with. There will also probably be unsettling ambiguities in the new code, and failures of prediction. The adaptive stage may involve some hostility toward non-participating people.

If the society actually transforms to a more adaptive mode and finds a new homeostasis, there should be a marked decline in pathological symptoms such as pain, violence, numbness and anomie. People will conform to the new cultural consensus without coercion. This accounts for the quality of brotherhood, or communitas, Victor Turner described in newly revitalized societies (1969). In the final stage of revitalization, the society may know a period of stability as organizations and institutions function well under routine administration (Wallace 1966: 160-162).

Not all successful revitalization movements have the same temporal-cultural orientation. Some are primarily oriented to the past, seeking to regain a golden age believed to have existed before the current time of stress. Others face toward the future, seeking an utopian goal. Others, especially societies under colonial domination, are assimilative, seeking to combine into one system the core of the old culture and aspects of an alien society, dissolving social boundaries between the two. Some are expropriative, as when cargo cults in the Pacific sought to import many customs of an alien group, combine them with indigenous cultural customs, but expel alien colonialists (Wallace 1966: 165-166).

No revitalization is purely of one type. If we were to think of the holistic movement as an incipient revitalization movement, to some extent it looks back to a pre-Cartesian, pre-dualistic golden age of holistic cultural stability. It is a culture of simultaneous magic, religion and science, not the evolutionary one-dimensional progression from magic through religion to science we have been congratulating ourselves for achieving. This view of a holistic past accounts, for example, in renewed interest in such people as Hildegard von Bingen. The movement also looks forward, to the frontiers of theoretical science. But the movement is primarily assimilative as people take to heart Sufi, Taoist, or Buddhist teachings about the way to proceed. Even an eminent Christian theologian has suggested that our European spiritual life would benefit from a Buddhist approach (Cupitt 1980), and Jesuit writers who have lived in the orient, from Teilhard de Chardin (1965) to William Johnston (1979), have helped Europeans understand Eastern concepts of unity consciousness.

A transformed world view in our times will inevitably be assimilative as improved communications enable world-wide sharing of insights. Never before have so many people been able to vicariously experience the lives of others, in the past and in other cultures, through biography, novels, films and records. Never before have so many people known the song of whales and the habits of badgers. Beyond electronic communication, might we Europeans cultivate a greater capacity to tune in on the kind of nested morphic fields Sheldrake has hypothesized, and *feel* people of another culture to be our 'mother's brother'; to *feel* raven and witchetty grub to be our ancestors? Perhaps in our new age many will know what 'primitives' knew.

## References


ALDRIDGE D. and PIETRONI P. 1987: Research Trials in General Practice *Family Practice* 4: 311-315.

BATESON G. 1936: *Naven.* Cambridge: Cambridge University Press.

— 1972: *Steps to an Ecology of Mind.* New York: Ballantine.

BOHM D. 1980: *Wholeness and the Implicate Order.* London: Ark.

— 1985: *Unfolding Meaning.* London: Ark.

BOHM D. and PEAT D.1987: *Science, Order and Creativity.* London: Bantam.

BRADLEY K. and CONSIDINE M. 1988: *The London Guide to Mind, Body and Spirit.* London: Brainwave.

BHMA (British Holistic Medical Association) 1988: London, Personal communication.

BMA (British Medical Association) 1988: London, Personal communication.

BURY J.B. 1932: *The Idea of Progress.* New York: Dover.

CAPRA F. 1975: *The Tao of Physics.* London: Flamingo.

-- 1982: *The Turning Point.* London: Flamingo.

CHASE H.D. 1988: Holistic Medicine: What is it and what are its Clinical Implications? *Maternal and Child Health,* October: 292.

CUPITT D. 1980: *Taking Leave of God.* London: SCM Press.

ENDACOTT M. 1988: Institute for Complementary Medicine. London, personal communication.

FERGUSON M. 1980: *The Aquarian Conspiracy.* London: Paladin.

FESTINGER L. 1957: *A Theory of Cognitive Dissonance.* Evanston, Ill.: Row, Peterson.

FRAZER Sir J. 1959 (1890): *The Golden Bough* (abridged). New York: Anchor.

HILDEGARD OF BINGEN 1985: *Illuminations.* Translated by Matthew Fox. Santa Fe: Bear and Co.

HUMPHREYS C. 1951: *Buddhism.* Harmondsworth: Penguin.

INSTITUTE FOR COMPLEMENTARY MEDICINE 1984: *Report on Trends in Complementary Medicine.* London: ICM.

ISHAM C. 1988: Professor of Theoretical Physics, Imperial College. London, personal communication.

JANTSCH E. 1980: *The Self-Organizing Universe.* Oxford: Pergamon.

JOHNSTON W. 1978: *The Inner Eye of Love: Mysticism and Religion.* London: Collins.

KROEBER A.L. 1948. *Anthropology.* New York: Harcourt Brace.

— 1963: *An Anthropologist Looks at History.* Berkeley: University of California Press.

KUHN T. 1962: *The Structure of Scientific Revolutions.* Chicago: University of Chicago Press.

LAO TZU 1963: *Tao Te Ching*. Translated by D.C. Lau. Harmondworth: Penguin.

LEVI-STRAUSS C. 1978: *Myth and Meaning*. London: Routledge and Kegan Paul.

LIFTON R.J. 1967: *Death in Life: Survivors of Hiroshima*. New York: Random House.

MACCORMACK C. 1980: Nature, Culture and Gender: A Critique. In: MACCORMACK C. and STRATHERN M. (eds.). *Nature, Culture and Gender*(pp. 1-24). Cambridge: Cambridge Univ. Press.

MARYLEBONE HEALTH CENTRE 1989: *Annual Report 1988/1989*. Wates Foundation Primary Health Care Research Project. London: Marylebone Health Centre.

MC DONAGH S. 1986: *To Care for the Earth: A Call to a New Theology*. London: Geoffrey Chapman.

*NATURE*
30 June 1988: 'When to Believe the Unbelievable', Editorial, 333: 787.
30 June 1988: Davenas E. et al.: 'Human Basophile Degranulation Triggered by Very Dilute Antiserum Against IgE', 333: 816-818.
28 July 1988: Maddox J., Randi J. and W. Stewart: 'High Dilution Experiment a Delusion', 334: 287-290. Correspondence following into volume 335.

PIETRONI P. 1986: The Case for a Real Alternative in the Treatment of Cancer *Cancer Topics* 6: 22-23.

– 1987a: The Meaning of Illness: Holism Dissected. *Journal of the Royal Society of Medicine* 80: 357-360.

– 1987b: Holistic Medicine: New Lessons to be Learned. *The Practitioner* 231: 1386-1390.

– 1990: *The Greening of Medicine* . London: Gollancz.

*OXFORD ENGLISH DICTIONARY* 1971: Ethos. Vol E., p. 314. Oxford: Oxford University Press.

PRICE R. 1986: *The Healing Within: Medicine, Health and Wholeness*. Monograph series, no.23. Tunbridge Wells: Institute for Cultural Research.

PRIGOGINE I. and STENGERS I. 1984: *Order Out of Chaos*. London: Heinemann.

SCHEPER-HUGHES N. and LOCK M.M. 1987: The Mindful Body: A Prolegomenon to FutureWork in Medical Anthropology. *Medical Anthropology* 1 (1): 6-41.

SCHWARTZ T. 1962: *The Paliau Movement in the Admirality Islands, 1946-1954*. Anthropological Papers of the American Museum of Natural History, Vol. 49, part 2. NewYork: American Museum.

SHELDRAKE R. 1985: *A New Science of Life: The Hypothesis of Formative Causation*, second edition. London: Anthony Blond.

-- 1988: *The Presence of the Past: Morphic Resonance and the Habits of Nature*. London: Collins.

— 1990: *The Rebirth of Nature*. London: Century.

SUMNER W.G. 1960 [1906]: *Folkways: A Study of the Sociological Importance of Usages, Manners, Customs, Mores and Morals*. New York: Mentor.

TEILHARD DE CHARDIN P. 1965: *The Phenomenon of Man*. New York: Harper and Row.

THOMPSON J. 1986: *Psychological Aspects of Nuclear War*. New York: John Wiley & Sons.

TURNER V. 1969: *The Ritual Process: Structure and Anti-Structure*. Chicago: University ofChicago Press.

TYLOR E.B. 1958 [1871]: *Religion in Primitive Culture*. New York: Harper and Row.

*UPANISHADS* 1965: Translated by Juan Mascaro. Harmondsworth: Penguin.

WALLACE A. 1966: *Religion: An Anthropological View.* New York: Random House.

WILBER K. 1979: *No Boundary.* Boston: New Science Library.

-- 1984: *Quantum Questions.* Boston: New Science Library.

WILSON G. 1988: New Life Design, Ltd. London, personal communication.

WHO (World Health Organization) 1983: *Apartheid and Health,* part II, pp. 59-258. Geneva: WHO.

ZUKAV G. 1979: *The Dancing Wu Li Masters.* London: Fontana.

## List of Contributors

**Gilles Bibeau**, Ph.D., Professor of Anthropology and Chair, Department of Anthropology, Université de Montréal, Départment d' Anthropologie, C.P. 6128, Succursale A, Montréal (Québec) H3C 3J7, Canada.

**René Devisch**, Ph.D., Professor of Anthropology, Research Unit on Symbol and Symptom, Katolieke Universiteit te Leuven, Centre for Social and Cultural Anthropology, Tiensestraat 102, B-3000 Leuven, Belgium

**Jutta Dornheim**, Ph.D., Professor, Fachbereich Sozialwissenschaften (LGW), Universität Osnabrück, An der Katharinenkirche 8b, Postfach 44 69, D-4500 Osnabrück, West-Germany.

**Atwood D. Gaines**, Ph.D., MPH, Professor, Case Western Reserve University, Department of Anthro-pology, Cleveland, Ohio 44106, U.S.A..

**Mary-Jo DelVecchio Good**, Ph.D., Associate Professor, Harvard Medical School, Department of Social Medicine, 25 Shattuck Street, Boston, Massachusetts 02115, U.S.A.

**Deborah R. Gordon**, Ph.D., Consultant for Research, Center for the Study and Prevention of Cancer, Viale Volta, 171, I-50131 Florence, Italy.

**Peter Hahn**, M.D., Professor, Medizinische Klinik Abt. 1.1.2. - Innere Medizin II, Ruprecht-Karls-Universität Heidelberg, Bergheimer Straße 58, D-6900 Heidelberg 1, Germany.

**Wolfgang Jacob**, M.D., Professor em., Institut für Sozialmedizin, Ruprecht-Karls-Universität Heidelberg, Im Neuenheimer Feld, D-6900 Heidelberg, Germany.

**Margaret Lock**, Ph.D., Professor and Chair Department of Humanities and Social Studies in Medicine, McGill University, 3655 Drummond Street, Montréal, Qué. H3G 1Y6, Canada.

**Carol MacCormack**, Ph.D., Professor, Department of Anthropology, Bryn Mawr College. Bryn Mawr, Pennsylvania 19010, U.S.A.

**Thomas Ots**, M.D., Institut für Ethnologie, Universität Hamburg, Rothenbaumchaussee 64a, D-2000 Hamburg 13, Germany

**Mariella Pandolfi**, Ph.D., Assistant, Department of Glottoanthropological Studies, University of Rome "La Sapienza", Piazza dei Ricci, 129, Roma, 00186, Italy.

**Beatrix Pfleiderer**, Ph.D., Professor, Institut für Ethnologie, Universität Hamburg, Rothenbaum-chaussee 64a, D-2000 Hamburg 13, Germany

**Thure von Uexküll**, M.D., Professor em., Sonnhalde 15, D-7800 Freiburg, Germany.

**Paul U. Unschuld**, Ph.D., Professor, Institut für Geschichte der Medizin, Ludwig-Maximilians-Universität, Lessingstr. 2, D-8000 München 2, Germany.

**Allan Young**, Ph.D., Professor, Department of Humanities and Social Studies in Medicine, McGill University, McIntyre Medical Sciences Building, 3655 Drummond Street, Montréal, Qué., H3G 1Y6, Canada.

**Francis Zimmermann**, Ph.D., Professor, Ecole des Hautes Etudes en Sciences Sociales, Centre d' Etudes de l' Inde et de l' Asie du Sud , 54, Boulevard Raspail, F.-75006 Paris, France.